AF412160

Advances in Experimental Medicine and Biology

Volume 1015

More information about this series at http://www.springer.com/series/5584

Rommy von Bernhardi • Jaime Eugenín
Kenneth J. Muller

Editors

The Plastic Brain

 Springer

Editors
Rommy von Bernhardi
Pontificia Universidad Católica de Chile
Santiago, Chile

Jaime Eugenín
Universidad de Santiago de Chile
Santiago, Chile

Kenneth J. Muller
University of Miami
Miami, USA

ISSN 0065-2598 ISSN 2214-8019 (electronic)
Advances in Experimental Medicine and Biology
ISBN 978-3-319-62815-8 ISBN 978-3-319-62817-2 (eBook)
DOI 10.1007/978-3-319-62817-2

Library of Congress Control Number: 2017952702

This Springer imprint is published by Springer Nature
The registered company is Springer International Publishing AG
The registered company address is: Gewerbestrasse 11, 6330 Cham, Switzerland

Contents

Part III   Neural Plasticity in the Respiratory Rhythm

Part IV   Damage-Triggered Neural Plasticity

Contributors

Marta C. Antonelli Instituto de Biología Celular y Neurociencia "Prof. Eduardo De Robertis", Facultad de Medicina, Universidad de Buenos Aires, Buenos Aires, Argentina

José Bargas División de Neurociencias, Instituto de Fisiología Celular, Universidad Nacional Autónoma de México, Mexico City, DF, Mexico

Janet Barroso-Flores División de Neurociencias, Instituto de Fisiología Celular, Universidad Nacional Autónoma de México, Mexico City, DF, Mexico

Yesser H. Belgacem INMED, Aix-Marseille Univ, INSERM, Marseille, France and Aix-Marseille Université, IMéRA, Marseille, France

Sebastián Beltrán-Castillo Laboratorio de Sistemas Neurales, Facultad de Química y Biología, Departamento de Biología, Universidad de Santiago de Chile, USACH, Santiago, Chile

Luis Beltran-Parrazal Centro de Investigaciones Cerebrales, Universidad Veracruzana, Xalapa, Mexico

Laura N. Borodinsky Department of Physiology & Membrane Biology and Institute for Pediatric Regenerative Medicine, University of California Davis School of Medicine and Shriners Hospital for Children Northern California, Sacramento, CA, USA

Karina Bravo Laboratorio de Sistemas Neurales, Facultad de Química y Biología, Departamento de Biología, Universidad de Santiago de Chile, USACH, Santiago, Chile

Marco Contreras Departamento de Fisiología, Facultad de Ciencias Biológicas, Pontificia Universidad Católica de Chile, Santiago, Chile

José M. Delgado-García Division of Neurosciences, Pablo de Olavide University, Seville, Spain

Jaime Eugenín Laboratorio de Sistemas Neurales, Facultad de Química y Biología, Departamento de Biología, Universidad de Santiago de Chile, USACH, Santiago, Chile

Laura Eugenín-von Bernhardi Servicio de Neurología, Hospital Sótero del Río, Santiago, Chile

Paula Farías Laboratorio de Sistemas Neurales, Departamento de Biología, Facultad de Química y Biología, Universidad de Santiago de Chile, Santiago, Chile

Elvira Galarraga División de Neurociencias, Instituto de Fisiología Celular, Universidad Nacional Autónoma de México, Mexico City, DF, Mexico

Karina Gómez Departamento de Fisiología, Facultad de Ciencias Biológicas, Pontificia Universidad Católica de Chile, Santiago, Chile

Alejandro González Departamento de Biología, Facultad de Química y Biología, and Millennium Nucleus of Ion Channels-Associated Diseases (MiNICAD), Universidad de Santiago de Chile, Santiago, Chile

Gabriela González-Mariscal Centro de Investigación en Reproducción Animal, CINVESTAV-Universidad Autónoma de Tlaxcala, Tlaxcala, Tlax, Mexico

Agnès Gruart Division of Neurosciences, Pablo de Olavide University, Seville, Spain

Gaspar Herrera Centro Interdisciplinario de Neurociencia de Valparaíso and Instituto de Neurociencia, Facultad de Ciencias, Universidad de Valparaíso, Valparaíso, Chile

Marco A. Herrera-Valdez Departamento de Matemáticas, Facultad de Ciencias, Universidad Nacional Autónoma de México, Mexico City, DF, Mexico

Isabel Llona Laboratorio de Sistemas Neurales, Departamento de Biología, Facultad de Química y Biología, Universidad de Santiago de Chile, Santiago, Chile

Carlos Madrid Centro de Fisiología Celular e Integrativa, Facultad de Medicina, Clínica Alemana - Universidad del Desarrollo, Concepción, Chile

Rodolfo Madrid Departamento de Biología, Facultad de Química y Biología, and Millennium Nucleus of Ion Channels-Associated Diseases (MiNICAD), Universidad de Santiago de Chile, Santiago, Chile

Nicolás Marichal Neurofisiología Celular y Molecular, Instituto de Investigaciones Biológicas Clemente Estable, Montevideo, Uruguay

Angel I. Melo Centro de Investigación en Reproducción Animal, CINVESTAV-Universidad Autónoma de Tlaxcala, Tlaxcala, Tlax, Mexico

Consuelo Morgado-Valle Centro de Investigaciones Cerebrales, Universidad Veracruzana, Xalapa, Mexico

Patricio Orio Centro Interdisciplinario de Neurociencia de Valparaíso and Instituto de Neurociencia, Facultad de Ciencias, Universidad de Valparaíso, Valparaíso, Chile

María Eugenia Pallarés Instituto de Biología Celular y Neurociencia "Prof. Eduardo De Robertis", Facultad de Medicina, Universidad de Buenos Aires, Buenos Aires, Argentina

Fernando Peña-Ortega Departamento de Neurobiología del Desarrollo y Neurofisiología, Instituto de Neurobiología, Universidad Nacional Autónoma de México, UNAM-Campus Juriquilla, Querétaro, Mexico

María Pertusa Departamento de Biología, Facultad de Química y Biología, and Millennium Nucleus of Ion Channels-Associated Diseases (MiNICAD), Universidad de Santiago de Chile, Santiago, Chile

Ricardo Piña Departamento de Biología, Facultad de Química y Biología, and Millennium Nucleus of Ion Channels-Associated Diseases (MiNICAD), Universidad de Santiago de Chile, Santiago, Chile

Cecilia Reali Neurofisiología Celular y Molecular, Instituto de Investigaciones Biológicas Clemente Estable, Montevideo, Uruguay

María Inés Rehermann Neurofisiología Celular y Molecular, Instituto de Investigaciones Biológicas Clemente Estable, Montevideo, Uruguay

Carlos Restrepo Departamento de Biología, Facultad de Química y Biología, and Millennium Nucleus of Ion Channels-Associated Diseases (MiNICAD), Universidad de Santiago de Chile, Santiago, Chile

Raúl E. Russo Neurofisiología Celular y Molecular, Instituto de Investigaciones Biológicas Clemente Estable, Montevideo, Uruguay

Fernando Torrealba Departamento de Fisiología, Facultad de Ciencias Biológicas, Pontificia Universidad Católica de Chile, Santiago, Chile

Jennifer L. Troc-Gajardo Laboratorio de Sistemas Neurales, Departamento de Biología, Facultad de Química y Biología, Universidad de Santiago de Chile, Santiago, Chile

Omar Trujillo-Cenóz Neurofisiología Celular y Molecular, Instituto de Investigaciones Biológicas Clemente Estable, Montevideo, Uruguay

Gonzalo Ugarte Departamento de Biología, Facultad de Química y Biología, and Millennium Nucleus of Ion Channels-Associated Diseases (MiNICAD), Universidad de Santiago de Chile, Santiago, Chile

Rommy von Bernhardi Laboratorio de Neurociencias, Facultad de Medicina, Departamento de Neurología, P. Universidad Católica de Chile, Santiago, Chile

Chapter 1
What Is Neural Plasticity?

Rommy von Bernhardi, Laura Eugenín-von Bernhardi, and Jaime Eugenín

Abstract "Neural plasticity" refers to the capacity of the nervous system to modify itself, functionally and structurally, in response to experience and injury. As the various chapters in this volume show, plasticity is a key component of neural development and normal functioning of the nervous system, as well as a response to the changing environment, aging, or pathological insult. This chapter discusses how plasticity is necessary not only for neural networks to acquire new functional properties, but also for them to remain robust and stable. The article also reviews the seminal proposals developed over the years that have driven experiments and strongly influenced concepts of neural plasticity.

Keywords Neuronal plasticity • Homeostatic plasticity • Hebbian plasticity

Abbreviations

AMPAR	α-amino-3-hydroxy-5-methyl-4-isoxazolepropionic acid receptor
CaMKII	Ca^{2+}/calmodulin-dependent Kinase II
CREB	3′, 5′-cyclic adenosine monophosphate (cAMP) responsive element binding protein
ECM	Extracellular matrix
LTD	Long-term depression
LTP	Long-term potentiation

R. von Bernhardi, MD, PhD (✉)
Laboratorio de Neurociencias, Facultad de Medicina, Departamento de Neurología,
P. Universidad Católica de Chile, Santiago, Chile
e-mail: rvonb@med.puc.cl

L. Eugenín-von Bernhardi, MD
Servicio de Neurología, Hospital Sótero del Río, Santiago, Chile

J. Eugenín, MD, PhD (✉)
Laboratorio de Sistemas Neurales, Facultad de Química y Biología, Departamento de
Biología, Universidad de Santiago de Chile, USACH, PO 9170022, Santiago, Chile
e-mail: jaime.eugenin@usach.cl; jeugenin@gmail.com

© Springer International Publishing AG 2017
R. von Bernhardi et al. (eds.), *The Plastic Brain*, Advances in Experimental
Medicine and Biology 1015, DOI 10.1007/978-3-319-62817-2_1

NMDA   N-methyl-D aspartate
NMDAR   N-methyl-D-aspartate receptor
PSD   Postsynaptic density

Introduction

The brain's architecture arises from a combination of genetic blueprint (design) and self-organization (Kaiser 2007). More than a century has elapsed since the word "plasticity" was first used to propose that the brain is inherently modifiable (Berlucchi and Buchtel 2009). Plasticity is contrary to what is known as localizationism (Duffau 2016; Latash et al. 2000), by which the human brain operates as the sum of distinct parts, each performing a single function. In contrast, plasticity endows the brain with a very different and distinctive feature: the ability to build or adapt itself to variable or persistent demands. With plasticity the brain can use its history and experience.

Neuroplasticity involves a broad spectrum of changes at different levels of the nervous system's organization. This is exemplified in this special edition of the Advances in Experimental Medicine and Biology volume 1015, in which activity-dependent, developmental, and injury-dependent neuronal plasticity are addressed. Here in Chap. 1 we review seminal ideas that led to the current view of neural plasticity. More extensive treatment of varieties and mechanisms of plasticity, their historical development, and clinical research supporting the notion that the brain is a plastic structure may be found elsewhere (Berlucchi and Buchtel 2009; Doidge 2009; Bence and Levelt 2005; Merzenich et al. 2013; Huganir and Nicoll 2013; Feldman 2009; Sweatt 2016; Lisman 2017).

As discussed in Chaps. 2 (Belgacem and Borodinsky 2017), 3 (Barroso-Flores et al. 2017), 4 (Torrealba et al. 2017), and 5 (Delgado-García and Gruart 2017), activity-dependent plasticity is crucial for our current understanding of learning and memory. Chapter 2 describes how activity-dependent plasticity influences the levels of CREB, a transcription factor regulating the expression of crucial target genes (Belgacem and Borodinsky 2017). Chapters 3 and 5 address how neuronal activity influences short-term (Barroso-Flores et al. 2017) and long-term (Delgado-García and Gruart 2017) synaptic plasticity. In addition, a prominent example of plasticity in learning and memory tasks is provided by the interoceptive system as part of the acquisition of conditioned taste aversion, drug addiction, neophobia, and aversive memory, as discussed in Chap. 4 (Torrealba et al. 2017).

Adult and young brains are plastic, and a remarkable example of concomitant adult-young brain plasticity is the result of the interaction between the mother and offspring, as discussed in Chap. 6 (González-Mariscal and Melo 2017). During maternal care, mother and young affect each other's brains, an interesting process likely involved in adaptation and evolution in mammals (González-Mariscal and Melo 2017).

The environment can cause phenotypic changes in the brain during development. The somatostatinergic system, addressed in Chap. 8 (Llona et al. 2017), illustrates how brainstem innervation varies day-to-day during development. The possibility that the timing of appearance of emergent function in the respiratory neural network may be initiated by external clues is discussed in Chap. 10 (Beltrán-Castillo et al. 2017). On the other hand, Chap. 7 (Pallarés and Antonelli 2017) describes how susceptible brains exposed during development to stress conditions can generate abnormal cognitive, behavioral and psychosocial outcomes in postnatal life. Likewise, gestational modification of serotonin levels, although compatible with a relative normal postnatal basal ventilatory function, can lead to functional impairments manifested only during extreme physiological demands, as described in Chap. 11 (Bravo et al. 2017). In Chap. 9 (Morgado-Valle and Beltran-Parrazal 2017), an interesting example of dynamic reconfiguration of a robust neural network is provided by the respiratory system exposed to hypoxia, a mechanisms for survival in extreme conditions, as further discussed in Chap. 12 (Peña-Ortega 2017).

In Chaps. 13 (Marichal et al. 2017) and 14 (González et al. 2017), two examples of plasticity associated with injury are addressed. In contrast to the peripheral nervous system, injury to the central nervous system, such as the mammalian spinal cord, is followed by limited endogenous neuronal repair (Marichal et al. 2017). Injury can also trigger neuronal changes leading to pathological conditions (González et al. 2017). Marichal et al. describe properties of spinal cord stem cells as a first step toward overcoming limitations to neuronal repair and regeneration, whereas González et al. describe how the expression of ionic channels can change in response to neuronal damage and possibly cause pathological pain.

The Meaning of Plasticity

In colloquial terms, plasticity is referred to as the property of a material of being physically malleable; by analogy with plastics, the organic polymers of high molecular mass most commonly derived from petrochemicals, the property of being plastic alludes to the attribute of being able to be shaped without breaking. Etymologically, plasticity derives from the Greek "*plassein*," "to mold," which has two meanings: the capacity to receive form or the capacity to give form (Malabou 2008). Thus, plasticity makes the brain modifiable, i.e. "formable," while being formed (Malabou 2008). In other words, neuroplasticity can be defined "*as the ability of the nervous system to respond to intrinsic or extrinsic stimuli by reorganizing its structure, connections, and function*" (Cramer et al. 2011).

Plasticity occurs at various organizational levels of the nervous system. Thus, we can talk of nervous tissue plasticity, neuronal or glial plasticity, synaptic plasticity, etc.

Historic Milestones in Our Understanding of Brain Plasticity

The history of the study of brain plasticity can be subdivided into the "age of intuition" and the "age of experimental bases" (Fig. 1.1). During the age of intuition, plasticity was inferred from changes in behavior rather than from morphological, biochemical-molecular or electrophysiological evidence (see Box 1.1).

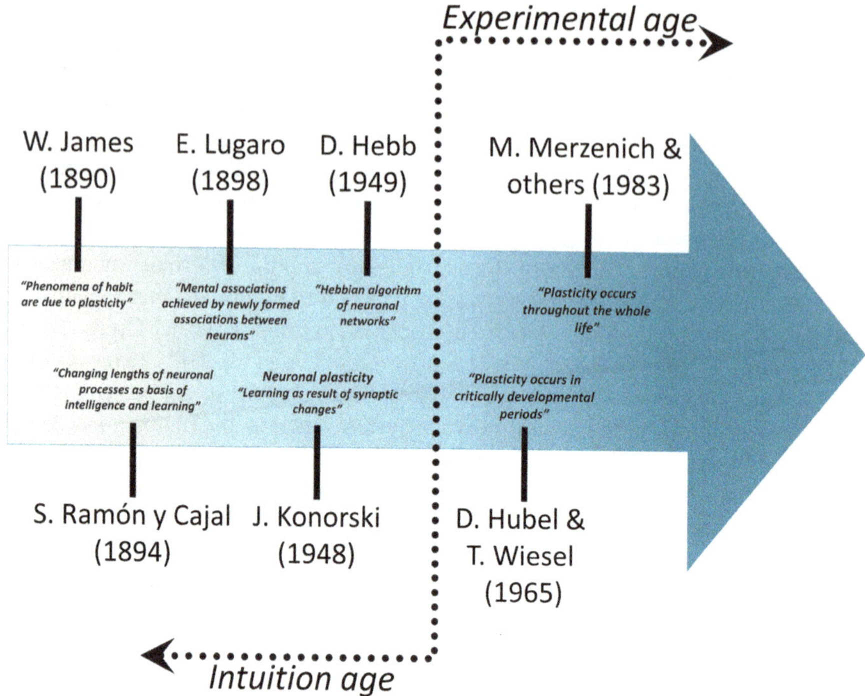

Fig. 1.1 Historic milestones in the concept of neuroplasticity. "*The age of intuition,*" during which brain plasticity was inferred from behavioral observations, preceded "*the age of experimental bases,*" that in addition used information from animal experiments, although the periods may overlap

Box 1.1 The Age of Intuition

The age of intuition started with William James (1842–1910), a North American physician, philosopher and psychologist. He is credited with being the first to express in his book *Principles of Psychology* (1890) the notion that the brain has plasticity: "*Organic matter, especially nervous tissue, seems endowed with a very ordinary degree of plasticity of this sort: so that we may without hesitation lay down as our first proposition the following, that the phenomena of habit in living beings are due to the plasticity of the organic materials of which their bodies are composed*" (Berlucchi and Buchtel 2009). He also proposed that plasticity emerges from the repeated use of neural paths, which leads to the acquisition of behavioral habits. Surprisingly, this was proposed when the "neuron doctrine," or "neuron theory", and the concept of synapse were not yet established (Berlucchi and Buchtel 2009).

Santiago Ramón y Cajal's work on the architecture of the nervous system was fundamental to the neuron theory (Azmitia 2007; Stahnisch and Nitsch

2002; Berlucchi and Buchtel 2009). Although Ramón y Cajal described neuronal changes that could be classified as examples of neuroplasticity, his concept of neuronal plasticity remained ambiguous in his manuscripts (Stahnisch and Nitsch 2002). In his initial writings, he speculated about the plasticity of brain neurons, which via the expansion and retraction of their processes can modify their inputs, as the basis of intelligence and the process of learning (1894). However, in later writings, he expressed the opposite view, favoring the idea of a static mature nervous system (Azmitia 2007).

The Italian neuropsychiatrist Eugenio Tanzi, influenced by Ramón y Cajal's work, proposed in 1893 that associative memories and practice-dependent motor skills may depend on a localized facilitation of neuronal contacts. Note that these "neuronal contacts" correspond to what several years later were experimentally supported and baptised as synapses by Charles Sherrington in his book *The Integrative Action of the Nervous System* (Levine 2007). Tanzi's disciple Ernesto Lugaro, another fervent admirer of Ramón y Cajal's work, expanded Tanzi's ideas, proposing in 1898 that mental associations may depend on newly formed associations between neurons, based on a coincidence of activity (Berlucchi and Buchtel 2009).

Jerzi Konorski, a Polish neurophysiologist, in his monograph *Conditioned Reflexes and Neuron Organization*, published in 1948, proposed that the central nervous system possesses two main properties, reactivity and plasticity (Berlucchi and Buchtel 2009). He inferred from his own studies of conditioned reflexes that learning is the result of morphological synaptic changes. He was the first to coin the term "neural plasticity" to denote the phenomenon through which the nervous system acquires new functions or behaviors as seen in conditioned reflexes. In his book, he proposed a set of remarkable hypotheses to explain how connections between neurons are modified. One of those proposals was that a prerequisite for establishing a conditioned reflex is the existence of potential or silent connections between cortical neurons. He proposed that synaptic changes occur when the neurons of an "emitting center" of "potential connections" are excited coincidentally with the neurons of a "receiving center". This coincidental activity would allow that "potential connections" be transformed into actual "excitatory connections" (Zielinski 2006). Later, Donald Hebb, a Canadian Psychologist, in his book *The Organization of Behavior: A Neurophysiological Theory* (1949) expressed his famous algorithm of learning in neural networks: "*When an axon of cell A is near enough to excite cell B and repeatedly or persistently takes part in firing it, some growth process or metabolic change takes place in one or both cells such that A's efficiency, as one of the cells firing B, is increased*" (Hebb 1949). Hebb also had the intuition that synaptic changes not only are expressed as reinforcement of weak connections, but they also can be expressed as synapse formation: "*When one cell repeatedly assists in firing another, the axon of the first cell develops synaptic knobs (or enlarges them if they already exist) in contact with the soma of the second cell*" (Hebb 1949).

Later, during the age of experimental bases, direct evidence of functional and structural plasticity in neuronal circuits was obtained (see Box 1.2).

Box 1.2 The Age of Experimental Bases

This era is represented by the work of many researchers. We will only mention some of the main contributions that had, in our opinion, a major pioneering influence in the field. David Hubel and Torsten Wiesel recorded the striate cortex activity maps in kittens after occluding only one eye for various periods from weeks to months of postnatal life (Hubel and Wiesel 1965). They found a clear decline in the number of cells in the kitten's visual cortex activated by visual stimulation of an eye that was previously closed for an early postnatal time (Hubel and Wiesel 1965). This result was well correlated with visual behavior. Kittens deprived from 1-to-4 months of age, when later forced to use only the deprived eye, behaved as if they were blind, whereas if they were allowed to use the non-deprived eye, they behaved as usual. Susceptibility to the monocular eye closure ranged from the fourth week to the end of the third month of life leading to the notion of a critical period (Constantine-Paton 2008). By contrast, monocular closure performed for over a year in an adult cat did not affect the number of striate cortex cells driven by that eye (Hubel and Wiesel 1970).

Until around the 1960–1970s, most of the neuroscientists considered the structure and function of the brain as being essentially fixed throughout adulthood. However, challenging results rattled the idea of a brain built on basis of a hardwired fixed circuitry. It was demonstrated that plasticity can occur beyond the critical period, including in adult brains (Kaas et al. 1983). After a lesion or silencing of sensory peripheral nerves, new patterns of somatosensory cortical activations can be recorded from adult mammals. Such reorganization depends on the somatotopic expansion of remaining, intact representations of body areas and, in some cases, a "nonsomatotopic" activation of the cortex from scattered receptive fields (Kaas et al. 1983). In the visual cortex, the onset of the critical period depends on the maturation of specific GABA circuits, and GABA pharmacological or genetic manipulation can initiate or delay the critical period (Le Magueresse and Monyer 2013). Conversely, during adulthood, expression of inhibitory transmission limits circuit rewiring (Takesian and Hensch 2013). Such limitations in the adult brain must be overridden in specific areas where the ability to incorporate newborn neurons into functional circuits is part of the normal physiology, as in the dentate gyrus of the hippocampus and the olfactory bulb (Kelsch et al. 2010).

Evidence supporting the notion that synaptic plasticity underlies learning was provided by clever experiments (Luco and Aranda 1964; Luco and Aranda 1966). During the 1970–1980s, the discovery of long-term potentiation (LTP) (Bliss and Lomo 1973; Huganir and Nicoll 2013) and long-term depression (LTD) (Ito et al. 1982; Collingridge et al. 2010) of excitatory synaptic transmission was the beginning of many important experimental and conceptual advances in understanding the mechanisms responsible for synaptic plasticity. For instance, hippocampal NMDAR-dependent LTP drew the attention of researchers because the LTP process appeared to be the biological correlate for Hebb's algorithm for the strengthening of functional network connections (Nicoll et al. 1988; Collingridge et al. 1983; Wigstrom et al. 1986; Malinow and Miller 1986; Malenka and Bear 2004; Mayer et al. 1984; Nowak et al. 1984; Jahr and Stevens 1987; Ascher and Nowak 1988; Lynch et al. 1983; Nicoll et al. 1988; Huganir and Nicoll 2013; Feldman 2009). Nowadays, at glutamatergic synapses, multiple plasticity processes are recognized beyond LTP and LTD (Lisman 2017).

Self-organizing Systems Use Modularity, Redundancy, and Degeneracy to Maintain Their Functional Integrity in a Changing and Unexpected Environment

Cornerstones of the structure of sensory and motor pathways are modularity, hierarchy, parallel and distributed processing. Brain structure must underlie such neural network properties as robustness (Kaiser 2007) and stability that make the brain's multiple functions reliable. Robustness is a critical attribute of functional networks, being defined as *"the ability to maintain performance in the face of perturbations and uncertainty"* (Stelling et al. 2004; Kitano 2007; Wagner 2005). Thus, a system increases in robustness when the effects of perturbations on performance become reduced.

Robustness can be achieved using such strategies as modularity, redundancy, or degeneracy (Gershenson 2012). *Modularity* is the organization of a system into fundamental building blocks or modules that are spatially or structurally isolated. Each module performs a discrete function. If one module is eliminated in a selective way, the remaining modules are enough to sustain the global function of the system, so this is not largely affected. It is thought that modules can be useful for preventing the propagation of damage in a system (Gershenson 2012). *Redundancy* infers that a same function is performed by multiple *identical* elements, endowing the brain with fail-safe or excess capacity for ensuring the persistence of a function even after the brain's partial destruction. *Degeneracy*, unlike redundancy, indicates that distinctly different elements may sustain the same function. Thus, despite the disappearance of one participating element, the function persists through the contribution of another.

It has been realized that robustness is applied to the maintenance of functions of a system, whereas stability and homeostasis are concerned with the maintenance of the system state (Kitano 2007; Wiggins 1990; Jen 2003). Figure 1.2 illustrates such distinction between both concepts..

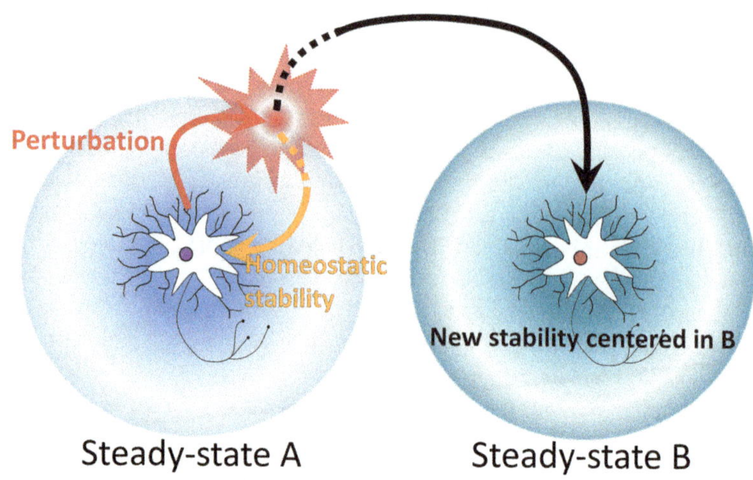

Fig. 1.2 Concept of stability and homeostasis. Let's assume that the initial state of a system is the steady state "*A*". When a perturbation (*red arrow*) drives the state of the system toward the boundary of *A*, the system has two alternatives: to return to the steady state "*A*", what is called stability or homeostasis (*orange arrows*) or to be derived to a new steady state "*B*", losing its stability around the attractor "*A*", and regaining stability around the new steady state "*B*". When the transition from steady state "*A*" towards to "*B*" is performed without loss of the system function, then the system is considered to be robust (Modified from Kitano 2007)

Self-organizing systems use combinations of modularity, redundancy, and degeneracy to work in a changing and unexpected environment. However, modularity, redundancy, and degeneracy *per se* are not enough to explain how the nervous system constantly adapts. Modifications of brain circuitry occur during development, after injury, or with changes in neural activity.

Plasticity and Fragility-Instability of a Neural Network

In general, brains can show plasticity in response to physiological demands, changes in neural activity, or damage of the nerve tissue. Additionally, plasticity is involved in formation of the network during development and the acquisition of new motor behaviors or learning throughout lifespan (Caroni et al. 2014).

The variety of biological processes involving neural plasticity includes neurogenesis, cell migration, changes in neuronal excitability and neurotransmission, the generation of new connections and modification of existing ones. Remodelling and refinement of connections use synapse formation and elimination, dendritic arborisation expansion and retraction, and axonal sprouting and pruning. In general, the biological processes perturb the neural network. Thus, an intriguing and crucial question is, how does a neural network remains functional while it is being remodelled? For example, neurogenesis, cell migration, and changes in ion channels, neurotransmitter receptors, or endogenous ligands could change a delicate balance and cause seizures or other malfunction during development. However, in infants under 5 years old, whose nervous system is under explosive growth and remodelling, the risk of seizure is rather small (Wong 2005). It has been argued that homeostatic processes occurring in individual cells of the network ensure network stability despite its plasticity (Perez-Otano and Ehlers 2005; Burrone and Murthy 2003; Marder and Prinz 2002; Marder and Prinz 2003; Turrigiano and Nelson 2000; Turrigiano et al. 1998; Davis and Bezprozvanny 2001; Davis and Goodman 1998; Davis 2006).

Structural and Functional Plasticity

Typically, the input-output relationship in neural networks is constantly regulated by activity-dependent processes. These processes include changes in the transmission efficacy of existing synapses or changes in the circuit connectivity by the formation and deletion of synapses. Thus, "transmission efficacy plasticity" may depend on adaptive changes in presynaptic, extracellular, or postsynaptic molecules. Thus, the plasticity may occur without modification of the number, site, distribution, density, or total area of synapses. Early long-term potentiation (Lisman 2017) and changes in the electrotonic properties due to geometrical changes of dendrites are clear examples of this kind of plasticity. In contrast, changes in the circuit connectivity involving formation, removal (Fauth and Tetzlaff 2016) or enlargement of synapses, as in late long-term potentiation (Lisman 2017), are called "structural or architectural plasticity" (Fauth and Tetzlaff 2016).

Hebbian and Homeostatic Plasticity

Transmission efficacy plasticity and the structural plasticity may be classified as Hebbian and homeostatic plasticity, respectively (Fauth and Tetzlaff 2016). Hebbian plasticity involves a change of the synaptic strength, either increasing or decreasing depending on the level of neuronal activity, on a time scale of seconds or minutes after stimulation onset. Early LTP is a typical example of Hebbian plasticity (Brown et al. 1990; Muller et al. 2002). In early LTP, a tetanic stimulus drives the coincident pre- and postsynaptic firing, inducing the increase in synaptic efficacy. This increase in the input drive will enhance potentiation. Thus, Hebbian synaptic plasticity may foster another Hebbian plastic event in a positive feedback loop. Without regulation, it would grow out of control. By contrast, homeostatic plasticity refers to the synaptic changes that counterbalance those induced by Hebbian plasticity. Homeostatic processes are slower (hours to days) than the Hebbian plasticity and can include modifications in ion channel density, transmitter release, or postsynaptic receptor sensitivity, triggered as responses to activity-dependent plastic changes (Marder and

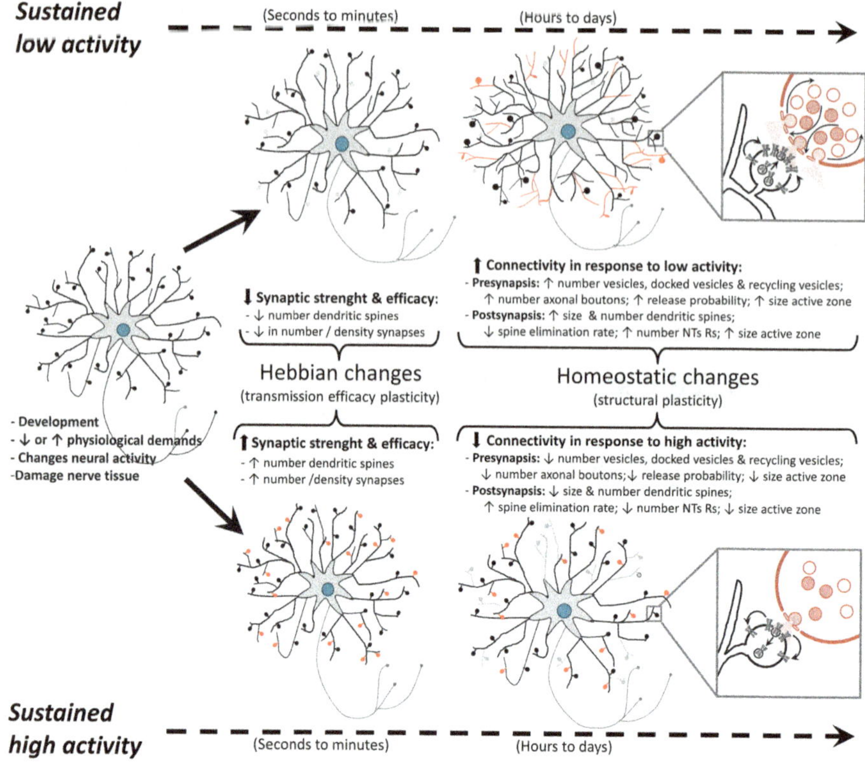

Fig. 1.3 Diagram of Hebbian and homeostatic plasticity. Increased neuronal activity strengthens synapses and can form new ones (Hebbian changes). If neuronal activity persists at a high level for longer, synapses may be removed and dendrites may retract, reducing the postsynaptic activity to lower basal values (homeostatic changes). In contrast, acute reduction of synaptic activity may reduce the number and extension of synapses (shrinkage) following Hebbian-type changes. Persistent low activity, may be compensated by increased number of synaptic sites and extension of the dendrite arborization (homeostatic changes) (Modified from Fauth and Tetzlaff 2016)

Prinz 2002; Davis and Bezprozvanny 2001; Davis 2006). In contrast to Hebbian plasticity, homeostatic plasticity constitutes a negative feedback loop. Thus, homeostatic dynamics decrease connectivity in response to high neuronal activity and increase in connectivity when activity drops (Turrigiano and Nelson 2000; Fauth and Tetzlaff 2016; Davis 2006) (Fig. 1.3). Synaptic scaling is one of the homeostatic mechanisms used by neurons (Lisman 2017), which detect changes in their own firing rates through molecular sensors such as calcium-dependent sensors (Turrigiano 2008; Turrigiano et al. 1998). The sensors can regulate receptor trafficking to scale the accumulation of neurotransmitter receptors at synaptic sites (Turrigiano 2008) and thus affect synaptic efficacy.

Transmission efficacy plasticity and structural plasticity are linked. Changes in neural activity can change the number or density of synapses. If the new level of activity persists, then homeostatic structural plasticity is activated to counterbalance initial functional and structural changes through addition or removal of synapses and by growth or retraction of the neurites (Fig. 1.3). Mechanisms underlying homeostatic plasticity can involve a variety of activity-dependent non-neural components. The brain extracellular matrix (ECM) appears to be a probable contributor. The ECM is a dynamic structure under constant activity-dependent remodeling, able to promote structural and functional plasticity (Acklin and Nicholls 1990; Frischknecht et al. 2014). In conjunction with neurons, glia is relevant for neuronal development and plasticity influencing neuronal survival, neuronal trophism, synapse remodeling and efficacy, among others contributions (von Bernhardi et al. 2016).

It has been proposed that Hebbian and homeostatic plasticities have different roles in terms of neural network functions. Hebbian plasticity is involved in life-long changes, storage capacity, and robustness of memory, whereas homeostatic plasticity self-organizes the connectivity of the neural network to avoid network instability (Fauth and Tetzlaff 2016).

Homeostatic plasticity involves synaptic and extra-synaptic mechanisms such as regulation of neuronal excitability, regulation of synapse formation, and stabilization of total synaptic strength and dendritic arborization (Yin and Yuan 2014; Turrigiano and Nelson 2004; Fauth and Tetzlaff 2016; Marder and Goaillard 2006). Such mechanisms should operate harmoniously with Hebbian mechanisms to enable experience to modify the properties of neuronal networks selectively (Turrigiano 1999).

Concluding Remarks

Neural plasticity can be observed during development of the nervous system and in adults. It emerges as an essential attribute that endows the brain with the ability to modify its structure and function in response to changes in neural activity and demand, as well as to acquire new capabilities as substrates for learning and memory or recovery of functionality after injury. Brain networks exhibit various forms of plasticity. Hebbian synaptic plasticity is a rapid response to changes in neural activity. Although Hebbian mechanisms are important for modifying neuronal circuitry selectively, they constitute a positive-feedback loop that destabilizes the activity of neuronal networks. For that reason, homeostatic plasticity plays a fundamental role in preserving robustness and stability in Hebbian processes. Homeostatic plasticity also stabilizes changes associated with developmental plasticity, learning, and injury.

Acknowledgment Grants Fondo Nacional de Desarrollo Científico y Tecnológico (FONDECYT) # 1171434 (JE) and 1171645 (RvB), DICYT-USACH (JE).

References

Acklin SE, Nicholls JG (1990) Intrinsic and extrinsic factors influencing properties and growth patterns of identified leech neurons in culture. J Neurosci 10(4):1082–1090

Ascher P, Nowak L (1988) The role of divalent cations in the N-methyl-D-aspartate responses of mouse central neurones in culture. J Physiol 399:247–266

Azmitia EC (2007) Cajal and brain plasticity: insights relevant to emerging concepts of mind. Brain Res Rev 55(2):395–405

Barroso-Flores J, Herrera-Valdez MA, Galarraga E, Bargas J (2017) Models of short term synaptic plasticity. Adv Exper Med Biol 1015

Belgacem YH, Borodinsky LN (2017) CREB at the crossroad of activity-dependent regulation of nervous system development and function. Adv Exper Med Biol 1015

Beltrán-Castillo S, Morgado-Valle C, Eugenín JL (2017) The onset of the fetal respiratory rhythm: an emergent property triggered by chemosensory drive? Adv Exper Med Biol 1015

Bence M, Levelt CN (2005) Structural plasticity in the developing visual system. Prog Brain Res 147:125–139. doi:10.1016/S0079-6123(04)47010-1

Berlucchi G, Buchtel HA (2009) Neuronal plasticity: historical roots and evolution of meaning. Exp Brain Res 192(3):307–319. doi:10.1007/s00221-008-1611-6

Bliss TV, Lomo T (1973) Long-lasting potentiation of synaptic transmission in the dentate area of the anaesthetized rabbit following stimulation of the perforant path. J Physiol 232(2):331–356

Bravo K, Eugenin J, Llona I (2017) Neurodevelopmental effects of serotonin on the brainstem respiratory network. Adv Exper Med Biol 1015

Brown TH, Kairiss EW, Keenan CL (1990) Hebbian synapses: biophysical mechanisms and algorithms. Annu Rev Neurosci 13:475–511. doi:10.1146/annurev.ne.13.030190.002355

Burrone J, Murthy VN (2003) Synaptic gain control and homeostasis. Curr Opin Neurobiol 13(5):560–567

Caroni P, Chowdhury A, Lahr M (2014) Synapse rearrangements upon learning: from divergent-sparse connectivity to dedicated sub-circuits. TINS 37(10):604–614. doi:10.1016/j.tins.2014.08.011

Collingridge GL, Kehl SJ, McLennan H (1983) Excitatory amino acids in synaptic transmission in the Schaffer collateral-commissural pathway of the rat hippocampus. J Physiol 334:33–46

Collingridge GL, Peineau S, Howland JG, Wang YT (2010) Long-term depression in the CNS. Nat Rev Neurosci 11(7):459–473. doi:10.1038/nrn2867

Constantine-Paton M (2008) Pioneers of cortical plasticity: six classic papers by Wiesel and Hubel. J Neurophysiol 99(6):2741–2744. doi:10.1152/jn.00061.2008

Cramer SC, Sur M, Dobkin BH, O'Brien C, Sanger TD, Trojanowski JQ, Rumsey JM, Hicks R, Cameron J, Chen D, Chen WG, Cohen LG, deCharms C, Duffy CJ, Eden GF, Fetz EE, Filart R, Freund M, Grant SJ, Haber S, Kalivas PW, Kolb B, Kramer AF, Lynch M, Mayberg HS, McQuillen PS, Nitkin R, Pascual-Leone A, Reuter-Lorenz P, Schiff N, Sharma A, Shekim L, Stryker M, Sullivan EV, Vinogradov S (2011) Harnessing neuroplasticity for clinical applications. Brain 134(Pt 6):1591–1609. doi:10.1093/brain/awr039

Davis GW (2006) Homeostatic control of neural activity: from phenomenology to molecular design. Annu Rev Neurosci 29:307–323. doi:10.1146/annurev.neuro.28.061604.135751

Davis GW, Bezprozvanny I (2001) Maintaining the stability of neural function: a homeostatic hypothesis. Annu Rev Physiol 63:847–869. doi:10.1146/annurev.physiol.63.1.847

Davis GW, Goodman CS (1998) Synapse-specific control of synaptic efficacy at the terminals of a single neuron. Nature 392(6671):82–86. doi:10.1038/32176

Delgado-García JM, Gruart A (2017) Learning as a functional state of the brain: studies in wild-type and transgenic animals. Adv Exper Med Biol 1015

Doidge N (2009) The brain that changes itself. Penguin Group, New York/Toronto/London/Dublin/Victoria/New Delhi/Johannesburg

Duffau H (2016) A two-level model of interindividual anatomo-functional variability of the brain and its implications for neurosurgery. Cortex 86:303–313. doi:10.1016/j.cortex.2015.12.009

Fauth M, Tetzlaff C (2016) Opposing effects of neuronal activity on structural plasticity. Front Neuroanat 10:75. doi:10.3389/fnana.2016.00075

Feldman DE (2009) Synaptic mechanisms for plasticity in neocortex. Annu Rev Neurosci 32:33–55. doi:10.1146/annurev.neuro.051508.135516

Frischknecht R, Chang KJ, Rasband MN, Seidenbecher CI (2014) Neural ECM molecules in axonal and synaptic homeostatic plasticity. Prog Brain Res 214:81–100. doi:10.1016/B978-0-444-63486-3.00004-9

Gershenson C (2012) Guiding the self-organization of random Boolean networks. Theor Biosci 131(3):181–191. doi:10.1007/s12064-011-0144-x

González A, Herrera G, Ugarte G, Piña R, Pertusa M, Orio P, Madrid R (2017) I_{KD} current in cold transduction and damage-triggered cold hypersensitivity. Adv Exper Med Biol 1015

González-Mariscal G, Melo AI (2017) Bidirectional effects of mother-young contact on the maternal and neonatal brains. Adv Exper Med Biol 1015

Hebb D (1949) The organization of behavior: a neurophysiological theory. Lawrence Erlbaum Associates, Mahwah/London

Hubel DH, Wiesel TN (1965) Binocular interaction in striate cortex of kittens reared with artificial squint. J Neurophysiol 28(6):1041–1059

Hubel DH, Wiesel TN (1970) The period of susceptibility to the physiological effects of unilateral eye closure in kittens. J Physiol 206(2):419–436

Huganir RL, Nicoll RA (2013) AMPARs and synaptic plasticity: the last 25 years. Neuron 80(3):704–717. doi:10.1016/j.neuron.2013.10.025

Ito M, Sakurai M, Tongroach P (1982) Climbing fibre induced depression of both mossy fibre responsiveness and glutamate sensitivity of cerebellar Purkinje cells. J Physiol 324:113–134

Jahr CE, Stevens CF (1987) Glutamate activates multiple single channel conductances in hippocampal neurons. Nature 325(6104):522–525. doi:10.1038/325522a0

Jen E (2003) Stable or robust? What's the difference? Complexity 8(3):12–18. doi:10.1002/cplx.10077

Kaas JH, Merzenich MM, Killackey HP (1983) The reorganization of somatosensory cortex following peripheral nerve damage in adult and developing mammals. Annu Rev Neurosci 6:325–356. doi:10.1146/annurev.ne.06.030183.001545

Kaiser M (2007) Brain architecture: a design for natural computation. Philos Trans R Soc A Mat Phys Eng 365(1861):3033–3045. doi:10.1098/rsta.2007.0007

Kelsch W, Sim S, Lois C (2010) Watching synaptogenesis in the adult brain. Annu Rev Neurosci 33:131–149. doi:10.1146/annurev-neuro-060909-153252

Kitano H (2007) Towards a theory of biological robustness. Mol Syst Biol 3:137. doi:10.1038/msb4100179

Latash LP, Latash ML, Meijer OG (2000) 30 years later: on the problem of the relation between structure and function in the brain from a contemporary viewpoint (1996), part II. Mot Control 4(2):125–149

Le Magueresse C, Monyer H (2013) GABAergic interneurons shape the functional maturation of the cortex. Neuron 77(3):388–405. doi:10.1016/j.neuron.2013.01.011

Levine DN (2007) Sherrington's "the integrative action of the nervous system": a centennial appraisal. J Neurol Sci 253(1–2):1–6. doi:10.1016/j.jns.2006.12.002

Lisman J (2017) Glutamatergic synapses are structurally and biochemically complex because of multiple plasticity processes: long-term potentiation, long-term depression, short-term potentiation and scaling. Philos Trans R Soc Lond Ser B Biol Sci 372(1715). doi:10.1098/rstb.2016.0260

Llona I, Farias P, Troc-Gajardo JL (2017) Early postnatal development of somastostatinergic systems in brainstem respiratory network. Adv Exper Med Biol 1015

Luco JV, Aranda LC (1964) An electrical correlate to the process of learning. Exp Blatta Orientalis. Nature 201:1330–1331

Luco JV, Aranda LC (1966) Reversibility of an electrical correlate to the process of learning. Nature 209:205–206

Lynch G, Larson J, Kelso S, Barrionuevo G, Schottler F (1983) Intracellular injections of EGTA block induction of hippocampal long term potentiation. Nature 305(5936):719–721

Malabou C (2008) What should we do with our brain? Fordham University Press, New York

Malenka RC, Bear MF (2004) LTP and LTD: an embarrassment of riches. Neuron 44(1):5–21. doi:10.1016/j.neuron.2004.09.012

Malinow R, Miller JP (1986) Postsynaptic hyperpolarization during conditioning reversibly blocks induction of long-term potentiation. Nature 320(6062):529–530. doi:10.1038/320529a0

Marder E, Goaillard JM (2006) Variability, compensation and homeostasis in neuron and network function. Nat Rev Neurosci 7(7):563–574. doi:10.1038/nrn1949

Marder E, Prinz AA (2002) Modeling stability in neuron and network function: the role of activity in homeostasis. BioEssays 24(12):1145–1154. doi:10.1002/bies.10185

Marder E, Prinz AA (2003) Current compensation in neuronal homeostasis. Neuron 37(1):2–4

Marichal N, Reali C, Rehermann MI, Trujillo-Cenóz O, Russo RE (2017) Progenitors in the ependyma of the spinal cord: a potential resource for self-repair after injury. Adv Exper Med Biol 1015

Mayer ML, Westbrook GL, Guthrie PB (1984) Voltage-dependent block by Mg2+ of NMDA responses in spinal cord neurones. Nature 309(5965):261–263

Merzenich MM, Nahum M, Van Vleet TM (2013) Neuroplasticity: introduction. Prog Brain Res 207:xxi–xxvi. doi:10.1016/B978-0-444-63327-9.10000-1

Morgado-Valle C, Beltran-Parrazal L (2017) Respiratory rhythm generation: the whole is greater than the sum of the parts. Adv Exper Med Biol 1015

Muller D, Nikonenko I, Jourdain P, Alberi S (2002) LTP, memory and structural plasticity. Curr Mol Med 2(7):605–611

Nicoll RA, Kauer JA, Malenka RC (1988) The current excitement in long-term potentiation. Neuron 1(2):97–103

Nowak L, Bregestovski P, Ascher P, Herbet A, Prochiantz A (1984) Magnesium gates glutamate-activated channels in mouse central neurones. Nature 307(5950):462–465

Pallarés ME, Antonelli MC (2017) Prenatal stress and neurodevelopmental plasticity: relevance to psychopathology. Adv Exper Med Biol 1015

Peña-Ortega F (2017) Neural network reconfigurations: the changes of the respiratory network by hypoxia as example. Adv Exper Med Biol 1015

Perez-Otano I, Ehlers MD (2005) Homeostatic plasticity and NMDA receptor trafficking. TINS 28(5):229–238. doi:10.1016/j.tins.2005.03.004

Stahnisch FW, Nitsch R (2002) Santiago Ramon y Cajal's concept of neuronal plasticity: the ambiguity lives on. TINS 25(11):589–591

Stelling J, Sauer U, Szallasi Z, Doyle FJ 3rd, Doyle J (2004) Robustness of cellular functions. Cell 118(6):675–685. doi:10.1016/j.cell.2004.09.008

Sweatt JD (2016) Neural plasticity & behavior – sixty years of conceptual advances. J Neurochem. doi:10.1111/jnc.13580

Takesian AE, Hensch TK (2013) Balancing plasticity/stability across brain development. Prog Brain Res 207:3–34. doi:10.1016/B978-0-444-63327-9.00001-1

Torrealba F, Madrid C, Contreras M, Gómez K (2017) Plasticity in the interoceptive system. Adv Exper Med Biol 1015

Turrigiano GG (1999) Homeostatic plasticity in neuronal networks: the more things change, the more they stay the same. TINS 22(5):221–227

Turrigiano GG (2008) The self-tuning neuron: synaptic scaling of excitatory synapses. Cell 135(3):422–435. doi:10.1016/j.cell.2008.10.008

Turrigiano GG, Nelson SB (2000) Hebb and homeostasis in neuronal plasticity. Curr Opin Neurobiol 10(3):358–364

Turrigiano GG, Nelson SB (2004) Homeostatic plasticity in the developing nervous system. Nat Rev Neurosci 5(2):97–107. doi:10.1038/nrn1327

Turrigiano GG, Leslie KR, Desai NS, Rutherford LC, Nelson SB (1998) Activity-dependent scaling of quantal amplitude in neocortical neurons. Nature 391(6670):892–896. doi:10.1038/36103 von Bernhardi R, Eugenin-von Bernhardi J, Flores B, Eugenin J (2016) Glial cells and integrity of the nervous system. Adv Exp Med Biol 949:1–24. doi:10.1007/978-3-319-40764-7_1

Wagner A (2005) Robustness and evolvability in living systems. In: Princeton Studies in Complexity, Levin SA, Strogatz SH (eds). Princeton University Press, Princeton

Wiggins S (1990) Introduction to applied nonlinear dynamical systems and chaos. Springer, New York. doi:10.1007/978-1-4757-4067-7

Wigstrom H, Gustafsson B, Huang YY, Abraham WC (1986) Hippocampal long-term potentiation is induced by pairing single afferent volleys with intracellularly injected depolarizing current pulses. Acta Physiol Scand 126(2):317–319. doi:10.1111/j.1748-1716.1986.tb07822.x

Wong M (2005) Advances in the pathophysiology of developmental epilepsies. Semin Pediat Neurol 12(2):72–87

Yin J, Yuan Q (2014) Structural homeostasis in the nervous system: a balancing act for wiring plasticity and stability. Front Cell Neurosci 8:439. doi:10.3389/fncel.2014.00439

Zielinski K (2006) Jerzy Konorski on brain associations. Acta Neurobiol Exp 66(1):75–84. discussion 85–90, 95–77

Part I

Neural Plasticity in Learning and Memory

Chapter 2
CREB at the Crossroads of Activity-Dependent Regulation of Nervous System Development and Function

Yesser H. Belgacem and Laura N. Borodinsky

Abstract The central nervous system is a highly plastic network of cells that constantly adjusts its functions to environmental stimuli throughout life. Transcription-dependent mechanisms modify neuronal properties to respond to external stimuli regulating numerous developmental functions, such as cell survival and differentiation, and physiological functions such as learning, memory, and circadian rhythmicity. The discovery and cloning of the cyclic adenosine monophosphate (cAMP) responsive element binding protein (CREB) constituted a big step toward deciphering the molecular mechanisms underlying neuronal plasticity. CREB was first discovered in learning and memory studies as a crucial mediator of activity-dependent changes in target gene expression that in turn impose long-lasting modifications of the structure and function of neurons. In this chapter, we review the molecular and signaling mechanisms of neural activity-dependent recruitment of CREB and its cofactors. We discuss the crosstalk between signaling pathways that imprints diverse spatiotemporal patterns of CREB activation allowing for the integration of a wide variety of stimuli.

Keywords Spatiotemporal integration • Crosstalk • Plasticity • Activity-dependent transcription factors • CREB cofactors

Y.H. Belgacem (✉)
INMED, Aix-Marseille Univ, INSERM, Marseille, France and Aix-Marseille Université, IMéRA, F-13000, Marseille, France
e-mail: yesser.belgacem-tellier@inserm.fr

L.N. Borodinsky
Department of Physiology & Membrane Biology and Institute for Pediatric Regenerative Medicine, University of California Davis School of Medicine and Shriners Hospital for Children Northern California, Sacramento, CA, USA

© Springer International Publishing AG 2017
R. von Bernhardi et al. (eds.), *The Plastic Brain*, Advances in Experimental Medicine and Biology 1015, DOI 10.1007/978-3-319-62817-2_2

Abbreviations

AC	Adenylate cyclase
ATF1	Activating transcription factor 1
BDNF	Brain-derived neurotrophic factor
b-zip	Basic leucine zipper domain
CaMKII/IV	Ca^{2+}/Calmodulin dependent Kinase II and IV
cAMP	$3',5'$-cyclic adenosine monophosphate
CaRE	Ca^{2+} Responsive Element
CBP	CREB Binding Protein
Cn	Calcineurin
CRE	cAMP-responsive elements
CREB	cAMP responsive element binding protein
CRTC	cAMP-regulated transcriptional coactivator
CREM	cAMP responsive element modulator ERK
DARPP	Dopamine and cAMP–regulated phosphoprotein
ERK	Extracellular signal-regulated kinase
IP3	Inositol 1,4,5-trisphosphate
KID	Kinase Inducible Domain
LTP	Long-term potentiation
MAPK	Mitogen-activated protein kinase
MSK-I	Mitogen/Stress Activated Kinase I
NGF	Nerve growth factor
NMDA	N-Methyl-D-Aspartate
NMDAR	N-Methyl-D-Aspartate ionotropic glutamate receptor
PDGF	Platelet-derived growth factor
PI3K	Phosphatidylinositol 3-kinase
PIP3	(3,4,5)-trisphosphate
PKA	cAMP-dependent Protein Kinase
PKB	Protein Kinase B
PKC	Protein Kinase C
pp90RSK	pp90 ribosomal S6 kinase
Shh	Sonic hedgehog
SIK1	Salt-inducible kinase 1
TORC	Transducer of regulated CREB
TRPC	Transient receptor potential canonical channel
VGCC	Voltage-gated Ca^{2+} channel

Introduction

During development of the nervous system, numerous cascades of transcription factors are the means of a genetic program governing a wide variety of critical events such as neural cell proliferation, migration, differentiation and synapse formation.

As mentioned in Chap. 1, until recently, the dogma has been that genetic programs determine the fate of different neural cells. However, emerging studies provided by multiple research groups indicate that the cell environment allows for plasticity in the process of neural cell specification. This phenomenon often involves activity-dependent mechanisms that add a homeostatic dimension to nervous system' development. Once development concludes, control of gene expression by activity-dependent mechanisms becomes a predominant feature of neurons. Interestingly, these mechanisms resemble those occurring during development.

In this chapter, we review prominent examples of activity-induced molecular mechanisms modulating gene expression in the developing and adult central nervous system. We particularly focus on 3′,5′-cyclic adenosine monophosphate (cAMP) responsive element binding protein (CREB) as the paradigmatic model of an activity-dependent transcription factor that participates in diverse processes of the developing and mature nervous system by regulating expression of crucial target genes.

We present some of the classical intracellular signaling cascades that transduce extracellular activity-dependent stimuli to CREB activation and the crosstalk among them. We also present the more recently discovered mechanisms controlling CREB expression and activity.

Molecular Structure of CREB

The CREB family of transcription factors is composed of several members, including CREB itself, the activating transcription factor 1 (ATF1), and the cAMP responsive element modulator CREM, among others (Altarejos and Montminy 2011; Flavell and Greenberg 2008; Mayr and Montminy 2001). There is a high level of redundancy between members of the CREB family that act as homo- or heterodimers to bind to cAMP-responsive elements (CRE) found in the regulatory regions of target genes (Altarejos and Montminy 2011; Flavell and Greenberg 2008; Mayr and Montminy 2001).

CREB is mainly activated or inactivated through phosphorylation or dephosphorylation of key serine amino acids, a mechanism characterized by a quick response rate leading to transcription of target genes, peaking 30 min to 1 h after stimulation (Michael et al. 2000). Several kinases have been shown to phosophorylate Ser133, such as cAMP-dependent Protein Kinase (PKA), Protein Kinase C (PKC), Akt, MAPKAP Kinase 2, Ca^{2+}/Calmodulin dependent Kinase II and IV (CaMKII/IV) and Mitogen/Stress Activated Kinase I (MSK-I) (Mayr and Montminy 2001). Ser133 is dephosphorylated by such phosphatases as serine/threonine phosphatases PP-1 (Bito et al. 1996; Alberts et al. 1994) and PP-2A (Wadzinski et al. 1993).

Once phosphorylated at Ser 133, CREB is able to recruit transcription coactivators such as the histone acetyltransferase CREB Binding Protein (CBP) and its paralogue p300; it thus upregulates transcription of target genes (Parker et al. 1996;

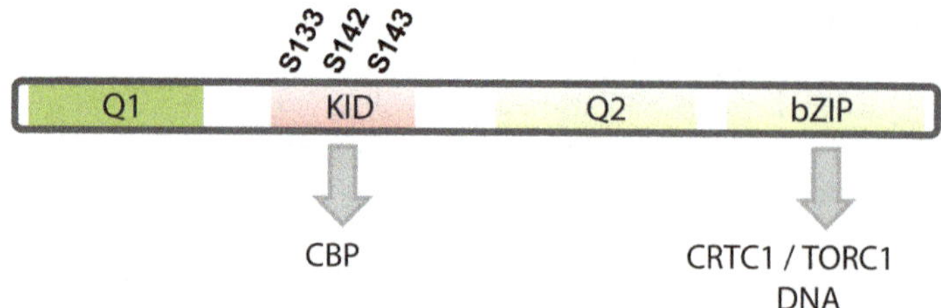

Fig. 2.1 CREB structure and function. Main domains of CREB are represented by *colored boxes*. The C-terminal basic leucine zipper domain (b-zip) is important for CREB homodimerization and for binding to the CRE element on target gene enhancer regions. The KID domain contains several serines (S) that, depending on their phosphorylation status, control the interaction between CREB and cofactors such as CBP. Two glutamine-rich domains (Q1 and Q2) are constitutive activators of transcription and interact with several co-activators and members of the transcriptional machinery. *Arrows* indicate the binding potential of KID and bZIP domains

Mayr and Montminy 2001). While Ser133 phosphorylation is the main way of activating CREB, other phosphorylation sites are important for the regulation of CREB activity (Kornhauser et al. 2002; Sakamoto et al. 2011) such as Ser142, which depending on the context, can activate or inhibit CREB's promotion of transcription (Wu and McMurray 2001; Gau et al. 2002).

CREB1 is a 43 kD soluble protein, which is constitutively expressed and is mainly found in the nucleus. It is a member of the basic leucine zipper domain (b-zip) transcription factor family (Mayr and Montminy 2001) (Fig. 2.1). The C-terminal b-zip domain is important for CREB homodimerization and for binding to the specific DNA palindromic consensus sequences of CRE "TGACGTCA" (Montminy and Bilezikjian 1987). However, CREB is capable, though with a five-fold reduced affinity, to bind on half CREs "CGTCA" (Fink et al. 1988; Craig et al. 2001). Phosphorylation of CREB does not seem to affect binding on CRE (Mayr and Montminy 2001). Methylation of the CpG present in CRE (Iguchi-Ariga and Schaffner 1989; Zhang et al. 2005) as well as the sequence of DNA surrounding CRE are important regulators of CREB binding on CREs (Connor and Marriott 2000; Mayr and Montminy 2001).

Upon binding to DNA and activation by phosphorylation on the Kinase Inducible Domain (KID), CREB recruits cofactors and members of the transcriptional machinery. The KID domain is surrounded by two glutamine-rich domains (Q1 and Q2) (Fig. 2.1) that are constitutive activators of transcription (Felinski et al. 2001; Brindle et al. 1993; Quinn 1993) and interact with several co-activators and members of the transcriptional machinery (Felinski and Quinn 1999).

While more than 750,000 potential binding sites for CREB have been identified in the human genome (Zhang et al. 2005), CREB stimulation leads to the expression of specific set of genes (Impey et al. 2004; Zhang et al. 2005), highlighting the importance of the regulation of CREB's action. This is done at multiple levels ranging from the type of ligand and signaling mechanism triggering CREB phosphory-lation/dephosphorylation, the cell type and subcellular localization of the transducing machinery, the presence of cofactors, and the methylation level of target DNA sequences.

Mechanisms of CREB Activation

Researchers in the field of learning and memory have been major contributors to the discovery of how electrical activity is translated into gene expression. Stimulation of sensory neurons activates specific neuronal networks through synaptic connections that, under specific conditions, are reinforced and stabilized in time (for reviews see Kandel et al. 2014; Flavell and Greenberg 2008).

The cAMP/PKA Axis

The first described signaling pathway involving CREB as playing a role in central nervous system physiology was discovered while studying learning and memory. The studies done on the mollusk *Aplysia californica* identified a model for simple forms of procedural memories such as habituation, dishabituation and sensitization (Kandel et al. 2014). Aplysia has a simple reflex called the gill withdrawal reflex: if the siphon of the animal is mechanically stimulated, sensory neurons innervating motor neurons trigger the withdrawal of the gill. Sensitization occurs if, before stimulating the siphon, an electric shock is applied to the tail of the animal, resulting in a stronger depolarization of motor neurons during the gill withdrawal reflex. Eric Kandel's group discovered that this sensitization is due to a release of serotonin by modulatory neurons, resulting in synthesis of cAMP in sensory neurons (Brunelli et al. 1976) thus activating PKA (Castellucci et al. 1980). Other groups also identified the cAMP signaling axis as a component in various forms of learning and memory in *Drosophila melanogaster* (Dudai et al. 1976; Byers et al. 1981). Interestingly, when a single shock is applied to the Aplysia's tail, the sensitization does not last more than few hours. However, if multiple shocks are delivered, the sensitization lasts several days, suggesting that a long-term memory has been acquired. This long term facilitation is dependent on mRNA and protein synthesis (Montarolo et al. 1986). The mechanism linking cAMP and activity to gene transcription in sensory neurons during long-term facilitation remained elusive until Montminy and colleagues' breakthrough while investigating the mechanisms by which cAMP regulates somatostatin gene transcription. By analyzing the regulatory sequences of the somatostatin gene, they found a short palindromic sequence (5′-TGACGTCA-3) responsive to cAMP (CRE) that is highly conserved in the regulatory regions of other genes for which transcription is controlled by cAMP (Montminy et al. 1986). Shortly after, Marc Montminy and Louise Bilezikjian identified a nuclear protein that binds to CRE and named it CREB (Montminy and Bilezikjian 1987). Using an elegant approach, Dash and colleagues (Dash et al. 1990) injected oligonucleotides coding for CRE and abolished long-term but not short-term facilitation in *Aplysia* sensory neurons, suggesting a prominent role for CREB in transducing activity into gene transcription necessary for long-term memory. In summary, the serotonin receptor, a G-protein-coupled membrane receptor recruits an adenylate cyclase (AC), leading to production of cAMP and activation of PKA that phosphorylates CREB on the KID domain, particularly at the Ser133, activating it (Figs. 2.1 and 2.2).

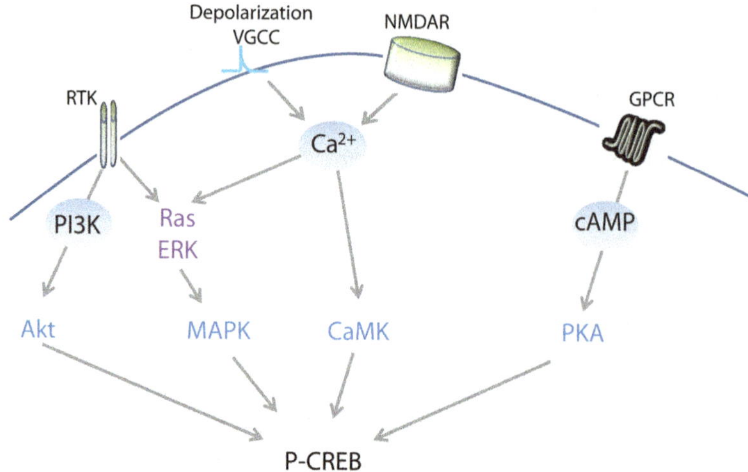

Fig. 2.2 Classical signaling pathways that lead to CREB activation. Membrane depolarization (voltage-gated Ca²⁺ channels, VGCC) and glutamate receptors (NMDAR) lead to cytosolic and nuclear Ca²⁺ elevations recruiting the CaMK or Ras/MAPK signaling pathways. G-protein-coupled receptors (GPCR) activate the cAMP/PKA axis, while Receptor Tyrosine Kinases (RTKs) generally recruit the Ras/MAPK or Akt signaling cascades

Calcium as Second Messenger

Ca^{2+} is a very important second messenger implicated in the transduction of numerous signaling pathways and playing a wide variety of roles in the central nervous system development and physiology (Bito and Takemoto-Kimura 2003; Flavell and Greenberg 2008; Ghosh et al. 1994; Carlezon et al. 2005; Lonze and Ginty 2002; Mayr and Montminy 2001; Sakamoto et al. 2011; Rosenberg and Spitzer 2011; Spitzer 2006). Increases in cytosolic Ca^{2+} concentration occur through a wide variety of mechanisms (Ghosh et al. 1994; Kornhauser et al. 1990; Averaimo and Nicol 2014; Brini et al. 2014), including voltage-gated Ca^{2+} channels (VGCCs) upon depolarization, and ligand-gated Ca^{2+} channels such as the N-Methyl-D-Aspartate (NMDA) ionotropic glutamate receptor (NMDAR), which upon glutamate binding allows Ca^{2+} influx (Fig. 2.2). Transient increases in Ca^{2+} are followed by recruitment of different signaling pathways, which will modify the activation status of CREB.

Ca²⁺/CaMK Axis

Long-term synaptic plasticity is triggered by depolarization of the postsynaptic membrane and, consequently, transcription of immediate early genes (Flavell and Greenberg 2008; Kandel et al. 2014; Mayr and Montminy 2001; Shaywitz and Greenberg 1999). The Greenberg and Kandel groups simultaneously discovered the mechanisms controlling the immediate early gene *c-fos* expression by membrane depolarization (Sheng et al. 1990; Dash et al. 1991; Sheng et al. 1991). Sheng and colleagues first found that in the rat pheochromocytoma cell line PC12, c-fos responds to potassium chloride-induced depolarization through a cis-regulating element present in the c-fos promoter (Sheng et al. 1990). This element, called CaRE (for $\underline{Ca}^{2+}$ $\underline{R}$esponsive $\underline{E}$lement), is responsive to depolarization and Ca^{2+} influx through voltage-gated Ca^{2+} channels. CaRE (-TGACGTTT-) is very similar to CRE (-TGACGTCA-) as it is recognized by cAMP/PKA-activated CREB (Sheng et al. 1990). In PC12 cells, depolarization and Ca^{2+} influx do not activate the cAMP/PKA axis, suggesting that another kinase transduces the depolarizing signal to CREB (Sheng et al. 1990). Instead, CREB activation is mediated by CaMKI and II (Sheng et al. 1991) (Fig. 2.2). This depolarization/Ca^{2+}/CaMK axis is independent but synergistic to the effects of the cAMP/PKA axis on CREB activation and subsequent c-fos transcription (Sheng et al. 1990, 1991).

Ca^{2+}/MAPK Axis

Another source of Ca^{2+} influx is the set of ligand-controlled Ca^{2+} channel receptors including the NMDAR. Ginty and collaborators (Ginty et al. 1993) discovered that these receptors are capable of activating CREB while studying the molecular basis of circadian rhythms in the suprachiasmatic neurons of the hypothalamus. The pacemaker cells in this structure regulate the circadian rhythms of the whole organism. Rhythms can be shifted by inputs of environmental information such as light: during the subjective night, a pulse of light activates retinal ganglionic cells that project axons to the suprachiasmatic nuclei establishing glutamatergic synapses. Interestingly, this signal triggers expression of immediate early genes such as c-fos, similarly to what is observed in long-term memory. Taking this analogy in consideration, Ginty et al. (1993) investigated the possibility that CREB transduces the signal of light pulses to the transcription of the early gene c-fos. They first isolated an antibody recognizing specifically phosphorylated CREB at the Ser133 and then, found, in vivo, that a pulse of light during the subjective night induces rapid CREB phosphorylation at Ser133 (Ginty et al. 1993). Trying to understand the mechanism responsible for the light-induced CREB phosphorylation, they found that in vitro, a 7-min NMDA incubation or depolarization induces a very strong phosphorylation of CREB at Ser133. Furthermore, the effect observed with NMDA but not with depolarization was prevented by APV, a potent NMDAR antagonist (Ginty et al. 1993). These results suggest that NMDARs are capable of activating CREB through a mechanism other than depolarization. It has been shown that an increase in cytosolic Ca^{2+} concentration is responsible for the recruitment of the $\underline{e}$xtracellular signal–$\underline{r}$elated protein $\underline{k}$inase/$\underline{m}$itogen-$\underline{a}$ctivated $\underline{p}$rotein $\underline{k}$inase (ERK/MAPK) pathway and subsequent CREB phosphorylation (Impey et al. 1998; Rosen et al. 1994). Obrietan et al. (1998) later showed that CREB activation by NMDAR in the suprachiasmatic nucleus, is, at least partially, mediated through the Ras/ERK/MAPK signaling pathway (Obrietan et al. 1998) (Fig. 2.2).

Receptor Tyrosine Kinases

Ras /MAPK

Neurotrophins and their tyrosine kinase receptors play important roles in the central nervous system development and homeostasis. Nerve growth factor (NGF) is a member of this family that induces expression of immediate early genes, particularly c-fos. Greenberg's group, using PC12 cells, investigated how NGF activates c-fos transcription (Ginty et al. 1994). They ruled out the implication of cAMP/PKA and Ca^{2+}/CamK axes. Instead, by using an inducible dominant-negative mutant of Ras, they identified a 105 kD CREB kinase (Ras-dependent p105 kinase) under the control of Ras that transduces the NGF-dependent phosphorylation of CREB on Ser 133 (Ginty et al. 1994) (Fig. 2.2). Xing et al. (1996) later identified the Ras-dependent p105 kinase as RSK2 (Xing et al. 1996), a member of the pp90 ribosomal S6 kinase (pp90RSK) family. Numerous extracellular signals mediated via receptor tyrosine kinases trigger MAP kinases and subsequently CREB kinases, such as MAPKAP-K2/3, MSK1 or MSK 1–3 (Shaywitz and Greenberg 1999; Flavell and Greenberg 2008; Carlezon et al. 2005; Lonze and Ginty 2002; Mayr and Montminy 2001; Sakamoto et al. 2011).

PI3K/Akt-PKB

The serine/threonine protein kinase Akt is mainly known for its anti-apoptotic and cell survival roles following growth factor stimulation. Upon binding to their receptors, growth factors trigger a signaling cascade involving the phosphatidylinositol 3-kinase (PI3K), which synthesizes the second messenger phosphatidylinositol (3,4,5)-trisphosphate (PIP3) at the plasma membrane. Akt (also called Protein kinase B (PKB)) is consequently recruited and docked to the membrane, where it is activated. Akt is then translocated to the cytosol and then to the nucleus to phosphorylate its targets (see Manning and Cantley 2007 for review). Du and Montminy demonstrated in 293 T cells that serum induces CREB phosphorylation at Ser 133 via recruitment of the PI3K/Akt axis, thus promoting cell survival (Du and Montminy 1998). Later, the PI3K/Akt/CREB signaling was also implicated in cell survival in the central nervous system (Chong et al. 2005) (Fig. 2.2).

Crosstalk Between CREB-Activating Pathways

Since the identification of the classical signaling cascades described previously in this chapter, numerous studies in multiple models discovered that combinations of these pathways work together, whether in parallel, synergistically or in an antagonistic manner to modulate CREB phosphorylation and activation.

CREB-induced immediate early genes are important for long-term potentiation (LTP), and both cAMP and Ca^{2+} second messengers play an important role in activating CREB (Impey et al. 1996). Moreover, the CRE element present in immediate early gene promoters is responsive to both Ca^{2+} and cAMP in a synergistic manner (Deutsch et al. 1987; Sheng et al. 1990; Impey et al. 1994). Late phase LTP (L-LTP) is a good example of the dual action of Ca^{2+} and cAMP (Impey et al. 1996). Indeed, Ca^{2+} induces phosphorylation of CREB at Ser133, an event necessary but not sufficient to promote sustained transcription of target genes (Brindle et al. 1995; Wagner et al. 2000; Ravnskjaer et al. 2007; Ginty et al. 1994; Impey et al. 1996). Impey and colleagues discovered that in PC12 cells and hippocampal neurons, Ca^{2+} influx recruits the ERK signaling cascade leading to CREB phosphorylation at Ser133. Remarkably, cAMP-induced PKA activation is mandatory for translocation of activated ERK in the nucleus where it promotes CREB phosphorylation. These studies demonstrate crosstalk between the cAMP-PKA and Ca^{2+}-MAPK signaling pathways in neurons important for L-LTP (Impey et al. 1998).

Another example of crosstalk between signaling pathways has come from a study on brain-derived neurotrophic factor (BDNF) signaling in neurons. In cultured cortical neurons and hippocampal slices, BDNF induces release of Ca^{2+} from intracellular stores and subsequent activation of CaMKIV. In parallel, BDNF triggers the Ras/ERK/RSK pathway. Both the Ca^{2+}/CaMKIV and Ras/ERK signaling lead to CREB phosphorylation at Ser133 (Finkbeiner et al. 1997). The study suggests that the two signaling pathways may induce different spatio-temporal dynamics for CREB activation given the distinct kinetics and localization of the Ras pathway and the IP3-induced mobilization of intracellular Ca^{2+} (Finkbeiner et al. 1997).

Platelet-derived growth factor (PDGF) promotes a neuroprotective action against toxicity induced by HIV-1 in primary midbrain neurons and in vivo on dopaminergic neurons of the *substantia nigra*, also by a crosstalk between pathways converging in CREB. PDGF activates the PI3K/Akt and Ca^{2+}/ERK signaling through IP3-mediated intracellular Ca^{2+} release and Ca^{2+} influx through transient receptor potential canonical (TRPC) channels leading to CREB phosphorylation (Yao et al. 2009). Whether both pathways are additive in CREB activation remains unclear.

Another example of signaling crosstalk converging on CREB activation comes from our recent study on the Sonic hedgehog (Shh) pathway in the developing *Xenopus laevis* spinal cord (Belgacem and Borodinsky 2015). Shh stimulation of embryonic spinal neurons elicits a non-canonical, Ca^{2+} spike-dependent pathway that regulates neurotransmitter specification (Belgacem and Borodinsky 2011). Shh activation of its coreceptor Smoothened, recruits a phospholipase C, inducing inositol 1,4,5-trisphosphate (IP3) oscillations in the primary cilium of embryonic spinal neurons and enhancing Ca^{2+} spike activity that relies on both IP3 receptor (IP3R)-regulated intracellular Ca^{2+} stores and Ca^{2+} influx through voltage-gated Ca^{2+} channels and TRPC1 (Belgacem and Borodinsky 2011) (Fig. 2.3). This Ca^{2+} spike-dependent, non-canonical Shh pathway results in PKA activation, which activates CREB that, in turn, represses gli1 transcription and contributes to the switch off of the Shh canonical, Gli transcription factor-mediated pathway during spinal cord development (Belgacem and Borodinsky 2015) (Fig. 2.3).

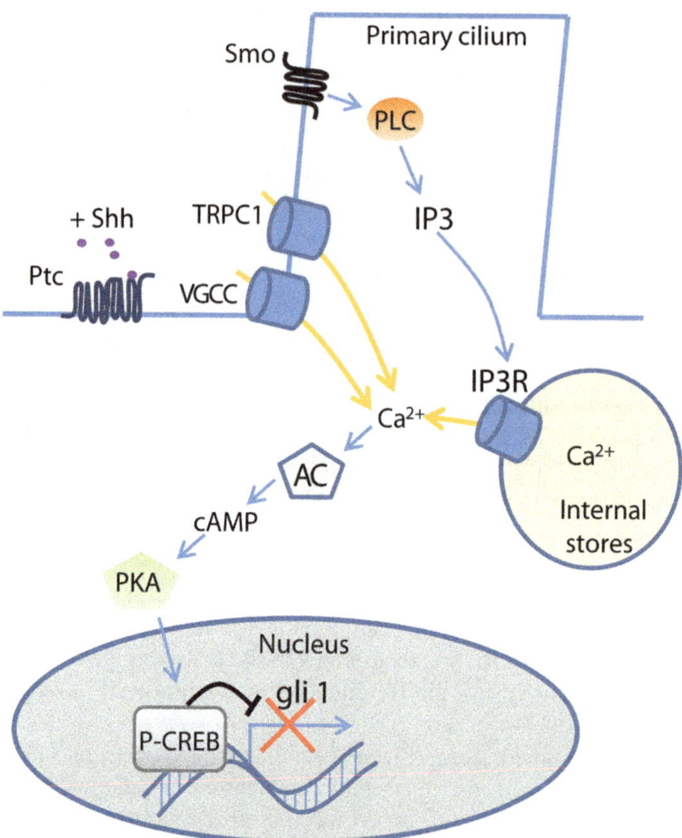

Fig. 2.3 CREB contributes to the Shh-calcium signaling axis-dependent switch off of the canonical Shh pathway in the embryonic spinal cord. In embryonic spinal neurons, Smo-dependent Shh signaling activates the synthesis of IP3 second messenger at the primary cilium. IP3 transients trigger the release of intracellular Ca^{2+} from intracellular stores that, in conjunction with Ca^{2+} influx through voltage-gated Ca^{2+} channels (VGCC) and transient receptor potential canonical (TRPC) channels lead to activation of Ca^{2+}-sensitive adenylate cyclase (AC) and an increase in cAMP levels. In turn, this leads to PKA-dependent CREB phosphorylation at serine 133. CREB represses expression of the canonical Shh transcription factor Gli1

Activation of CREB Cofactors by Activity Dependent Signaling Pathways: Consequences for CREB Function

CREB phosphorylation on key amino acids is a crucial step in its activation and recruitment of cofactors allowing for specific gene transcription. Activity-dependent signaling controls all the steps from CREB phosphorylation to recruitment and activation of the cofactors. In this section, we will focus particularly on two important CREB cofactors: CBP and TORC/CRTC1.

CBP

The most studied site of phosphorylation by serine/threonine kinases is the Ser133 that controls the binding of CREB to CBP/p300 (Gonzalez et al. 1989; Gonzalez and Montminy 1989; Altarejos and Montminy 2011; Parker et al. 1996). However, it has been observed that growth factor and stress pathways, while promoting CREB phosphorylation at Ser133, do not promote gene transcription as efficiently as the cAMP-induced Ser133 phosphorylation does (Mayr and Montminy 2001; Hardingham et al. 2001; Chawla et al. 1998; Altarejos and Montminy 2011; Brindle et al. 1995; Ravnskjaer et al. 2007; Bito et al. 1996; Mayr et al. 2001). This difference could be due to subcellular events such as the presence of nuclear Ca^{2+} waves as observed in hippocampal neurons (Hardingham et al. 2001) or the activation of other factors that could act as positive or negative regulators and that are preferentially activated by Ras or cAMP (discussed in (Mayr and Montminy 2001)). For instance, Chawla et al. (1998) in mouse pituitary cell line AtT20 show that CBP is activated by nuclear Ca^{2+}, CaMKIV and cAMP (Chawla et al. 1998) (Fig. 2.4). Receptor tyrosine kinase-mediated growth factors and stress pathways that do not elevate intracellular Ca^{2+} and cAMP might then fail to efficiently recruit CBP and, consequently, prevent the phosphorylated CREB from promoting gene transcription.

Activity-dependent signaling can negatively or positively regulate CREB activity by recruiting CaMKII or CaMKIV, respectively (Matthews et al. 1994). Sun et al. (1994) investigated the mechanisms responsible for this difference. They demonstrated, in vitro in GH3 pituitary tumor cells using mutagenesis studies and phosphopeptide mapping analysis, that, while both CaMKII and IV phosphorylate CREB at Ser133, CaMKII also phosphorylates Ser142 (Fig. 2.1) (Sun et al. 1994). Additionally, they showed that Ser 142 has an inhibitory effect on CREB-induced gene transcription. Wu and McMurray (2001) later showed in human neuroblastoma cells (SK-N-MC) and African green monkey kidney cells (CV-1) that phosphorylation of Ser142 by CaMKII prevents CREB dimerization and binding to CBP (Wu and McMurray 2001), explaining its negative effect on CREB transcriptional capability (Sun et al. 1994). These results have been confirmed in vivo (Flavell and Greenberg 2008; Carlezon et al. 2005; Lonze and Ginty 2002; Mayr and Montminy 2001; Sakamoto et al. 2011), highlighting the importance of CREB phosphorylation status, as well as the fact that Ca^{2+} can be a positive or a negative regulator of CREB activity depending on the subtype of CaMK recruited in the cell.

Although in general Ser142 phosphorylation inhibits CREB-dependent gene transcription, Kornhauser and colleagues showed that in rat cortical neurons in vitro and in vivo, Ca^{2+} influx through VGCCs or NMDAR triggers CREB triple phosphorylation at Ser133, 142 and 143, promoting CREB transcriptional action (Kornhauser et al. 2002). Strikingly, this triple phosphorylation disrupts CBP binding on CREB, suggesting that CBP is not always necessary to transduce CREB-dependent signaling.

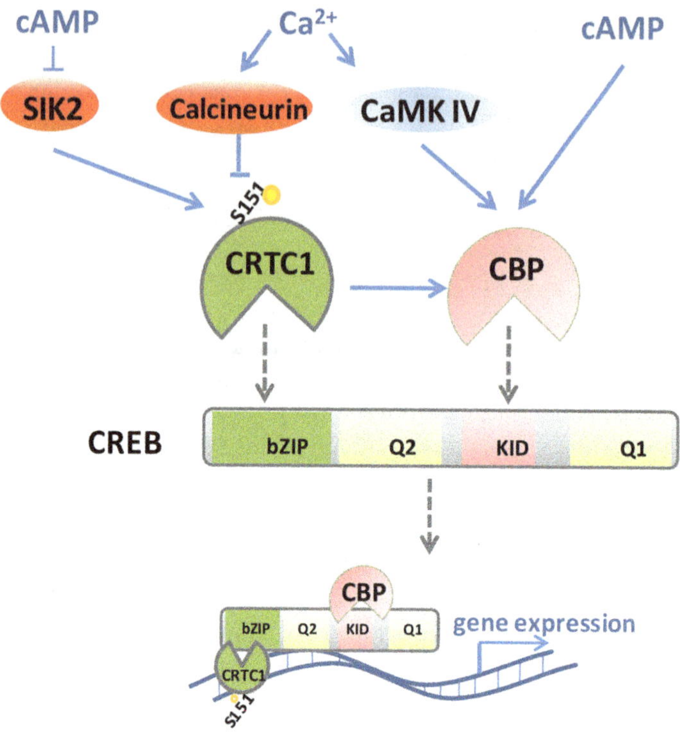

Fig. 2.4 Model of activity-dependent recruitment of CREB co-activators. Binding of cofactors to CREB and DNA is modulated by Ca^{2+} and cAMP second messengers. Activity-induced signaling leads to increases in cAMP and Ca^{2+}, which inhibit Salt-induced kinase 2 (SIK2) and activate calcineurin. This results in dephosphorylation of serine 151 (S151) on the cofactor CRTC1 and its translocation to the nucleus, where it promotes binding of CREB to the TFIID complex. In parallel, Ca^{2+}-activated CaMKIV and cAMP lead to CBP recruitment and a reciprocal synergism between CBP and CRTC1

TORC/CRTC1

Another important CREB coactivator called cAMP-regulated transcriptional coactivator (CRTC) or transducer of regulated CREB (TORC) was discovered simultaneously by two groups (Iourgenko et al. 2003; Conkright et al. 2003). Three genes code for this evolutionarily conserved cofactor: CRTC1–3, with CRTC1 as the prevalent isoform found in the brain. Contrary to CBP, CRTC binds to the bZIP domain of CREB (Altarejos and Montminy 2011; Xue et al. 2015). Under basal conditions, TORC1 phosphorylated at Ser 151 is maintained in the cytosol by 14–3-3 proteins (Screaton et al. 2004). Upon Ca^{2+} influx in the cell, calcineurin dephosphorylates TORC and, thus, allows its translocation to the nucleus, where it promotes binding of CREB to the TFIID complex, enhancing CREB DNA binding activity independently of CREB's phosphorylation status (Bittinger et al. 2004; Screaton et al. 2004). cAMP elevation also leads to CRTC dephosphorylation and nuclear translocation by inhibiting the TORC2 kinase SIK2 (Screaton et al. 2004) (Fig. 2.4).

TORC is essential for CRE-dependent transcription of numerous genes triggered by Ca^{2+} or cAMP elevation (Altarejos and Montminy 2011; Xue et al. 2015). For instance, TORC plays a crucial role in long term memory: Zhou et al. (2006) showed that, in hippocampus, neuronal activity elicits CRTC1 translocation to the nucleus, leading to CRE-mediated gene expression necessary for L-LTP (Zhou et al. 2006). Interestingly, Kovacs et al. (2007) later showed that CRTC1 was acting as an integrator of neuronal activity by detecting the coincidence of Ca^{2+} and cAMP increases in hippocampal neurons, thus leading to activation of the genetic machinery responsible of L-LTP (Kovacs et al. 2007) (Fig. 2.4). These studies illustrate the role of signaling pathway crosstalk in CREB function via regulating its cofactors.

CBP and TORC Interaction

CBP and TORC2 interact in an activity-dependent synergistic way. cAMP signaling that leads to TORC dephosphorylation and translocation to the nucleus also promotes its association with CBP/p300 and subsequently increases CBP occupancy on promoters of target genes (Ravnskjaer et al. 2007). Interestingly, CBP/p300 has a reciprocal effect on TORC2 recruitment. This interaction participates in the specificity of CRE-driven gene expression by favoring cAMP-dependent signaling over other pathways such as the stress cascade (Ravnskjaer et al. 2007) (Fig. 2.4).

Spatiotemporal Patterns of CREB Activation

Regulation of gene expression through CREB is fast and transitory; for instance, in PC12 cells cAMP-induced somatostatin expression peaks 15–30 min after stimulation and goes back to baseline after 4 h (Hagiwara et al. 1992). The decay in CREB activity is controlled by specific serine/threonine protein phosphatases such as protein phosphatase PP1 and PP2A that dephosphorylate CREB at Ser133 (Flavell and Greenberg 2008). The duration of CREB activation is key for the efficiency of target gene expression. Bito et al. (1996) showed in hippocampal neurons that a short stimulation of synaptic activity was translated into a transient CREB phosphorylation that was not sufficient to cause gene expression (Bito et al. 1996). This might be due to activation of CaMKIV, which phosphorylates CREB, followed by calcineurin (Cn) or PP2B, a Ca^{2+}/calmodulin-dependent serine/threonine phosphatase, and PP1-mediated CREB dephosphorylation (Fig. 2.5a). However, when synaptic stimulation is prolonged, the effects of Cn/PP1 are strongly reduced, allowing for a

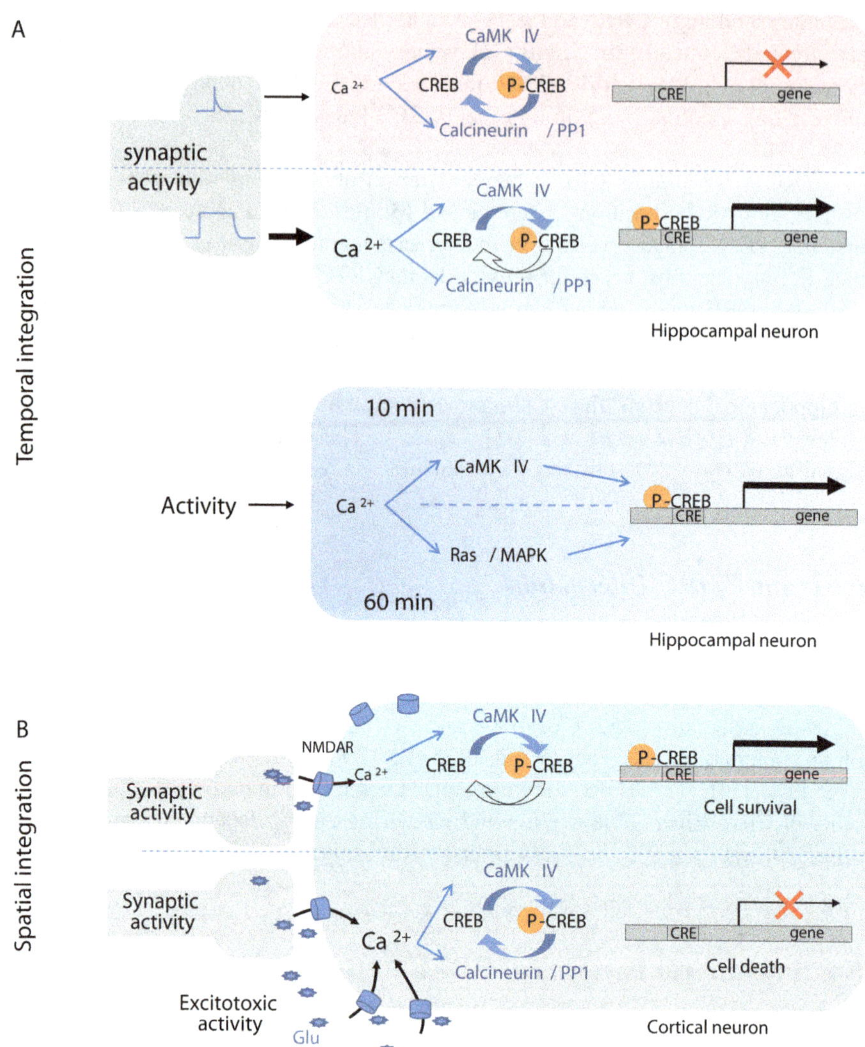

Fig. 2.5 Spatiotemporal control of CREB activation. (**a**) Temporal integration of activity through CREB. *Upper panel*: The duration of synaptic activity influences the duration of CREB phosphorylation. Short stimulation induces a low increase in cytosolic [Ca^{2+}] and transient CREB phosphorylation that does not lead to expression of target genes. Longer synaptic stimulation triggers a large increase in cytosolic [Ca^{2+}] and sustained CREB activation, enhancing expression of target genes. *Lower panel*: In hippocampal neurons, synaptic activity triggers a fast acting signaling that recruits CaMKIV and CREB phosphorylation within 10 min. This signal is then replaced by a slower and long lasting Ras/MAPK-dependent pathway that dominates 60 min after the initial stimulation, thus both fast and long-lasting CREB activation are ensured. (**b**) Spatial integration of Ca^{2+}-mediated activity through CREB. In cortical neurons, glutamatergic synaptic activity triggers CREB phosphorylation through recruitment of CaMKIV leading to expression of target genes and cell survival. Excitotoxic levels of glutamate (Glu), also lead to recruitment of phosphatases such as calcineurin, with consequent CREB dephosphorylation, inactivation and cell death

sustained activation of CREB and expression of target genes (Mayr and Montminy 2001). Moreover, Cn can be activated and inactivated by Ca^{2+} and Calmodulin (Stemmer et al. 1995), which might result in varied temporal patterns of CREB activation.

Another example of dynamic temporal control of CREB activity comes from the work of Wu et al. (2001), who showed in hippocampal neurons in vitro that within 10 min of stimulation, the Ca^{2+}-induced CaMKIV signaling phosphorylates CREB at Ser133. Then, the slower intracellular Ca^{2+}-sensitive Ras/MAPK pathway follows and predominates after 60 min of stimulation, extending CREB phosphorylation and promoting gene expression (Wu et al. 2001) (Fig. 2.5a). The fast CaMKIV signaling may convey acute and precise information to the nucleus, while MAPK signaling might code information about the duration of the stimulation.

The expression of phosphatases is cell specific, allowing for different responses in CREB-mediated gene expression depending on the neuronal subtype. Liu and Graybiel (1996) compared the convergence of Ca^{2+} and cAMP signaling on CREB phosphorylation by activating the D1/D5 dopamine receptor (D1/D5R) and L-type VGCC at the same time in two populations of striatal neurons in organotypic slices (Liu and Graybiel 1996). The two neuronal populations differ in their expression of two phosphatases: while neurons present in the striosomes express the phosphatase dopamine and cAMP–regulated phosphoprotein (DARPP-32) and Ca^{2+}/PP2B, which are induced by cAMP and Ca^{2+} respectively, neurons found in the striatal matrix do not. While both D1/D5R and L-type VGCC were able to transiently stimulate CREB phosphorylation, only D1/D5R promoted sustained CREB phosphorylation in DARPP-32-expressing cells, while L-type VGCC did it in DARPP-32-lacking cells (Liu and Graybiel 1996). This is interesting because only sustained CREB phosphorylation leads to c-fos expression, highlighting the importance of the spatiotemporal control of CREB activation. Thus, the duration of CREB phosphorylation is not only controlled by the coincidence of the synaptic stimulation, but also by the cell-specific presence of CREB phosphatases.

Strong activation of the NMDAR can lead to an excitotoxic cell death signal, while exerting a cell survival role or mediating synaptic plasticity when activated within the synapse. This divergent effect seems to be due, at least partially, to the duration of CREB phosphorylation following NMDAR activation. In cortical neurons, synaptic NMDAR stimulation leads to a sustained phosphorylation of CREB at the Ser133 that lasts 3 h, while excitotoxic activation of NMDAR leads to a transient, shorter activation of CREB. This difference on the effects of NMDAR activation on CREB phosphorylation is due to the recruitment of Cn after excitotoxic activation of NMDAR and therefore CREB dephosphorylation (Lee et al. 2005). This effect could be, at least partially, due to the activation of extrasynaptic NMDAR during excitotoxic glutamatergic stimulation (Fig. 2.5b).

Dynamic control of CREB activity also occurs at the CREB coactivator level. For instance, in rat developing cortical neurons, CRTC1/TORC is dephosphorylated via calcineurin following depolarization-induced Ca^{2+} influx, leading to CRTC1 nuclear translocation. This promotes CREB activation and expression of genes such as the salt-inducible kinase 1 (SIK1). In turn, SIK1 phosphorylates CRTC1 in response to persistent depolarization and therefore inactivates CRTC1 and CREB-driven transcription (Li et al. 2009).

Examples of spatially restricted CREB activation include the regulation of CREB in axons. In embryonic dorsal root ganglia cultures, CREB mRNA is translated in axons under the control of NGF in an activity-dependent manner; CREB is then retrogradely transported to the nucleus by endosomes containing the NGF receptor TrkA and phosphorylated in endosomes by this receptor through a MEK5-ERK5 signaling. CREB is finally translocated in the nucleus where it promotes cell survival (Cox et al. 2008).

Additional Regulatory Mechanisms of CREB Activation/ Recruitment

CREB can be regulated independently of phosphorylation through such mechanisms as ubiquitination (Comerford et al. 2003) and glycosylation (Lamarre-Vincent and Hsieh-Wilson 2003), and also through activity-dependent epigenetic modifications of CREB mRNA. Rajasethupathy et al. (2009) showed that in Aplysia the small RNA miR-124 inhibits CREB translation. In turn, serotonin signaling in presynaptic sensory neurons relieves this negative effect enabling long-term facilitation (Rajasethupathy et al. 2009). Similarly, miR-134 exerts a negative effect on CREB translation and synaptic plasticity (Gao et al. 2010). These studies present alternative mechanisms for activity-dependent regulation of CREB in the central nervous system.

Concluding Remarks

Long-term plasticity of the central nervous system relies on a molecular machinery that in neurons transduces extracellular stimuli into long lasting structural and functional modifications. In the context of a living organism, changes in the environment lead to the activation of numerous intracellular signaling cascades that need to be integrated to achieve the appropriate neuronal response. The crosstalk among different signaling pathways allows for spatiotemporal integration of multiple signals converging at CREB and its cofactors. In addition to regulation of the level of phosphorylation of CREB, several other mechanisms, such as activity-dependent epigenetic modifications of CREB mRNA, are emerging as crucial in controlling CREB activity and target gene transcription. Future studies will likely reveal novel activity-dependent mechanisms that utilize CREB to integrate diverse stimuli that generate complex neuronal responses.

Acknowledgements Research in the laboratory has been supported by the Basil O'Connor Starter Scholar Research Award Grant 5-FY09-131 from the March of Dimes Foundation, Klingenstein Foundation Award in Neuroscience, NSF 1120796, NIH-NINDS R01NS073055 and

Shriners Hospital for Children 86500-NCA and 85220-NCA grants to LNB and Shriners Hospital for Children Fellowship, Brain and Behavior foundation NARSAD grant #25356, IMERA grant and fellowship with the support of INSERM / Labex RFIEA / ANR (Investissements d'avenir), and Prestige / Marie Curie fellowship #PRESTIGE-2016-3-0025 to YHB.

References

Alberts AS, Montminy M, Shenolikar S, Feramisco JR (1994) Expression of a peptide inhibitor of protein phosphatase 1 increases phosphorylation and activity of CREB in NIH 3T3 fibroblasts. Mol Cell Biol 14(7):4398–4407

Altarejos JY, Montminy M (2011) CREB and the CRTC co-activators: sensors for hormonal and metabolic signals. Nat Rev Mol Cell Biol 12(3):141–151. doi:10.1038/nrm3072

Averaimo S, Nicol X (2014) Intermingled cAMP, cGMP and calcium spatiotemporal dynamics in developing neuronal circuits. Front Cell Neurosci 8:376. doi:10.3389/fncel.2014.00376

Belgacem YH, Borodinsky LN (2011) Sonic hedgehog signaling is decoded by calcium spike activity in the developing spinal cord. Proc Natl Acad Sci U S A 108(11):4482–4487

Belgacem YH, Borodinsky LN (2015) Inversion of sonic hedgehog action on its canonical pathway by electrical activity. Proc Natl Acad Sci U S A 112(13):4140–4145. doi:10.1073/pnas.1419690112

Bito H, Takemoto-Kimura S (2003) Ca(2+)/CREB/CBP-dependent gene regulation: a shared mechanism critical in long-term synaptic plasticity and neuronal survival. Cell Calcium 34(4–5):425–430

Bito H, Deisseroth K, Tsien RW (1996) CREB phosphorylation and dephosphorylation: a Ca(2+)- and stimulus duration-dependent switch for hippocampal gene expression. Cell 87(7):1203–1214

Bittinger MA, McWhinnie E, Meltzer J, Iourgenko V, Latario B, Liu X, Chen CH, Song C, Garza D, Labow M (2004) Activation of cAMP response element-mediated gene expression by regulated nuclear transport of TORC proteins. Curr Biol 14(23):2156–2161. doi:10.1016/j.cub.2004.11.002

Brindle P, Linke S, Montminy M (1993) Protein-kinase-A-dependent activator in transcription factor CREB reveals new role for CREM repressors. Nature 364(6440):821–824. doi:10.1038/364821a0

Brindle P, Nakajima T, Montminy M (1995) Multiple protein kinase A-regulated events are required for transcriptional induction by cAMP. Proc Natl Acad Sci U S A 92(23):10521–10525

Brini M, Cali T, Ottolini D, Carafoli E (2014) Neuronal calcium signaling: function and dysfunction. Cell Mol Life Sci 71(15):2787–2814. doi:10.1007/s00018-013-1550-7

Brunelli M, Castellucci V, Kandel ER (1976) Synaptic facilitation and behavioral sensitization in Aplysia: possible role of serotonin and cyclic AMP. Science 194(4270):1178–1181

Byers D, Davis RL, Kiger JA Jr (1981) Defect in cyclic AMP phosphodiesterase due to the dunce mutation of learning in *Drosophila melanogaster*. Nature 289(5793):79–81

Carlezon WA Jr, Duman RS, Nestler EJ (2005) The many faces of CREB. Trends Neurosci 28(8):436–445. doi:10.1016/j.tins.2005.06.005

Castellucci VF, Kandel ER, Schwartz JH, Wilson FD, Nairn AC, Greengard P (1980) Intracellular injection of t he catalytic subunit of cyclic AMP-dependent protein kinase simulates facilitation of transmitter release underlying behavioral sensitization in Aplysia. Proc Natl Acad Sci U S A 77(12):7492–7496

Chawla S, Hardingham GE, Quinn DR, Bading H (1998) CBP: a signal-regulated transcriptional coactivator controlled by nuclear calcium and CaM kinase IV. Science 281(5382):1505–1509

Chong ZZ, Li F, Maiese K (2005) Activating Akt and the brain's resources to drive cellular survival and prevent inflammatory injury. Histol Histopathol 20(1):299–315

Comerford KM, Leonard MO, Karhausen J, Carey R, Colgan SP, Taylor CT (2003) Small ubiquitin-related modifier-1 modification mediates resolution of CREB-dependent responses to hypoxia. Proc Natl Acad Sci U S A 100(3):986–991. doi:10.1073/pnas.0337412100

Conkright MD, Canettieri G, Screaton R, Guzman E, Miraglia L, Hogenesch JB, Montminy M (2003) TORCs: transducers of regulated CREB activity. Mol Cell 12(2):413–423

Connor LM, Marriott SJ (2000) Sequences flanking the cAMP responsive core of the HTLV-I tax response elements influence CREB protease sensitivity. Virology 270(2):328–336. doi:10.1006/viro.2000.0262

Cox LJ, Hengst U, Gurskaya NG, Lukyanov KA, Jaffrey SR (2008) Intra-axonal translation and retrograde trafficking of CREB promotes neuronal survival. Nat Cell Biol 10(2):149–159. doi:10.1038/ncb1677

Craig JC, Schumacher MA, Mansoor SE, Farrens DL, Brennan RG, Goodman RH (2001) Consensus and variant cAMP-regulated enhancers have distinct CREB-binding properties. J Biol Chem 276(15):11719–11728. doi:10.1074/jbc.M010263200

Dash PK, Hochner B, Kandel ER (1990) Injection of the cAMP-responsive element into the nucleus of Aplysia sensory neurons blocks long-term facilitation. Nature 345(6277):718–721. doi:10.1038/345718a0

Dash PK, Karl KA, Colicos MA, Prywes R, Kandel ER (1991) cAMP response element-binding protein is activated by Ca2+/calmodulin- as well as cAMP-dependent protein kinase. Proc Natl Acad Sci U S A 88(11):5061–5065

Deutsch PJ, Jameson JL, Habener JF (1987) Cyclic AMP responsiveness of human gonadotropin-alpha gene transcription is directed by a repeated 18-base pair enhancer. Alpha-promoter receptivity to the enhancer confers cell-preferential expression. J Biol Chem 262(25):12169–12174

Du K, Montminy M (1998) CREB is a regulatory target for the protein kinase Akt/PKB. J Biol Chem 273(49):32377–32379

Dudai Y, Jan YN, Byers D, Quinn WG, Benzer S (1976) Dunce, a mutant of drosophila deficient in learning. Proc Natl Acad Sci U S A 73(5):1684–1688

Felinski EA, Quinn PG (1999) The CREB constitutive activation domain interacts with TATA-binding protein-associated factor 110 (TAF110) through specific hydrophobic residues in one of the three subdomains required for both activation and TAF110 binding. J Biol Chem 274(17):11672–11678

Felinski EA, Kim J, Lu J, Quinn PG (2001) Recruitment of an RNA polymerase II complex is mediated by the constitutive activation domain in CREB, independently of CREB phosphorylation. Mol Cell Biol 21(4):1001–1010. doi:10.1128/mcb.21.4.1001-1010.2001

Fink JS, Verhave M, Kasper S, Tsukada T, Mandel G, Goodman RH (1988) The CGTCA sequence motif is essential for biological activity of the vasoactive intestinal peptide gene cAMP-regulated enhancer. Proc Natl Acad Sci U S A 85(18):6662–6666

Finkbeiner S, Tavazoie SF, Maloratsky A, Jacobs KM, Harris KM, Greenberg ME (1997) CREB: a major mediator of neuronal neurotrophin responses. Neuron 19(5):1031–1047

Flavell SW, Greenberg ME (2008) Signaling mechanisms linking neuronal activity to gene expression and plasticity of the nervous system. Annu Rev Neurosci 31:563–590. doi:10.1146/annurev.neuro.31.060407.125631

Gao J, Wang WY, Mao YW, Graff J, Guan JS, Pan L, Mak G, Kim D, Su SC, Tsai LH (2010) A novel pathway regulates memory and plasticity via SIRT1 and miR-134. Nature 466(7310):1105–1109. doi:10.1038/nature09271

Gau D, Lemberger T, von Gall C, Kretz O, Le Minh N, Gass P, Schmid W, Schibler U, Korf HW, Schutz G (2002) Phosphorylation of CREB Ser142 regulates light-induced phase shifts of the circadian clock. Neuron 34(2):245–253

Ghosh A, Ginty DD, Bading H, Greenberg ME (1994) Calcium regulation of gene expression in neuronal cells. J Neurobiol 25(3):294–303. doi:10.1002/neu.480250309

Ginty DD, Kornhauser JM, Thompson MA, Bading H, Mayo KE, Takahashi JS, Greenberg ME (1993) Regulation of CREB phosphorylation in the suprachiasmatic nucleus by light and a circadian clock. Science 260(5105):238–241

Ginty DD, Bonni A, Greenberg ME (1994) Nerve growth factor activates a Ras-dependent protein kinase that stimulates c-fos transcription via phosphorylation of CREB. Cell 77(5):713–725

Gonzalez GA, Montminy MR (1989) Cyclic AMP stimulates somatostatin gene transcription by phosphorylation of CREB at serine 133. Cell 59(4):675–680

Gonzalez GA, Yamamoto KK, Fischer WH, Karr D, Menzel P, Biggs W 3rd, Vale WW, Montminy MR (1989) A cluster of phosphorylation sites on the cyclic AMP-regulated nuclear factor CREB predicted by its sequence. Nature 337(6209):749–752. doi:10.1038/337749a0

Hagiwara M, Alberts A, Brindle P, Meinkoth J, Feramisco J, Deng T, Karin M, Shenolikar S, Montminy M (1992) Transcriptional attenuation following cAMP induction requires PP-1-mediated dephosphorylation of CREB. Cell 70(1):105–113. doi:10.1016/0092-8674(92)90537-M

Hardingham GE, Arnold FJ, Bading H (2001) Nuclear calcium signaling controls CREB-mediated gene expression triggered by synaptic activity. Nat Neurosci 4(3):261–267. doi:10.1038/85109

Iguchi-Ariga SM, Schaffner W (1989) CpG methylation of the cAMP-responsive enhancer/promoter sequence TGACGTCA abolishes specific factor binding as well as transcriptional activation. Genes Dev 3(5):612–619

Impey S, Wayman G, Wu Z, Storm DR (1994) Type I adenylyl cyclase functions as a coincidence detector for control of cyclic AMP response element-mediated transcription: synergistic regulation of transcription by Ca2+ and isoproterenol. Mol Cell Biol 14(12):8272–8281

Impey S, Mark M, Villacres EC, Poser S, Chavkin C, Storm DR (1996) Induction of CRE-mediated gene expression by stimuli that generate long-lasting LTP in area CA1 of the hippocampus. Neuron 16(5):973–982

Impey S, Obrietan K, Wong ST, Poser S, Yano S, Wayman G, Deloulme JC, Chan G, Storm DR (1998) Cross talk between ERK and PKA is required for Ca2+ stimulation of CREB-dependent transcription and ERK nuclear translocation. Neuron 21(4):869–883

Impey S, McCorkle SR, Cha-Molstad H, Dwyer JM, Yochum GS, Boss JM, McWeeney S, Dunn JJ, Mandel G, Goodman RH (2004) Defining the CREB regulon: a genome-wide analysis of transcription factor regulatory regions. Cell 119(7):1041–1054. doi:10.1016/j.cell.2004.10.032

Iourgenko V, Zhang W, Mickanin C, Daly I, Jiang C, Hexham JM, Orth AP, Miraglia L, Meltzer J, Garza D, Chirn GW, McWhinnie E, Cohen D, Skelton J, Terry R, Yu Y, Bodian D, Buxton FP, Zhu J, Song C, Labow MA (2003) Identification of a family of cAMP response element-binding protein coactivators by genome-scale functional analysis in mammalian cells. Proc Natl Acad Sci U S A 100(21):12147–12152. doi:10.1073/pnas.1932773100

Kandel ER, Dudai Y, Mayford MR (2014) The molecular and systems biology of memory. Cell 157(1):163–186. doi:10.1016/j.cell.2014.03.001

Kornhauser JM, Nelson DE, Mayo KE, Takahashi JS (1990) Photic and circadian regulation of c-fos gene expression in the hamster suprachiasmatic nucleus. Neuron 5(2):127–134

Kornhauser JM, Cowan CW, Shaywitz AJ, Dolmetsch RE, Griffith EC, Hu LS, Haddad C, Xia Z, Greenberg ME (2002) CREB transcriptional activity in neurons is regulated by multiple, calcium-specific phosphorylation events. Neuron 34(2):221–233

Kovacs KA, Steullet P, Steinmann M, Do KQ, Magistretti PJ, Halfon O, Cardinaux JR (2007) TORC1 is a calcium- and cAMP-sensitive coincidence detector involved in hippocampal long-term synaptic plasticity. Proc Natl Acad Sci U S A 104(11):4700–4705. doi:10.1073/pnas.0607524104

Lamarre-Vincent N, Hsieh-Wilson LC (2003) Dynamic glycosylation of the transcription factor CREB: a potential role in gene regulation. J Am Chem Soc 125(22):6612–6613. doi:10.1021/ja028200t

Lee B, Butcher GQ, Hoyt KR, Impey S, Obrietan K (2005) Activity-dependent neuroprotection and cAMP response element-binding protein (CREB): kinase coupling, stimulus intensity, and temporal regulation of CREB phosphorylation at serine 133. J Neurosci 25(5):1137–1148. doi:10.1523/jneurosci.4288-04.2005

Li S, Zhang C, Takemori H, Zhou Y, Xiong ZQ (2009) TORC1 regulates activity-dependent CREB-target gene transcription and dendritic growth of developing cortical neurons. J Neurosci 29(8):2334–2343. doi:10.1523/jneurosci.2296-08.2009

Liu F-C, Graybiel AM (1996) Spatiotemporal dynamics of CREB phosphorylation: transient versus sustained phosphorylation in the developing striatum. Neuron 17(6):1133–1144. doi:10.1016/S0896-6273(00)80245-7

Lonze BE, Ginty DD (2002) Function and regulation of CREB family transcription factors in the nervous system. Neuron 35(4):605–623. doi:10.1016/S0896-6273(02)00828-0

Manning BD, Cantley LC (2007) AKT/PKB signaling: navigating downstream. Cell 129(7):1261–1274. doi:10.1016/j.cell.2007.06.009

Matthews RP, Guthrie CR, Wailes LM, Zhao X, Means AR, McKnight GS (1994) Calcium/calmodulin-dependent protein kinase types II and IV differentially regulate CREB-dependent gene expression. Mol Cell Biol 14(9):6107–6116

Mayr B, Montminy M (2001) Transcriptional regulation by the phosphorylation-dependent factor CREB. Nat Rev Mol Cell Biol 2(8):599–609. doi:10.1038/35085068

Mayr BM, Canettieri G, Montminy MR (2001) Distinct effects of cAMP and mitogenic signals on CREB-binding protein recruitment impart specificity to target gene activation via CREB. Proc Natl Acad Sci U S A 98(19):10936–10941

Michael LF, Asahara H, Shulman AI, Kraus WL, Montminy M (2000) The phosphorylation status of a cyclic AMP-responsive activator is modulated via a chromatin-dependent mechanism. Mol Cell Biol 20(5):1596–1603

Montarolo PG, Goelet P, Castellucci VF, Morgan J, Kandel ER, Schacher S (1986) A critical period for macromolecular synthesis in long-term heterosynaptic facilitation in Aplysia. Science 234(4781):1249–1254

Montminy MR, Bilezikjian LM (1987) Binding of a nuclear protein to the cyclic-AMP response element of the somatostatin gene. Nature 328(6126):175–178. doi:10.1038/328175a0

Montminy MR, Sevarino KA, Wagner JA, Mandel G, Goodman RH (1986) Identification of a cyclic-AMP-responsive element within the rat somatostatin gene. Proc Natl Acad Sci U S A 83(18):6682–6686

Obrietan K, Impey S, Storm DR (1998) Light and circadian rhythmicity regulate MAP kinase activation in the suprachiasmatic nuclei. Nat Neurosci 1(8):693–700. doi:10.1038/3695

Parker D, Ferreri K, Nakajima T, LaMorte VJ, Evans R, Koerber SC, Hoeger C, Montminy MR (1996) Phosphorylation of CREB at Ser-133 induces complex formation with CREB-binding protein via a direct mechanism. Mol Cell Biol 16(2):694–703

Quinn PG (1993) Distinct activation domains within cAMP response element-binding protein (CREB) mediate basal and cAMP-stimulated transcription. J Biol Chem 268(23):16999–17009

Rajasethupathy P, Fiumara F, Sheridan R, Betel D, Puthanveettil SV, Russo JJ, Sander C, Tuschl T, Kandel E (2009) Characterization of small RNAs in Aplysia reveals a role for miR-124 in constraining synaptic plasticity through CREB. Neuron 63(6):803–817. doi:10.1016/j.neuron.2009.05.029

Ravnskjaer K, Kester H, Liu Y, Zhang X, Lee D, Yates JR 3rd, Montminy M (2007) Cooperative interactions between CBP and TORC2 confer selectivity to CREB target gene expression. EMBO J 26(12):2880–2889. doi:10.1038/sj.emboj.7601715

Rosen LB, Ginty DD, Weber MJ, Greenberg ME (1994) Membrane depolarization and calcium influx stimulate MEK and MAP kinase via activation of Ras. Neuron 12(6):1207–1221

Rosenberg SS, Spitzer NC (2011) Calcium signaling in neuronal development. Cold Spring Harb Perspect Biol 3(10):a004259. doi:10.1101/cshperspect.a004259

Sakamoto K, Karelina K, Obrietan K (2011) CREB: a multifaceted regulator of neuronal plasticity and protection. J Neurochem 116(1):1–9. doi:10.1111/j.1471-4159.2010.07080.x

Screaton RA, Conkright MD, Katoh Y, Best JL, Canettieri G, Jeffries S, Guzman E, Niessen S, Yates JR 3rd, Takemori H, Okamoto M, Montminy M (2004) The CREB coactivator TORC2 functions as a calcium- and cAMP-sensitive coincidence detector. Cell 119(1):61–74. doi:10.1016/j.cell.2004.09.015

Shaywitz AJ, Greenberg ME (1999) CREB: a stimulus-induced transcription factor activated by a diverse array of extracellular signals. Annu Rev Biochem 68:821–861. doi:10.1146/annurev.biochem.68.1.821

Sheng M, McFadden G, Greenberg ME (1990) Membrane depolarization and calcium induce c-fos transcription via phosphorylation of transcription factor CREB. Neuron 4(4):571–582

Sheng M, Thompson MA, Greenberg ME (1991) CREB: a Ca(2+)-regulated transcription factor phosphorylated by calmodulin-dependent kinases. Science 252(5011):1427–1430

Spitzer NC (2006) Electrical activity in early neuronal development. Nature 444(7120):707–712. doi:10.1038/nature05300

Stemmer PM, Wang X, Krinks MH, Klee CB (1995) Factors responsible for the Ca2+-dependent inactivation of calcineurin in brain. FEBS Lett 374(2):237–240. doi:10.1016/0014-5793(95)01095-V

Sun P, Enslen H, Myung PS, Maurer RA (1994) Differential activation of CREB by Ca2+/calmodulin-dependent protein kinases type II and type IV involves phosphorylation of a site that negatively regulates activity. Genes Dev 8(21):2527–2539

Wadzinski BE, Wheat WH, Jaspers S, Peruski LF Jr, Lickteig RL, Johnson GL, Klemm DJ (1993) Nuclear protein phosphatase 2A dephosphorylates protein kinase A-phosphorylated CREB and regulates CREB transcriptional stimulation. Mol Cell Biol 13(5):2822–2834

Wagner BL, Bauer A, Schutz G, Montminy M (2000) Stimulus-specific interaction between activator-coactivator cognates revealed with a novel complex-specific antiserum. J Biol Chem 275(12):8263–8266

Wu X, McMurray CT (2001) Calmodulin kinase II attenuation of gene transcription by preventing cAMP response element-binding protein (CREB) dimerization and binding of the CREB-binding protein. J Biol Chem 276(3):1735–1741. doi:10.1074/jbc.M006727200

Wu GY, Deisseroth K, Tsien RW (2001) Activity-dependent CREB phosphorylation: convergence of a fast, sensitive calmodulin kinase pathway and a slow, less sensitive mitogen-activated protein kinase pathway. Proc Natl Acad Sci U S A 98(5):2808–2813. doi:10.1073/pnas.051634198

Xing J, Ginty DD, Greenberg ME (1996) Coupling of the RAS-MAPK pathway to gene activation by RSK2, a growth factor-regulated CREB kinase. Science 273(5277):959–963

Xue ZC, Wang C, Wang QW, Zhang JF (2015) CREB-regulated transcription coactivator 1: important roles in neurodegenerative disorders. Sheng Li Xue Bao 67(2):155–162

Yao H, Peng F, Fan Y, Zhu X, Hu G, Buch SJ (2009) TRPC channel-mediated neuroprotection by PDGF involves Pyk2/ERK/CREB pathway. Cell Death Differ 16(12):1681–1693. doi:10.1038/cdd.2009.108

Zhang X, Odom DT, Koo SH, Conkright MD, Canettieri G, Best J, Chen H, Jenner R, Herbolsheimer E, Jacobsen E, Kadam S, Ecker JR, Emerson B, Hogenesch JB, Unterman T, Young RA, Montminy M (2005) Genome-wide analysis of cAMP-response element binding protein occupancy, phosphorylation, and target gene activation in human tissues. Proc Natl Acad Sci U S A 102(12):4459–4464. doi:10.1073/pnas.0501076102

Zhou Y, Wu H, Li S, Chen Q, Cheng XW, Zheng J, Takemori H, Xiong ZQ (2006) Requirement of TORC1 for late-phase long-term potentiation in the hippocampus. PLoS One 1:e16. doi:10.1371/journal.pone.0000016

Chapter 3
Models of Short-Term Synaptic Plasticity

Janet Barroso-Flores, Marco A. Herrera-Valdez, Elvira Galarraga, and José Bargas

Abstract We focus on dynamical descriptions of short-term synaptic plasticity. Instead of focusing on the molecular machinery that has been reviewed recently by several authors, we concentrate on the dynamics and functional significance of synaptic plasticity, and review some mathematical models that reproduce different properties of the dynamics of short term synaptic plasticity that have been observed experimentally. The complexity and shortcomings of these models point to the need of simple, yet physiologically meaningful models. We propose a simplified model to be tested in synapses displaying different types of short-term plasticity.

Keywords Synapse • Synaptic facilitation • Synaptic depression • Short-term synaptic plasticity • Synapses • GABA • Striatum

Abbreviations

$[Ca^{2+}]_i$	Intracellular Ca^{2+} concentration
ACh	Acetylcholine
CB	Calbindin
CCK	Cholecystokinin
DA	Dopamine
EPSC	Excitatory postsynaptic current
EPSP	Excitatory postsynaptic potential
GABA	Gamma-aminobutyric acid

J. Barroso-Flores (✉) • E. Galarraga • J. Bargas
División de Neurociencias, Instituto de Fisiología Celular, Universidad Nacional Autónoma de México, Mexico City, DF 04510, Mexico
e-mail: janet.barroso@zi-mannheim.de; egalarraga@ifc.unam.mx; jbargas@ifc.unam.mx

M.A. Herrera-Valdez (✉)
Departamento de Matemáticas, Facultad de Ciencias, Universidad Nacional Autónoma de México, Mexico City, DF 04510, Mexico
e-mail: marcoh@ciencias.unam.mx

© Springer International Publishing AG 2017
R. von Bernhardi et al. (eds.), *The Plastic Brain*, Advances in Experimental Medicine and Biology 1015, DOI 10.1007/978-3-319-62817-2_3

IPSC Inhibitory postsynaptic current
IPSP Inhibitory postsynaptic potential
LTS   Low threshold spiking
LTSP Long terms synaptic plasticity
mEPSP Miniature excitatory postsynaptic potential
PV   Parvalbumin SOM somatostatin
STB   Short term biphasic plasticity
STD   Short term depression
STF   Short term facilitation
STSP Short term synaptic plasticity

Introduction

Synaptic efficacy is typically regarded as the ability to evoke postsynaptic events upon the release of neurotransmitter by the presynaptic terminal. Synaptic plasticity is the capability of changing synaptic efficacy as result of previous activity in the synapse. Synaptic efficacy changes dynamically during the on-going function of the synapse. As already mentioned in Chap. 1, plasticity involves processes that occur at various time scales. Changes in synaptic efficacy that occur within milliseconds to minutes are regarded as short-term synaptic plasticity (STSP) (Zucker and Regehr 2002). In turn, changes in efficacy that take place over tens of minutes to hours or even longer, are regarded as long-term synaptic plasticity (LTSP) (Morris 2003) and tend to be maintained for long time. In the present work, we focus on short-term synaptic plasticity. This review focuses on the dynamics of the short-term synaptic plasticity.

The arrival of an action potential at the presynaptic terminal opens voltage-dependent Ca^{2+} channels, raising intracellular calcium ($[Ca^{2+}]_i$) in the presynaptic terminal, triggering the activation of presynaptic machinery that results in the exo-cytosis of readily releasable vesicles containing transmitter molecules (Katz and Miledi 1968). Transmitter released into the synaptic cleft diffuses and binds to postsynaptic receptors. The activation of receptors triggers an electrical response in the postsynaptic neuron that involves the activation of a postsynaptic current, which in turn causes a local change: the postsynaptic membrane potential. Several reviews deal with the chain of events that take place during synaptic activation, including, but not limited to, vesicle cycling, docking and priming due to the SNARE complex, *exo-* and endocytosis, synaptic receptor regulation (Dutta Roy et al. 2014; Neher 2010; Gandhi and Stevens 2003; Triller and Choquet 2005).

Postsynaptic events are seen as postsynaptic currents when recorded with the voltage-clamp technique. The postsynaptic current is excitatory if increases the probability of postsynaptic spiking and inhibitory if it decreases the probability of postsynaptic spiking. A postsynaptic current is regarded as excitatory (EPSC) if it causes a positive deflection in the postsynaptic membrane potential, called excitatory postsynaptic potentials (EPSPs). Similarly, inhibitory postsynaptic currents (IPSCs) are those that produce negative deflections of the postsynaptic membrane, called inhibitory postsynaptic potentials (IPSPs). The arrival of several action potentials at the presynaptic terminals triggers sequences of pulses in the intra-terminal $[Ca^{2+}]_i$ that may add sublinearly, supralinearly, or both. Depending on the type of synaptic terminal and on the context, neurotransmitter release may be enhanced or decreased as action potentials arrive (Bollmann and Sakmann 2005; Dutta Roy et al. 2014; Tecuapetla et al. 2007). The postsynaptic currents triggered by the released neurotransmitter then change the postsynaptic potential in a way that often reflects on the sublinear or supralinear summation of the presynaptic $[Ca^{2+}]_i$ transients. Factors that affect the dynamics of such processes include the type of postsynaptic target; the presence of modulatory transmitters such as dopamine (DA) or acetylcholine (ACh), which modify presynaptic efficacy (Guzmán et al. 2003; Tecuapetla et al. 2007; Dehorter et al. 2009; Mansvelder et al. 2009) chelating proteins such as parvalbumin (PV) or calbindin (CB) that modify Ca^{2+} dynamics inside the terminal (Mochida et al. 2008); vesicle recycling including clathrin mediated endocytosis (Granseth et al. 2006; Lopez-Murcia et al. 2014; Kavalali 2007); and a variety of other factors that make synaptic plasticity highly dynamic.

Types of Short-Term Synaptic Plasticity

Short-Term Depression

Following a train of presynaptic stimuli, the synaptic efficacy of some terminals tends to decrease in time scales of a few milliseconds to seconds (Barroso-Flores et al. 2015). That is, postsynaptic events decrease in amplitude, possibly due to less transmitter release or less activation of postsynaptic receptors. This is called short-term depression (STD, Fig. 3.1a). Functionally, depression has been linked to habituation, sensory adaptation and an initial high probability of release (Zucker 1989; Fisher et al. 1997; O'Donovan and Rinzel 1997). Several mechanisms have been proposed to account for STD. One is the depletion of the readily releasable pool of synaptic vesicles (Del Castillo and Katz 1954; Rosenmund and Stevens 1996). If the frequency of action potentials is high enough, the fraction of vesicles in the readily releasable pool progressively declines after each action potential resulting in a decrease in synaptic efficacy over time. In such conditions, the incoming stimulus has to stop to allow replenishment of the vesicles. High stimulus frequency can therefore contribute to depression, as high presynaptic release frequencies lead to more rapid vesicle depletion and saturation of the mechanisms that mediate recovery of the presynaptic machinery and vesicle refilling. Another possible explanation for synaptic depression is decreased Ca^{2+} entry due to inactivation of Ca^{2+} channels

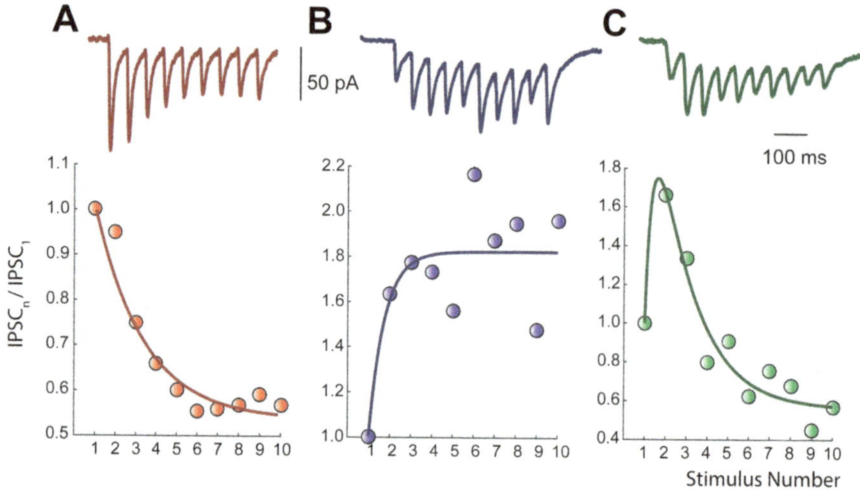

Fig. 3.1 Examples of inhibitory inputs onto striatal projection neurons with different short-term synaptic plasticity profiles. (**a**) Synapse with depression. (**b**) Facilitating synapse. (**c**) Biphasic plasticity

(Forsythe et al. 1998; Xu and Wu 2005). Other mechanisms may depend on modulators, as for example endocannabinoids (Freund and Hájos 2003; Freund et al. 2003) released by the presynaptic (Straiker and Mackie 2009) or postsynaptic elements (Fukudome et al. 2004; Uchigashima et al. 2007; Ohno-Shosaku and Kano 2014). In the extreme, some synapses arising from the same presynaptic neuron may display synaptic facilitation in one target and synaptic depression in another target (Muller and Nicholls 1974; Markram et al. 1998; Thomson et al. 2002a, b) depending on local $[Ca^{2+}]i$ and modulatory messengers. Nevertheless, if the postsynaptic target, the extracellular conditions and the presynaptic stimulus remain constant, the short-term synaptic plasticity can be thought of as entirely dependent on the presynaptic element and the presynaptic dynamics becomes a reliable indicator for the dynamics of the whole synapse (Gupta et al. 2000; Tecuapetla et al. 2007; Savanthrapadian et al. 2014; Sedlacek and Brenowitz 2014; Beierlein et al. 2003; Blackman et al. 2013; Wang et al. 2006). For instance, a common finding is that basket cells or neurons containing parvalbumin (PV-immunoreactive), especially the fast-spiking interneurons, exhibit short-term depression (Massi et al. 2012; Tecuapetla et al. 2007; Gittis et al. 2010; Planert et al. 2010), as well as the climbing fibers on Purkinje neurons (Dittman et al. 2000; Ohtsuki et al. 2009), the collateral projections between striatal projection neurons (Hoffman et al. 2003; Tecuapetla et al. 2007), and various synapses in the cochlear nucleus (MacLeod et al. 2007; Wang et al. 2011; MacLeod 2011).

Short-Term Facilitation

Short-term facilitation or STF occurs when synaptic efficacy increases for brief periods and involves processes that are typically independent of protein expression (Fig. 3.1b) (Barroso-Flores et al. 2015). Besides STF there are different kinds of facilitation, which are distinguished by the time scales in which they occur, and represent different physiological processes. Sensitization is a process in which repeated trains of action potentials lead to a progressive increase of the response (Marinesco et al. 2004; Fulton et al. 2008; Nikolaev et al. 2013). Another kind of facilitation, called synaptic enhancement, occurs in two phases: the first one lasting tens of milliseconds, followed by a slower phase lasting hundreds of milliseconds (McNaughton et al. 1978). In a longer time scale, augmentation lasts several seconds and post-tetanic potentiation may last from 30 s to several minutes (Zucker and Regehr 2002).

STF may be explained by diverse presynaptic phenomena. Quantal analysis has revealed that synaptic enhancement may be due to an increment in the probability of release (p) or to an increase in the number of release sites (*n*) (Clements 2003; Clements and Silver 2000). Residual Ca^{2+} has been proposed repeatedly to account for these changes, the hypothesis being that Ca^{2+} accumulation in the presynaptic terminal after an action potential increases the amount of released neurotransmitter if subsequent action potentials occur close enough in time (Katz and Miledi 1968; Ravin et al. 1999; Van der Kloot and Molgo 1993). Importantly, buffering presynaptic Ca^{2+} reduces both facilitation and augmentation (Magleby 1979; Regehr et al. 1994; Stevens and Wesseling 1999; Salin et al. 1996). Nevertheless, some synapses are remarkably fast in replenishing the readily releasable pool (Trommershäuser et al. 2003; Neher 2010; Watanabe et al. 2013), as is the case for parallel fibers synapsing onto Purkinje cells (Dittman et al. 2000), terminals of cholecystokinin (CCK) containing and somatostatin (SOM) containing interneurons (Blackman et al. 2013; Savanthrapadian et al. 2014), regular spiking pyramidal neurons synapsing onto somatostatin-expressing low threshold-spiking (LTS) neurons in the neocortex (Thomson et al. 1993; Markram et al. 1998; Reyes et al. 1998; Beierlein et al. 2003; Hayut et al. 2011), Schaffer collaterals onto SOM interneurons (Sun et al. 2009), and striatal projection neurons onto target cells in other basal ganglia nuclei (Sims et al. 2008; Miguelez et al. 2012; Hernandez-Martinez et al. 2015), among others.

Short-Term Biphasic Plasticity

During a train of presynaptic action potentials, some synapses may undergo a combination of facilitation followed by depression, a phenomenon called short-term biphasic plasticity or STB (Markram et al. 1998; Wang et al. 2006; Savanthrapadian et al. 2014, Barroso-Flores et al. 2015) (Fig. 3.1c). This is the case for the terminals from SOM containing interneurons of the hippocampus (Savanthrapadian et al.

2014). The mechanisms behind this type of plasticity have not been studied in depth, but basic explanations have been proposed through the use of mathematical models (Tsodyks and Markram 1997; Hennig 2013). A diverse collection of microcircuits capable of displaying multiple states, arguably fulfilling different functions, has been revealed (Traub et al. 2004; Klausberger and Somogyi 2008; Carrillo-Reid et al. 2009; Kopell et al. 2011). Such a multifunctional capability can be explained by combining the dynamics of synaptic plasticity mentioned above with the intrinsic properties of neurons, all subject to neuromodulation. We now concentrate on explaining how these synaptic dynamics combine from a functional perspective.

Short-Term Synaptic Plasticity as a Temporal Filter in Brain Microcircuits

As already pointed out, different classes of synapses behave in diverse ways with respect to time. The classic role of synapses is to mediate the induction of postsynaptic responses by presynaptic action potentials. Therefore, a most invoked hypothesis has been that they act as temporal filters for incoming input (Fortune and Rose 2000; Buonomano 2000; Zucker and Regehr 2002; George et al. 2011; Dittman et al. 2000; Wang et al. 2006; Abbott and Regehr 2004).

Each type of short-term synaptic plasticity mentioned above can be thought of as an electrical frequency filter, as revealed by experiments exploring the functional relationship between the magnitude of postsynaptic responses induced by a stimulus train at given stimulus frequencies (bode-plots) (Markram et al. 1998; Dittman et al. 2000; Izhikevich et al. 2003). In general, synapses displaying short-term depression (STD) show progressively decreased responses to high frequency inputs and tend to conserve low frequency inputs intact. In contrast, synapses that exhibit short-term facilitation (STF) amplify high frequency inputs and tend to produce responses of constant amplitude for low-enough input frequencies. For these reasons, synapses exhibiting STD, STF, and STB profiles have been described as equivalent to low, high, and band pass filters, respectively (Markram et al. 1998; Dittman et al. 2000; Fortune and Rose 2000; Wu et al. 2001; Izhikevich et al. 2003; Abbott and Regehr 2004; Lange-Asschenfeldt et al. 2007). However, it is worth remarking that synapses that display STF are not quite analogous to a high-pass filter because they tend to amplify high-frequency inputs. By extension, the commonly accepted analogy between band-pass filters and synapses exhibiting STB profiles is also not quite correct. Nevertheless, an important consequence of the observations above is that, synaptic function is not limited to excitation and inhibition and includes temporal filtering as an important function that has been proposed to play a role in adaptation, gain control, spatial detection, and rhythm generation (Wang and Buzsáki 1996; Whittington and Traub 2003; Sirota et al. 2008; Cardin et al. 2009). For instance, STF has been postulated as a mechanism for the generation of low frequency oscillations in the theta range (Whittington and Traub 2003; Sirota et al. 2008). STB could be playing a role in synchronizing neuronal ensembles in the beta frequency band, a main oscillatory behavior observed in Parkinson disease (Hammond et al. 2007; te Woerd et al. 2014). STD helps in determining sensory habituation; responses showing a greater efficacy at the onset of an input and at low frequencies (Abbott and Regehr 2004). Depressing synapses have been proposed to produce more reliable transmission in response to new stimuli and may contribute to the removal of redundant correlated inputs, allowing information to flow more efficiently (Goldman et al. 2002). Because of the multiple functions mentioned above, there is an increasing interest in incorporating the models of short-term synaptic plasticity into larger circuit models.

Some Mathematical Models of Short-Term Synaptic Plasticity

Although the postsynaptic terminal contributes to STSP (e.g. receptor desensitization and retrograde messengers), the presynaptic processes involving the dynamics of transmitter release and the replenishment of the vesicle pool are arguably among the most important determinants of short-term synaptic plasticity (Körber and Kuner 2016). Each of these processes depends on ion channels, pumps, enzymes, vesicle recycling and other factors (Trommershäuser et al. 2003; Neher 2010; Dutta Roy et al. 2014). Ideally, models try to lump all these pieces into the least possible number of variables and parameters while capturing the essence of the biophysical processes involved. The existing mathematical models of neurotransmitter release based on experimental data try to support and predict biophysical and behavioral aspects of synaptic transmission at different levels of detail (Trommershäuser et al. 2003; Pan and Zucker 2009; Neher 2010), including priming, vesicle-fusion (Magleby and Zengel 1982; Blank et al. 2001; Millar et al. 2005; Taschenberger et al. 2005) and endocytosis (Balaji and Ryan 2007; Granseth et al. 2006). We next review three of the most widely used models of short-term synaptic plasticity, and later we propose a new generalized model capable of reproducing a vast repertoire of synaptic plasticity dynamics that can be tested for different synapses. The main variables in models of synaptic plasticity typically include the amount of neurotransmitter being released, $N_T(t)$, detected by passive postsynaptic receptors of the membrane. Let $N(t)$ be the number of vesicles available for release, each vesicle assumed to contain the same amount of neurotransmitter and let $p(t)$ the release probability. If it is assumed that any two vesicles are released independently of each other, the average release N_T can then be thought of as proportional to $N(t)p(t)$.

Assuming that the released neurotransmitter is instantly detected, the postsynaptic conductance $g(t)$, can be assumed to be proportional to the amount of neurotransmitter released is $g(t) = gN_T(t)$, where g is the maximal conductance of the postsynaptic element. For simplicity, g is assumed to be a constant. However, to model synaptic desensitization or hypersensitization, g can be assumed to change in time. For instance, g can be a function of the amount of neurotransmitter concentration during depression ($g \rightarrow 0$). One model originally developed by (Liley and North 1953) for the neuromuscular junction explains synaptic depression by vesicle depletion during tetanic stimulation of the rat neuromuscular junction. This process was described by a first order, non-autonomous differential equation of the form

$$\partial_t n = \frac{1-n(t)}{\tau_n} - \sum_n^{i=1} \delta\left(t-t_i\right)\cdot p\cdot n(t),$$

(3.1)

where $n(t)$ is the proportion of occupancy of the readily releasable pool and p is a parameter representing the probability of release. The first part of Eq. (3.1) corresponds to the total replenishment of the release pool with a time constant τn. The second part corresponds to instantaneous release of neurotransmitter assuming action potentials arriving at times t_i, with function $\delta(s) = 1$ for $s = 0$ and 0 otherwise. This model accurately fits the behavior of a number of depressing synapses with a variety of time constants (Liley and North 1953; Tsodyks and Markram 1997). Note that replenishment and probability of release are part of the same equation. A more detailed model (Tsodyks and Markram 1997) characterized synaptic connections by describing the proportion of vesicles in three active states, effective (E), inactive (I), and recovered (R), respectively, with dynamics given by

$$\partial_t R = \frac{I}{\tau_{rec}} - U_{SE}\cdot R\cdot\delta\left(t-t_i\right),$$

$$\partial_t E = -\frac{E}{\tau_{inac}} + U_{SE}\cdot R\cdot\delta\left(t-t_i\right),$$

(3.2)

$$I \propto 1-R-E.$$

It is assumed that action potentials arrive at the terminal at times t_i, $i = 1, ...,n$, instantaneously activating a fraction of resources at each of those times. This is controlled by a parameter U_{SE} that controls the utilization of synaptic efficacy. The recovery process begins as soon as the utilization of resources does. The net post-synaptic current is assumed to be proportional to the fraction of resources in the effective state, which inactivates with a time constant τ_{inact} and recovers with a time constant τ_{rec}. However, synaptic facilitation could not be fitted until the model included a facilitating mechanism (Markram et al. 1998). Thus, short-term depression and facilitation had to be approximated with two independent variables, R and u, respectively, and time-dependent changes given by

$$\partial_t R = \frac{1-R}{D} - u\cdot R\cdot\delta\left(t-t_i\right),$$

$$\partial_t u = \frac{U-u}{\tau_{inac}} + U_{SE}\cdot\left(1-u\right)\cdot\delta\left(t-t_i\right)\cdot$$

(3.3)

The first equation models the exponential recovery (or replenishment) to the value $R = 1$ with a rate constant D^{-1}, minus the instantaneous release of neurotransmitter assuming action potentials arriving at times t_i. Notice the similarity with the Liley and North (1953) model. The second equation describes the time-dependent evolution of release after the ith stimulus. Again, there is an exponential recovery of u to its resting value U in the absence of inputs (synaptic efficacy during the first action potential), and an increase proportional to $U (1 - u)$ when inputs arrive. This model explains the kinetics of a variety of synapses (Markram et al. 1998; Izhikevich et al. 2003; Gupta et al. 2000) and reproduces the bode-plots of facilitating and depressing synapses shown therein and in other publications. However, the biphasic synapse, at least in the striatum, is not reproduced. Thus, a more detailed model of STSP was proposed (Dittman et al. 2000) with very different synaptic dynamics depending on the contributions of various calcium-dependent mechanisms. The two main dynamic variables in this model are facilitation and refractory depression, emphasizing the role of residual calcium (Ca_{res}). An end-plate synaptic current

$$EPC = \alpha \cdot N_T \cdot F \cdot D \tag{3.4}$$

where N_T is the number of release sites and the fraction of released sites is divided into two parts, facilitation (F) and depression (D), respectively, and α is a scaling factor for the averaged miniature end-plate postsynaptic current EPC (or mini EPSC) amplitude. Enhancement of release was assumed to be a calcium dependent increase from an initial value F_1 dependent on a calcium-bound molecule CaX_F with dissociation constant K_F

$$F(t) = \frac{CaX_F}{CaX_F + K_F} \tag{3.5}$$

where CaX_F was modeled with first order kinetics and a time constant τ_F after a jump of size ΔF during the action potential arriving at time t_i

$$\partial_t CaX_F = -\frac{CaX_F}{\tau_F} + \Delta_F \cdot \delta(t - t_i) \tag{3.6}$$

If CaX_F is allowed to decay to 0, a correction to Eq. (3.6) has to be made with the addition of a baseline release probability F_1 when $CaX_F = 0$

$$F(t) = F_1 + (1 - F_1)\frac{CaX_F}{CaX_F + K_F}. \tag{3.7}$$

Now, $F(t)$ ranges from F_1 to 1 as CaX_F increases from 0. After the arrival of an action potential, CaX_F increases to Δ_F and $F(t)$ increases to

$$F_2 = F_1 + (1 - F_1)\frac{\Delta_F}{\Delta_F + K_F}. \tag{3.8}$$

If a second stimulus arrives when no recovery from the refractory state has occurred, then the second EPC will be determined by the increased release probability F_2 and the remaining number of release sites as follows:

$$EPC_2 = \alpha \cdot N_T \cdot \left(1 - D_1 \cdot F_1\right) \cdot F_2. \tag{3.9}$$

Since $F_2 \leq 1$, the initial probability of release has as upper bound

$$F_1 = \frac{1}{1 + \rho} \tag{3.10}$$

where

$$\rho = \frac{EPC_2}{EPC_1} = \frac{\left(1 - F_1\right) F_2}{F_1} \tag{3.11}$$

is the facilitation ratio. The depression mechanism assumes three active states: R is a refractory state, T is a transitional state and N the readily releasable sites. A total of release sites N_T is determined as N_T. Similar to the dynamics of facilitation, recovery rate was assumed to be depend on the concentration of a Ca-bound molecule CaX_D with unbinding time constant τ_D defining the dynamics

$$\partial_t CaX_D = -\frac{CaX_D}{\tau_F} + \Delta_D \cdot \delta\left(t - t_i\right), \tag{3.12}$$

in which bound Ca^{2+} is assumed to jump, instantaneously by ΔD units after an action potential arrives at time t_i. The release probability is described by a depression variable $D = N/N_T$ (number available sites / total number of sites) with

$$\partial_t D = \left(1 - D\right) \cdot k_{recov} CaX_D - D \cdot F \cdot \delta\left(t - t_i\right), \tag{3.13}$$

where

$$k_{recov}\left(CaX_D\right) = \left(k_{max} - k_i\right) \frac{CaX_D}{CaX_D - K_D} + k_i. \tag{3.14}$$

After the first action potential, N_T, F_1, and D_1 sites have released transmitter and pass to a refractory state, leaving $N_T(1 - F_1 D_1)$ sites available to release. Then $D_2 = 1 - F_1 \cdot D_1$ For simplicity, $D_1 = 1$ if all sites are available to release, which means $D_2 = 1 - F_1$.

This model also captures a spectrum of synaptic dynamics of STSP, including the dynamics during irregular stimulus trains (Dittman et al. 2000). Nevertheless, the ideal to lump the various physiological phenomena into a minimal, simple and intuitive set of variables is partially lost, in particular, because extra assumptions about calcium management have to be made. In the next section we propose a simplified model that generalizes some of the ideas previously mentioned, which we have been testing in a set of GABAergic striatal synapses. The proposed model aims to recover the ideal of using the least number of variables and assumptions. This does not mean that more detailed models with experimentally testable variables are not needed.

A New Model for Short-Term Synaptic Plasticity

The present computational model was generated by characterizing short-term synaptic plasticity with two main dynamic variables: occupancy of vesicles in the readily releasable pool $x(t)$ (between 0 and 1) and the probability of release $p(t)$ (Barroso-Flores et al. 2015). Assume a presynaptic neuron firing a train of action potentials at times t_k, $k = 0,1,...,n$. The release dynamics in the presynaptic terminal can be described by the product xp with rules of evolution given by

$$\partial_t x = x^\alpha \left(\frac{x_\infty - x}{\tau_x} \right) - px \sum_{k=1}^{n} \phi(t - t_k)$$

$$\partial_t p = p^\beta \left(\frac{p_\infty - p}{\tau_p} \right) - (1 - p) x \sum_{k=1}^{n} q(h_k; a, \tau_q) \phi(t - t_k)$$
(3.15)

where

$$\phi\left(t; t_k, \tau_\phi \right) = \left(t - t_k \right) \exp \left[1 - \left(\frac{t - t_k}{\tau_\phi} \right) \right]$$
(3.16)

represents the activation time course of the presynaptic release machinery and

$$q\left(h; a, \tau_q \right) = ah \exp \left(1 - \frac{h}{\tau_q} \right)$$
(3.17)

is a function that models the dependence of the increase in the probability of release on the interspike interval h, caused by a presynaptic action potential. The parameter τ_q is a time constant after which the increase in p starts to decay. The parameters α and β allow better adjustment of experimental data. For instance, $(\alpha,\beta) = (1,0)$ describe logistic and linear dynamics for x and p, respectively, in the absence of inputs. The parameters and represent the steady state occupancy of the readily releasable pool and the probability of release, respectively. τ_x and τ_p are time constants for the recovery of the releasable pool from depression and the recovery from

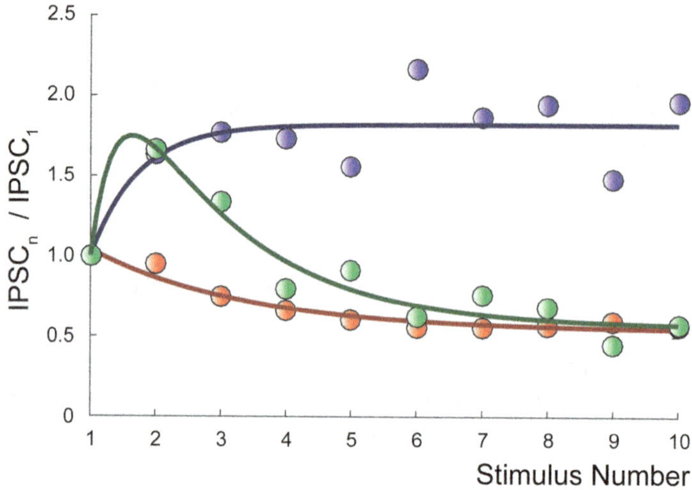

Fig. 3.2 Fitting of the proposed model to different types of short-term synaptic plasticity recorded from spiny neurons in the rat striatum. *Dots* represent IPSCs amplitudes in Fig. 3.2 and *lines* represent curves obtained by fitting the normalized response by the model in equations (3.15)–(3.17)

facilitation, respectively. The function $\varphi(t)$ represents the shape of the input (e.g. the shape of the Ca^{2+} transients resulting from an afferent volley). Increase in $p(t)$, denoted by $q(h)$ can be assumed to depend on the presynaptic interspike intervals (h) with a maximum increase a in the probability of release, reached at a time τ_q. In most experiments we have used trains of 20 Hz with regular interspike intervals. Up to now, this model has been able to reproduce, within reasonable approximation, the dynamics of a variety of inhibitory inputs onto striatal projection neurons (Körber and Kuner 2016) (Fig. 3.2). The present contribution was published in its present form in order to be tested by other researchers and be able to receive feedback.

Conclusion

The anatomical connections of neuronal networks, the intrinsic properties of neurons and the nature of their synapses, excitatory or inhibitory, are not sufficient to construct a meaningful model of a microcircuit. One reason is STSP, in which synapses change as a function of time in different ways. STSP plays several roles that are important for information processing and coding within circuits. The various synaptic dynamics supply the circuit with a variety of frequency filters that modulate the computational and the informational transfer capabilities of the network. Although existing models that approximate experimentally recorded short-term synaptic plasticity can be very complex there is still a need for simple models based on only a few variables that can be used as a first probe to explore the variety of synapses found in many circuits. The general and yet simple model proposed here approximates all the various short-term synaptic plasticity found in the literature. In addition, synapses do not work in "voltage-clamp" mode. When recorded in current clamp, the decay of synaptic events has to include the membrane time constant which induces different kinds of temporal and spatial summations which will modify the filtering of the presynaptic activity performed by the synapses. In turn, the membrane time constant can change due to postsynaptic intrinsic currents that can activate or inactivate depending on, for example, modulatory transmitters that activate G-protein coupled receptors. These last phenomena increase the richness and possibilities of neuronal circuits dynamics.

Acknowledgments We thank Antonio Laville, Dagoberto Tapia for technical support and advice and to Dr. Claudia Rivera for animal care. Acquisition software was made by Dr. Luis Carrillo-Reid and María Itze Lemus. Experiments were supported by the Consejo Nacional de Ciencia y Tecnología (CONACyT-México) Grant Fronteras de la Ciencia 57 (to J. Bargas and Marco Arieli Herrera-Valdez), the fellowship 290847 (to M.A. Herrera-Valdez), by Dirección General de Asuntos del Personal Académico, UNAM Grants IN-201914 and IN-205610 (to E. Galarraga and J. Bargas, respectively) Marco Arieli Herrera-Valdez thanks the Mathematics Institute at UNAM for the support provided during the initial drafting stages of this manuscript.

References

Abbott L, Regehr WG (2004) Synaptic computation. Nature 431(7010):796–803

Balaji J, Ryan T (2007) Single-vesicle imaging reveals that synaptic vesicle exocytosis and endocytosis are coupled by a single stochastic mode. Proc Natl Acad Sci U S A 104(51):20576–20581

Barroso-Flores J, Herrera-Valdez MA, Lopez-Huerta VG, Galarraga E, Bargas J (2015) Diverse short-term dynamics of inhibitory synapses converging on striatal projection neurons: differential changes in a rodent model of parkinson's disease. Neural Plast 2015:573543. doi:10.1155/2015/573543

Beierlein M, Gibson JR, Connors BW (2003) Two dynamically distinct inhibitory networks in layer 4 of the neocortex. J Neurophysiol 90(5):2987–3000

Blackman AV, Abrahamsson T, Costa RP, Lalanne T, Sjöström PJ (2013) Target cell-specific short-term plasticity in local circuits. Front Synaptic Neurosci 5:11. doi:10.3389/fnsyn.2013.00011

Blank PS, Vogel SS, Malley JD, Zimmerberg J (2001) A kinetic analysis of calcium-triggered exocytosis. J Gen Physiol 118(2):145–156

Bollmann JH, Sakmann B (2005) Control of synaptic strength and timing by the release-site Ca^{2+} signal. Nat Neurosci 8(4):426–434

Buonomano DV (2000) Decoding temporal information: a model based on short-term synaptic plasticity. J Neurosci 20(3):1129–1141

Cardin JA, Carlén M, Meletis K, Knoblich U, Zhang F, Deisseroth K, Tsai LH, Moore CI (2009) Driving fast-spiking cells induces gamma rhythm and controls sensory responses. Nature 459(7247):663–667

Carrillo-Reid L, Tecuapetla F, Ibáñez-Sandoval O, Hernández-Cruz A, Galarraga E, Bargas J (2009) Activation of the cholinergic system endows compositional properties to striatal cell assemblies. J Neurophysiol 101(2):737–749

Clements JD (2003) Variance–mean analysis: a simple and reliable approach for investigating synaptic transmission and modulation. J Neurosci Methods 130(2):115–125

Clements JD, Silver RA (2000) Unveiling synaptic plasticity: a new graphical and analytical approach. Trends Neurosci 23(3):105–113

Dehorter N, Guigoni C, Lopez C, Hirsch J, Eusebio A, Ben-Ari Y, Hammond C (2009) Dopamine-deprived striatal gabaergic interneurons burst and generate repetitive gigantic ipscs in medium spiny neurons. J Neurosci 29(24):7776–7787

Del Castillo J, Katz B (1954) Quantal components of the end-plate potential. J Physiol 124(3):560–573

Dittman JS, Kreitzer AC, Regehr WG (2000) Interplay between facilitation, depression, and residual calcium at three presynaptic terminals. J Neurosci 20(4):1374–1385

Dutta Roy R, Stefan MI, Rosenmund C (2014) Biophysical properties of presynaptic short-term plasticity in hippocampal neurons: insights from electrophysiology, imaging and mechanistic models. Front Cell Neurosci 8:141. doi:10.3389/fncel.2014.00141

Fisher SA, Fischer TM, Carew TJ (1997) Multiple overlapping processes underlying short-term synaptic enhancement. Trends Neurosci 20(4):170–177

Forsythe ID, Tsujimoto T, Barnes-Davies M, Cuttle MF, Takahashi T (1998) Inactivation of presynaptic calcium current contributes to synaptic depression at a fast central synapse. Neuron 20(4):797–807

Fortune ES, Rose GJ (2000) Short-term synaptic plasticity contributes to the temporal filtering of electrosensory information. J Neurosci 20(18):7122–7130

Freund TF, Hájos N (2003) Excitement reduces inhibition via endocannabinoids. Neuron 38(3):362–365

Freund TF, Katona I, Piomelli D (2003) Role of endogenous cannabinoids in synaptic signaling. Physiol Rev 83(3):1017–1066

Fukudome Y, Ohno-Shosaku T, Matsui M, Omori Y, Fukaya M, Tsubokawa H, Taketo MM, Watanabe M, Manabe T, Kano M (2004) Two distinct classes of muscarinic action on hippocampal inhibitory synapses: M2-mediated direct suppression and m1/m3-mediated indirect suppression through endocannabinoid signalling. Eur J Neurosci 19(10):2682–2692

Fulton D, Condro MC, Pearce K, Glanzman DL (2008) The potential role of postsynaptic phospholipase c activity in synaptic facilitation and behavioral sensitization in aplysia. J Neurophysiol 100(1):108–116

Gandhi SP, Stevens CF (2003) Three modes of synaptic vesicular recycling revealed by single-vesicle imaging. Nature 423(6940):607–613

George AA, Lyons-Warren AM, Ma X, Carlson BA (2011) A diversity of synaptic filters are created by temporal summation of excitation and inhibition. J Neurosci 31(41):14721–14734

Gittis AH, Nelson AB, Thwin MT, Palop JJ, Kreitzer AC (2010) Distinct roles of gabaergic interneurons in the regulation of striatal output pathways. J Neurosci 30(6):2223–2234

Goldman MS, Maldonado P, Abbott L (2002) Redundancy reduction and sustained firing with stochastic depressing synapses. J Neurosci 22(2):584–591

Granseth B, Odermatt B, Royle SJ, Lagnado L (2006) Clathrin-mediated endocytosis is the dominant mechanism of vesicle retrieval at hippocampal synapses. Neuron 51(6):773–786

Gupta A, Wang Y, Markram H (2000) Organizing principles for a diversity of gabaergic interneurons and synapses in the neocortex. Science 287(5451):273–278

Guzmán JN, Hernández A, Galarraga E, Tapia D, Laville A, Vergara R, Aceves J, Bargas J (2003) Dopaminergic modulation of axon collaterals interconnecting spiny neurons of the rat striatum. J Neurosci 23(26):8931–8940

Hammond C, Bergman H, Brown P (2007) Pathological synchronization in Parkinson's disease: networks, models and treatments. Trends Neurosci 30(7):357–364

Hayut I, Fanselow EE, Connors BW, Golomb D (2011) Lts and fs inhibitory interneurons, short-term synaptic plasticity, and cortical circuit dynamics. PLoS Comput Biol 7(10):e1002248. doi:10.1371/journal.pcbi.1002248

Hennig MH (2013) Theoretical models of synaptic short term plasticity. Front Comput Neurosci 7:154. doi:10.3389/fncom.2013.00154

Hernández-Martínez R, Aceves JJ, Rueda-Orozco PE, Hernández-Flores T, Hernández-González O, Tapia D, Galarraga E, Bargas J (2015) Muscarinic presynaptic modulation in gabaergic pallidal synapses of the rat. J Neurophysiol 113(3):796–807

Hoffman AF, Oz M, Caulder T, Lupica CR (2003) Functional tolerance and blockade of long-term depression at synapses in the nucleus accumbens after chronic cannabinoid exposure. J Neurosci 23(12):4815–4820

Izhikevich EM, Desai NS, Walcott EC, Hoppensteadt FC (2003) Bursts as a unit of neural information: selective communication via resonance. Trends Neurosci 26(3):161–167

Katz B, Miledi R (1968) The role of calcium in neuromuscular facilitation. J Physiol 195(2):481

Kavalali ET (2007) Multiple vesicle recycling pathways in central synapses and their impact on neurotransmission. J Physiol 585(3):669–679

Klausberger T, Somogyi P (2008) Neuronal diversity and temporal dynamics: the unity of hippocampal circuit operations. Science 321(5885):53–57

Kopell N, Whittington M, Kramer M (2011) Neuronal assembly dynamics in the beta1 frequency range permits short-term memory. Proc Natl Acad Sci U S A 108(9):3779–3784

Körber C, Kuner T (2016) Molecular machines regulating the release probability of synaptic vesicles at the active zone. Front Synaptic Neurosci 8:5. doi:10.3389/fnsyn.2016.00005

Lange-Asschenfeldt C, Lohmann P, Riepe MW (2007) Hippocampal synaptic depression following spatial learning in a complex maze. Exp Neurol 203(2):481–485

Liley A, North K (1953) An electrical investigation of effects of repetitive stimulation on mammalian neuromuscular junction. J Neurophysiol 16(5):509–527

López-Murcia FJ, Royle SJ, Llobet A (2014) Presynaptic clathrin levels are a limiting factor for synaptic transmission. J Neurosci 34(25):8618–8629

MacLeod KM (2011) Short-term synaptic plasticity and intensity coding. Hear Res 279(1):13–21

MacLeod KM, Horiuchi TK, Carr CE (2007) A role for short-term synaptic facilitation and depression in the processing of intensity information in the auditory brain stem. J Neurophysiol 97(4):2863–2874

Magleby K (1979) Facilitation, augmentation, and potentiation of transmitter release. Prog Brain Res 49:175–182

Magleby K, Zengel J (1982) A quantitative description of stimulation-induced changes in transmitter release at the frog neuromuscular junction. J Gen Physiol 80(4):613–638

Mansvelder HD, Mertz M, Role LW (2009) Nicotinic modulation of synaptic transmission and plasticity in cortico-limbic circuits. Semin Cell Dev Biol 20(4):432–440. doi:10.1016/j.semcdb.2009.01.007

Marinesco S, Kolkman KE, Carew TJ (2004) Serotonergic modulation in aplysia. I Distributed serotonergic network persistently activated by sensitizing stimuli. J Neurophysiol 92(4):2468–2486

Markram H, Wang Y, Tsodyks M (1998) Differential signaling via the same axon of neocortical pyramidal neurons. Proc Natl Acad Sci U S A 95(9):5323–5328

Massi L, Lagler M, Hartwich K, Borhegyi Z, Somogyi P, Klausberger T (2012) Temporal dynamics of parvalbumin-expressing axo-axonic and basket cells in the rat medial prefrontal cortex in vivo. J Neurosci 32(46):16496–16502

McNaughton BL, Douglas R, Goddard GV (1978) Synaptic enhancement in fascia dentata: cooperativity among coactive afferents. Brain Res 157(2):277–293

Miguelez C, Morin S, Martinez A, Goillandeau M, Bezard E, Bioulac B, Baufreton J (2012) Altered pallido-pallidal synaptic transmission leads to aberrant firing of globus pallidus neurons in a rat model of parkinson's disease. J Physiol 590(22):5861–5875

Millar AG, Zucker RS, Ellis-Davies GC, Charlton MP, Atwood HL (2005) Calcium sensitivity of neurotransmitter release differs at phasic and tonic synapses. J Neurosci 25(12):3113–3125

Mochida S, Few AP, Scheuer T, Catterall WA (2008) Regulation of presynaptic Ca(V)2.1 channels by Ca^{2+} sensor proteins mediates short-term synaptic plasticity. Neuron 57(2):210–216

Morris RG (2003) Long-term potentiation and memory. Philos Trans R Soc B: Biol Sci 358(1432):643–647

Muller K, Nicholls J (1974) Different properties of synapses between a single sensory neurone and two different motor cells in the leech CNS. J Physiol 238(2):357–369

Neher E (2010) What is rate-limiting during sustained synaptic activity: vesicle supply or the availability of release sites. Front Synaptic Neurosci 2:144. doi:10.3389/fnsyn.2010.00144

Nikolaev A, Leung KM, Odermatt B, Lagnado L (2013) Synaptic mechanisms of adaptation and sensitization in the retina. Nat Neurosci 16(7):934–941

O'Donovan MJ, Rinzel J (1997) Synaptic depression: a dynamic regulator of synaptic communication with varied functional roles. Trends Neurosci 20(10):431–433

Ohno-Shosaku T, Kano M (2014) Endocannabinoid-mediated retrograde modulation of synaptic transmission. Curr Opin Neurobiol 29:1–8

Ohtsuki G, Piochon C, Hansel C (2009) Climbing fiber signaling and cerebellar gain control. Front Cell Neurosci 3:4. doi:10.3389/neuro.03.004.2009

Pan B, Zucker RS (2009) A general model of synaptic transmission and short-term plasticity. Neuron 62(4):539–554

Planert H, Szydlowski SN, Hjorth JJ, Grillner S, Silberberg G (2010) Dynamics of synaptic transmission between fast-spiking interneurons and striatal projection neurons of the direct and indirect pathways. J Neurosci 30(9):3499–3507

Ravin R, Parnas H, Spira M, Volfovsky N, Parnas I (1999) Simultaneous measurement of evoked release and $[Ca^{2+}]i$ in a crayfish release bouton reveals high affinity of release to Ca^{2+}. J Neurophysiol 81(2):634–642

Regehr WG, Delaney KR, Tank DW (1994) The role of presynaptic calcium in short-term enhancement at the hippocampal mossy fiber synapse. J Neurosci 14(2):523–537

Reyes A, Lujan R, Rozov A, Burnashev N, Somogyi P, Sakmann B (1998) Target-cell specific facilitation and depression in neocortical circuits. Nat Neurosci 1(4):279–285

Rosenmund C, Stevens CF (1996) Definition of the readily releasable pool of vesicles at hippocampal synapses. Neuron 16(6):1197–1207

Salin PA, Scanziani M, Malenka RC, Nicoll RA (1996) Distinct short-term plasticity at two excitatory synapses in the hippocampus. Proc Natl Acad Sci U S A 93(23):13304–13309

Savanthrapadian S, Meyer T, Elgueta C, Booker SA, Vida I, Bartos M (2014) Synaptic properties of som-and cck-expressing cells in dentate gyrus interneuron networks. J Neurosci 34(24):8197–8209

Sedlacek M, Brenowitz SD (2014) Cell-type specific short-term plasticity at auditory nerve synapses controls feed-forward inhibition in the dorsal cochlear nucleus. Front Neural Circuits 8:78. doi:10.3389/fncir.2014.00078

Sims RE, Woodhall GL, Wilson CL, Stanford IM (2008) Functional characterization of gabaergic pallido-pallidal and striatopallidal synapses in the rat globus pallidus in vitro. Eur J Neurosci 28(12):2401–2408

Sirota A, Montgomery S, Fujisawa S, Isomura Y, Zugaro M, Buzsáki G (2008) Entrainment of neocortical neurons and gamma oscillations by the hippocampal theta rhythm. Neuron 60(4):683–697

Stevens CF, Wesseling JF (1999) Augmentation is a potentiation of the exocytotic process. Neuron 22(1):139–146

Straiker A, Mackie K (2009) Cannabinoid signaling in inhibitory autaptic hippocampal neurons. Neuroscience 163(1):190–201

Sun HY, Bartley AF, Dobrunz LE (2009) Calcium-permeable presynaptic kainate receptors involved in excitatory short-term facilitation onto somatostatin interneurons during natural stimulus patterns. J Neurophysiol 101(2):1043–1055

Taschenberger H, Scheuss V, Neher E (2005) Release kinetics, quantal parameters and their modulation during short-term depression at a developing synapse in the rat CNS. J Physiol 568(2):513–537

Tecuapetla F, Carrillo-Reid L, Bargas J, Galarraga E (2007) Dopaminergic modulation of short-term synaptic plasticity at striatal inhibitory synapses. Proc Natl Acad Sci U S A 104(24):10,258–10,263

Thomson A, Deuchars J, West D (1993) Single axon excitatory postsynaptic potentials in neocortical interneurons exhibit pronounced paired pulse facilitation. Neuroscience 54(2):347–360

Thomson AM, Bannister AP, Mercer A, Morris OT (2002a) Target and temporal pattern selection at neocortical synapses. Philos Trans R Soc Lond Ser B Biol Sci 357(1428):1781–1791

Thomson AM, West DC, Wang Y, Bannister AP (2002b) Synaptic connections and small circuits involving excitatory and inhibitory neurons in layers 2–5 of adult rat and cat neocortex: triple intracellular recordings and biocytin labelling in vitro. Cereb Cortex 12(9):936–953

Traub RD, Bibbig A, LeBeau FE, Buhl EH, Whittington MA (2004) Cellular mechanisms of neuronal population oscillations in the hippocampus in vitro. Annu Rev Neurosci 27:247–278

Triller A, Choquet D (2005) Surface trafficking of receptors between synaptic and extrasynaptic membranes: and yet they do move! Trends Neurosci 28(3):133–139

Trommershaüser J, Schneggenburger R, Zippelius A, Neher E (2003) Heterogeneous presynaptic release probabilities: functional relevance for short-term plasticity. Biophys J 84(3):1563–1579

Tsodyks MV, Markram H (1997) The neural code between neocortical pyramidal neurons depends on neurotransmitter release probability. Proc Natl Acad Sci U S A 94(2):719–723

Uchigashima M, Narushima M, Fukaya M, Katona I, Kano M, Watanabe M (2007) Subcellular arrangement of molecules for 2-arachidonoyl-glycerol-mediated retrograde signaling and its physiological contribution to synaptic modulation in the striatum. J Neurosci 27(14):3663–3676

Van der Kloot W, Molgo J (1993) Facilitation and delayed release at about 0 degree c at the frog neuromuscular junction: effects of calcium chelators, calcium transport inhibitors, and okadaic acid. J Neurophysiol 69(3):717–729

Wang XJ, Buzsaki G (1996) Gamma oscillation by synaptic inhibition in a hippocampal interneuronal network model. J Neurosci 16(20):6402–6413

Wang Y, Markram H, Goodman PH, Berger TK, Ma J, Goldman-Rakic PS (2006) Heterogeneity in the pyramidal network of the medial prefrontal cortex. Nat Neurosci 9(4):534–542

Wang Y, O'Donohue H, Manis P (2011) Short-term plasticity and auditory processing in the ventral cochlear nucleus of normal and hearing-impaired animals. Hear Res 279(1):131–139

Watanabe S, Rost BR, Camacho-Pérez M, Davis MW, Sohl-Kielczynski B, Rosenmund C, Jorgensen EM (2013) Ultrafast endocytosis at mouse hippocampal synapses. Nature 504(7479):242–247

Whittington MA, Traub RD (2003) Interneuron diversity series: inhibitory interneurons and network oscillations in vitro. Trends Neurosci 26(12):676–682

te Woerd ES, Oostenveld R, de Lange FP, Praamstra P (2014) A shift from prospective to reactive modulation of beta-band oscillations in parkinson's disease. NeuroImage 100:507–519

Wu N, Hsiao CF, Chandler SH (2001) Membrane resonance and subthreshold membrane oscillations in mesencephalic v neurons: participants in burst generation. J Neurosci 21(11):3729–3739

Xu J, Wu LG (2005) The decrease in the presynaptic calcium current is a major cause of short-term depression at a calyx-type synapse. Neuron 46(4):633–645

Zucker RS (1989) Short-term synaptic plasticity. Annu Rev Neurosci 12(1):13–31

Zucker RS, Regehr WG (2002) Short-term synaptic plasticity. Annu Rev Physiol 64(1):355–405

Chapter 4
Plasticity in the Interoceptive System

Fernando Torrealba, Carlos Madrid, Marco Contreras, and Karina Gómez

Abstract The most outstanding manifestations of the plastic capacities of brain circuits and their neuronal and synaptic components in the adult CNS are learning and memory. A reduced number of basic plastic mechanisms underlie learning capacities at many levels and regions of the brain. The interoceptive system is no exception, and some of the most studied behavioral changes that involve learning and memory engage the interoceptive pathways at many levels of their anatomical and functional organization.

In this chapter, we will review four examples of learning, mostly in rats, where the interoceptive system has a role. In the case of conditioned taste aversion, the interoceptive system is of outstanding importance. In drug addiction, the role of the insular cortex – the highest level of the interoceptive system– is unusual and complex, as many forebrain regions are engaged by the process of addiction. In the third example, neophobia, the gustatory region of the insular cortex plays a major role. Finally, the role of different areas of the insular cortex in different processes of aversive memory, particularly fear conditioning, will be reviewed.

Keywords Conditioned taste aversion • Drug addiction • Fear learning • Insular cortex • Interoception • Neophobia

F. Torrealba (✉) • M. Contreras • K. Gómez
Departamento de Fisiología, Facultad de Ciencias Biológicas, Pontificia Universidad Católica de Chile, Santiago, Chile
e-mail: ftorrealba@bio.puc.cl

C. Madrid
Centro de Fisiología Celular e Integrativa, Facultad de Medicina, Clínica Alemana – Universidad del Desarrollo, Concepción, Chile

© Springer International Publishing AG 2017
R. von Bernhardi et al. (eds.), *The Plastic Brain*, Advances in Experimental Medicine and Biology 1015, DOI 10.1007/978-3-319-62817-2_4

Abbreviations

BLA   Basolateral amygdala
CTA   Conditioned taste aversion
ERK   Extracellular signal-regulated kinases
IC   Insular cortex
MAPK   Mitogen-activated protein kinases
NMDA   N-Methyl-D-aspartate
NTS   Nucleus of the solitary tract
pIC   Posterior insular cortex
RAIC   Rostral agranular insular cortex
TRN   Thalamic reticular nucleus
VPLpc   Ventroposterolateral parvicellular thalamic nucleus
VPMpc   Ventroposteromedial parvicellular thalamic nucleus

Overview of the Interoceptive System

It is important to have an idea about the functional organization of the interoceptive system as a prerequisite to understand the specific learning capabilities that give rise to the four examples of plasticity presented above. As mentioned in Chap. 1, dozens of primary sensory receptors located in the internal organs (Iggo 1986), muscles, and skin (Craig 2002), or within the CNS provide information on the physiological state of the body (Craig 2004). The cell bodies of these sensory receptors may be located in the ganglia of cranial nerves (Claps and Torrealba 1988; Torrealba and Claps 1988) or in dorsal root ganglia or in the circumventricular organs of the brain. The peripheral interoceptors have thin axons, most of them umyelinated C fibers (Iggo 1986; Widdicombe 1998).

The Lamina I Subsystem

Those receptors with cell bodies in the dorsal root ganglia project centrally to lamina I neurons (Andrew and Craig 2001) located below the dorsal and lateral surface of the spinal cord and the medulla. The axons from these second order neurons cross the midline to form the spino-thalamic pathway, which makes synaptic contacts with dorsal thalamic neurons and gives off collaterals to the nucleus of the solitary tract in the caudal medulla and the parabrachial nucleus of the pons (Craig 2004). This set of neurons forms the lamina I interoceptive subsystem, which shares thalamic and cortical relays with the visceral interoceptive subsystem.

The Visceral Interoceptive Subsystem

A large variety of sensory receptors innervate the internal organs (Iggo 1986; Kalia and Richter 1988) through the glossopharyngeal, the vagus and the facial nerves with thin, mostly unmyelinated axons. These axons project to the second order neurons of this subsystem, located in the caudal two-thirds of the nucleus of the solitary tract (NTS). The majority of the peripheral axons make synaptic contacts with ipsilateral neurons of the NTS, but a considerable number project to the contralateral NTS.

About half of the NTS neurons send ascending projections to the ipsilateral parabrachial nucleus; they form the perceptual branch of the visceral subsystem (Acuña-Goycolea et al. 2000). The other half of the NTS neurons project locally to neurons located in the ventral and lateral medulla region (within what was known as the reticular formation of the medulla); these NTS second–order neurons contribute to the reflex branch of the interoceptive system (Blessing 1997; Acuña-Goycolea et al. 2000).

Interoceptive Thalamic Nuclei

The two interoceptive subsystems, the visceral and the lamina I subsystems, converge on four thalamic nuclei (Allen et al. 1991; Gauriau and Bernard 2002, 2004). They send projections through the lateral and the medial parabrachial subnuclei to the interoceptive thalamus and, in addition, to these thalamic nuclei via direct projections from lamina I neurons in the spinal cord and medulla (Gauriau and Bernard 2004). The principal dorsal thalamic nuclei involved in interoception, including nociception, are the ventroposterolateral parvicellular (VPLpc), the ventroposteromedial parvicellular (VPMpc), the posterior and the posterior triangular thalamic nuclei (Gauriau and Bernard 2002). These thalamic nuclei send interoceptive, including nociceptive, information to the insular cortex (IC) and some of them to the secondary somatosensory cortex.

At the thalamic level, the interoceptive system forms a network that appears to direct attention to the corresponding sensory system, much as do other sensory regions of the dorsal thalamus. Many, if not all, thalamic nuclei receive GABAergic projections from a limited and specific part of the thalamic reticular nucleus (TRN).

The TRN forms a shell of GABAergic neurons that surrounds the thalamus. A particular region of the TRN in turn receives excitatory connections from the thalamic nucleus to which it projects back and, also, excitatory projections from the cortical area connected to that particular thalamic nucleus. The interoceptive thalamus is no exception, and both VPLpc and VPMpc receive GABAergic projections from what may be called the interoceptive TRN (Stehberg et al. 2001).

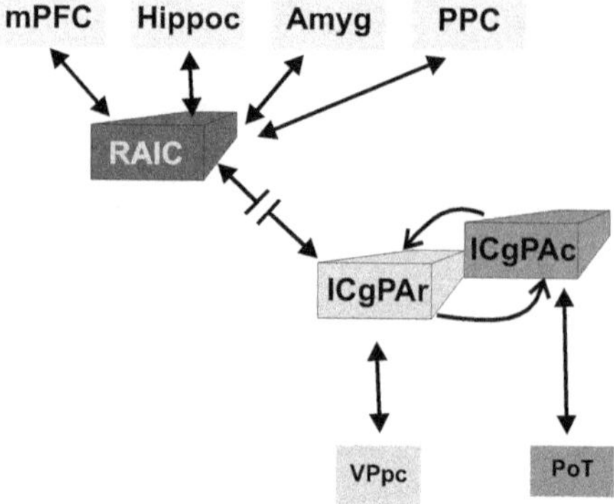

Fig. 4.1 Schematic drawing of the principal neural connections of the insular cortex (IC). At the center we placed the primary interoceptive cortex (ICgPAr in Shi and Cassell nomenclature), which receives direct thalamic input from the principal interoceptive thalamic nucleus (VPpc) as well as reciprocal connections from a more caudal IC area (ICgPAc, Parietal caudal IC in Shi and Cassell nomenclature), in turn the target of the posterior thalamic nucleus (PoT). The ICgPAr cortex is reciprocally (but indirectly) with the rostral agranular IC (RAIC), considered a higher order cortex. The RAIC together with ICgPAc establish neural connections with other forebrain regions including the medial prefrontal cortex (mPFC); the hippocampal formation (Hippoc); the amygdala complex (Amyg) and the posterior parietal cortex (PPC)

The Interoceptive Cortex

The IC of rats can be subdivided, based on cytoarchitectonics, into several areas that may also differ by both their neural connections and function. The thalamic input from the VPLpc defines a primary interoceptive cortex located in an intermediate (in the antero-posterior axis) part of the granular region of the IC. This region is adjacent to the dysgranular IC, which is recipient of gustatory and interoceptive input from the VPMpc thalamic nucleus. Between the dysgranular cortex above and the pyriform cortex below lies the posterior extension of the agranular IC. The agranular IC forms part of the rat prefrontal cortex because it shares thalamic inputs from the mediodorsal thalamic nucleus and lacks a granular layer IV. The rostral agranular IC (RAIC) is a site of convergence of interoceptive and limbic systems inputs (Saper 1982; Jasmin et al. 2004). Relevant to central networks involved in learning are the neural connections of the RAIC with the amygdaloid complex and the hippocampal formation. The existence of several subdivisions of the IC is important when considering the mechanisms of learning and memory that involve the insula (Fig. 4.1).

Conditioned Taste Aversion

A cardinal problem faced by animals is to decide whether to eat something that looks edible and has no aversive smell. Garcia et al. (1974), during a series of insights and experiments, proposed a behavioral mechanism by which animals have solved this problem. They ingest a small amount of the potential foodstuff; take a neural picture of the characteristics of that food and its location to return afterwards. If the food caused no malaise, nausea or harm, they go and get it and now have a larger, more substantial portion (Garcia et al. 1974). This learning that associates a taste with an aversive interoceptive perception/feeling—conditioned taste aversion (CTA) – establishes long-term memories with a single trial, and as such has been a very productive model for learning and memory.

In principle, the convergence of taste and malaise information may take place at several levels of the respective pathways. In fact, both sensory modalities are represented at the NTS, parabrachial nucleus, thalamus and the IC. However, at least up to the respective primary sensory cortices in the intermediate IC, taste and interoception have adjacent but not overlapping neural representations (Cechetto and Saper 1987). So far, recordings from single or few neurons have revealed that in the primary taste or interoceptive cortices neurons respond to either a taste or a visceral stimulus. Nevertheless, the single unit and multi-unit recordings obtained from naïve animals (that is, with no previous CTA) may not reveal a convergence of taste and interoceptive information if that convergence is the result of learning.

In spite of the uncertainties about the precise site of convergence of taste and gastrointestinal illness, it is clear that somewhere in the IC both signals interact to give rise to CTA. Bilateral lesioning of the IC at the level of the primary gustatory cortex disrupts CTA (Bermúdez-Rattoni and McGaugh 1991; Nerad et al. 1996). The lesions need to be placed at about +1.7 mm relative to bregma to be effective. More rostral or caudal IC lesions fail to prevent CTA acquisition (Nerad et al. 1996). No CTA is induced by a unilateral IC lesion (Schafe and Bernstein 1998) at bregma level +2.0–0 mm.

Other work has shown the importance of the IC in storage of CTA memories: the protein synthesis inhibitor anisomycin infused to the anterior IC impairs taste-aversion memory (Rosenblum et al. 1993). The anterior insular cortex has reciprocal connections with the amygdala (Mesulam and Mufson 1982; Shi and Cassell 1998a), part of a set of structures that seem to participate in the perception and organization of autonomic responses to aversive or threatening stimuli (Öngur and Price 2000). It has been suggested that the connections make a loop between insula and amygdala that is responsible for the storage of aversive memories. The cellular changes are a part of the switch from a safe to an aversive taste memory trace (Bermúdez-Rattoni et al. 2004).

Lesions made at different anterior-posterior levels of the IC that included a large portion of the anterior IC from +5.2 to +2.7 mm from bregma had no effect on acquisition of CTA or spatial learning (Nerad et al. 1996). However, lesions that included more caudal areas of the anterior IC (+3.2 to −0.3 mm from bregma) disrupt CTA acquisition and impair the retention of an inhibitory avoidance task (Bermúdez-Rattoni et all. 1997). Neither study distinguishes an important dorsal-ventral organization of the IC with different connections and therefore separate functions (Shi and Cassell 1998a, b).

Neophobia

Rats are careful when they are facing a novel food; when a new flavor reaches their mouth they ingest a limited quantity of the novel food, thus preventing negative post-ingestive consequences. This behavior is named taste neophobia. As it was described above, there is a region of IC that contains the gustatory cortex. This region, still poorly defined, apparently sits rostral to medial IC.

The role of IC in neophobia was revealed initially in the absence of a well identified gustatory cortex. One early study by Kiefer et al. (1982) described that neophobia for flavored solutions was absent in rats with IC gustatory ablation (Kiefer et al. 1982). Cell-specific lesions of gustatory IC with ibotenic acid attenuated the reaction to the novel taste of saccharin in a familiar environment (Dunn and Everitt 1988). Recently, it was shown that lesions +2.28 to 0.00 mm relative to bregma, gustatory cortex (involving agranular, dysgranular and slightly granular IC) impair taste neophobia (Lin et al. 2015). Furthermore, the evaluation of activity in this area reveals that, relative to a familiar taste, a novel taste induces higher levels of c-Fos expression principally in gustatory insular cortex (Lin et al. 2012).

The mechanisms that involve IC in neophobia have been described partially (Rosenblum et al. 1997; Berman et al. 1998; Bermúdez-Rattoni et al. 2004). It should be noted that vast majority of the published work, possibly all of it, involved the same region of the insular cortex, 1.2 mm anterior to bregma, without distinction between granular, dysgranular and agranular influence. Briefly, it has been suggested that an increase in the release of ACh in IC is induced by the first presentation of a new taste (i.e. saccharin or quinine). Therefore, cholinergic activity seems to signal the novelty of a stimulus during the initial stages of taste memory formation. The differential activation of mitogen-activated protein kinases (MAPKs), specifically the extracellular signal-regulated kinases 1–2 (ERK1–2), via activation mediated by muscarinic Ach receptors in IC, further supports a role for cholinergic transmission in neophobia. In rats, the first presentation of a new taste (i.e. 0.5% saccharin) produces a robust neophobic response. Although when saccharin is presented again the response is smaller, if these animals receive scopolamine before or after the first saccharin taste, they show the same strong neophobic response as they did to the first taste (Gutiérrez et al. 2003). Other results showed that acute intracortical microinfusion of the noncompetitive antagonist of the N-Methyl-D-aspartate (NMDA) receptor MK-801 in the same region of the IC, and given 20 min before the presentation of 0.5% saccharin solution prevents the attenuation of gustatory neophobia, indicating that this process also is an NMDA receptor-dependent phenomenon (Figueroa-Guzmán et al. 2006).

The formation of a safety memory process involves IC (Christianson et al. 2008, 2011) and requires IC plasticity, in a way not yet completely understood, as well as IC areas and circuitry not well identified. According to mapping evidence, the classic area of intervention on memory taste and neophobia is connected with the interoceptive insular cortex and with amygdaloidal nuclei. The integration of interoceptive information characteristic of food such as its temperature, texture, and astringency with its flavor should participate in the perception of a foodstuff; but whether this integration is necessary to identify a food as new is under discussion. This integration implies a region of gustatory and interoceptive convergence, which remains unidentified so far.

Another important brain area involved in neophobia is the amygdala. The role of amygdala-insula connectivity in neophobia has not been addressed in a direct way. One study based on lesions of the basolateral amygdala (BLA) showed participation of that region in neophobia (Reilly and Bornovalova 2005). Other studies have evaluated the role of NMDA receptors in the BLA and, similarly to studies of IC, they found an amygdala contribution to gustatory neophobia (Figueroa-Guzmán and Reilly 2008). So, if we consider the radical differences in terms of connectivity and general function of cortical and subcortical areas such as the amygdala, it is evident that their contributions should be different. Experiments with crossed-disconnection strategy suggest that the neophobic reaction to a new tastant depends, in part, upon an interaction between the BLA and IC (Lin and Reilly 2012). It appears that the joint activity of the two structures is critical for memory, as described above, and also for some behaviors that unfold in response to a stimulus with high saliency, as in neophobia.

Fear Memory and Insular Cortex

The neural substrates of the fear response include those associated with acquisition, storage and execution of defensive behaviors, and the set of physiological responses that are expressed in the presence of an actual or apparent danger (LeDoux 2000). It has been reported that expression of acquired fear occurs when animals have made an association between an innocuous cue and an aversive stimulus; now the cue predicts potential harm and induces fear responses, with the subject displaying various defensive behaviors in order to avoid damage. The exteroceptive sensory systems are involved in detecting stimuli that signal the presence of a predator (such as sounds or smells), providing information about the location and/or identity of an actual or potential danger (LeDoux 1996). Together with the neural processing of the external cues, the peripheral/visceral body changes are registered by the interoceptive sensory system (Saper 1982, 1995; Cechetto and Saper 1987, 1990; Allen et al. 1991), facilitating perceptual and evaluative processes related to fear and threat.

In the posterior parietal IC of rats, according to the nomenclature of Shi and Cassell (Shi and Cassell 1998a), the role of IC has been evaluated for emotional memories using pavlovian fear conditioning. Lesions of this parietal IC region by suction (Romanski and LeDoux 1992) or bilateral electrolytic damage (Romanski and LeDoux 1992; Rosen et al. 1992) have shown blockade of fear-potentiated startle, but excitotoxic lesions of the same area (Campeau and Davis 1995) found no sign of an involvement of this most caudal IC in fear conditioning. The interpretation of these results is limited due to the known effect of different methods of lesions on fibers of passage that might influence other areas, and so these studies do not necessarily indicate an involvement of IC in fear learning. Recent studies showed that a pre-training lesion (Christianson et al. 2008) or reversible inhibition during training, but not during testing (Christianson et al. 2011), prevents the effect of a safety signal on subsequent fear/anxiety-like behavior. The effect of safety signals would occur as consequence of reduced activity of basolateral amygdala and bed nucleus stria terminalis, nuclei responsible to sustain the fear response and that receive connections of the sensory IC, but the mechanism remains unclear. Therefore, the role of the more caudal IC seems relevant in integration of different cues, but the differential connectivity of these IC regions is still unclear, underlining the importance of understanding the interoceptive system connectivity and function.

The region of the IC from 0.95 to −1.4 mm rostral to bregma, the posterior insular cortex (pIC) in the nomenclature of Shi and Cassell (Shi and Cassell 1998a), and the region immediately rostral to somatosensory IC, are less explored. Functional studies on taste memory, explained earlier (Nerad et al. 1996), also revealed the heterogeneity of the IC in rats. Cell-specific lesion to intermediate IC (around bregma +1.7 mm) disrupted CTA learning while central or more caudal (−0.3 mm from bregma) lesions affected the acquisition of Morris water maze learning. We reported that the activity of the pIC is important for the long-term retention of auditory conditioned fear and that there is an involvement of the pIC in the consolidation of fear memory (Casanova et al. 2016). These findings suggest that fear-related interoceptive information is important for the encoding of auditory fear memories. The long-term inactivation of the pIC caused a prolonged reduction in expression of conditioned fear, suggesting a long-term storage of learned fear in pIC. This is further supported by the increased expression in pIC of early genes involved in neural plasticity that accompanies fear memory reactivation (Casanova et al. 2016).

The literature on the anterior area of IC (RAIC) is also limited. The anatomical organization of the RAIC and its connections with other insular cortices (Shi and Cassell 1998a), medial prefrontal cortex (Sesack et al. 1989; Condé et al. 1995; Gabbott et al. 2003; Vertes 2004; Hoover and Vertes 2007), dorsomedial thalamic nucleus (Krettek and Price 1977; Guldin and Markowitsch 1983; Ray and Price 1992; Jasmin et al. 2004) and the amygdala (Mcdonald et al. 1996; Reynolds and Zahm 2005) make it a plausible site for the integration of autonomic responses, contributing to fear learning and implementation of emotional behaviors related to fear memory. Bilateral lesion of the lateral prefrontal cortex (including the anterior IC), suggests that the anterior IC participates in the acquisition or expression of contextual fear (Morgan and LeDoux 1999). Another study where anisomycin was infused into the anterior IC revealed no effects on consolidation of fear extinction (Santini et al. 2004), suggesting that this area of the IC is not a site for storage of long-term memory for fear extinction. However, the above finding does not rule out the possibility that the anterior IC has another role in memory processes. Alves et al. (2013) found that reversible inhibition of anterior IC attenuated expression of fear of an aversive context (Alves et al. 2013). In contrast, lesions of the anterior IC increase fear learning (Lacroix et al. 2000). Therefore, the actual evidence suggests a role of the anterior IC in emotional memories.

We proposed the participation of the anterior IC, particularly the RAIC, in reconsolidation. To explore this idea, we used a fear conditioning paradigm to evaluate whether the impaired expression of zif268 in the RAIC during reconsolidation modifies the expression of learned fear. Briefly, rats with chronic bilateral cannulae in the RAIC received a 2 day-set of sessions to make them fear-conditioned to a tone. The rats were placed in the operant chambers and were exposed to five CS–US pairings (120 ± 30 s interstimulus interval). The CS was an auditory tone (5 KHz, 80 dB, 20 s) and the US a mild electric footshock (0.5 mA, 0.5 s). After conditioning, the protocol continued as indicated in Fig. 4.2a: Ninety minutes before the first test, one group received a bilateral infusion of zif268 antisense oligonucleotide (ASO group $n = 6$) while the other group received zif268 missense oligonucleotide (MSO group $n = 6$). These experiments were performed as described previously by Lee [46]. In the first exposure to the tone cue (20 s, 80 db, 5 KHz.) in a completely different room, all rats showed an elevated freezing (Fig. 4.2b). However, in the following test only ASO group exhibited a significant drop in freezing behavior. When we evaluated the zif268 expression in the IC in reactivated memory (Fig. 4.2c, black bars), we found pronounced differences in relation to non-reactivated fear memory (Fig. 4.2c, white bars), supporting the idea that area RAIC (Fig. 4.2d, e) is relevant for the reconsolidation of this fear memory.

It is known that the expression of zif268 increases in several brain structures after reactivation of consolidated fear memories (Hall et al. 2001; Bozon et al. 2003). This activation was described in the basolateral amygdala (Lee et al. 2006) and hippocampus of rats (Lee et al. 2004), in a demonstration that the expression of Zif268 is highly correlated with memory reactivation by a cue contingent on learning. So, it is well accepted that access to stored information, which makes this information momentarily labile, requires activation of inducible immediate early genes encoding regulatory transcription factors that interact with promoter regulatory elements zif268. Therefore, as in other brain areas, zif268 expression in IC can lead to long-term synaptic changes, activating signaling pathways through its inducible transcription and subsequent protein synthesis. These results support a role of the RAIC in reconsolidation of emotional memories and maybe in other associative learning processes such as consolidation, and it emphasizes the importance of knowing the hierarchical organization of the insular cortex to understand its specific roles in different learning processes and memory storage.

Drug Addiction and the Insula

The IC is a relatively new player in understanding addiction. So far, its main established role has been as a brain site relevant to drug craving. We will briefly state the main characteristics of addiction, as they are relevant to the role of the IC in learning and plasticity.

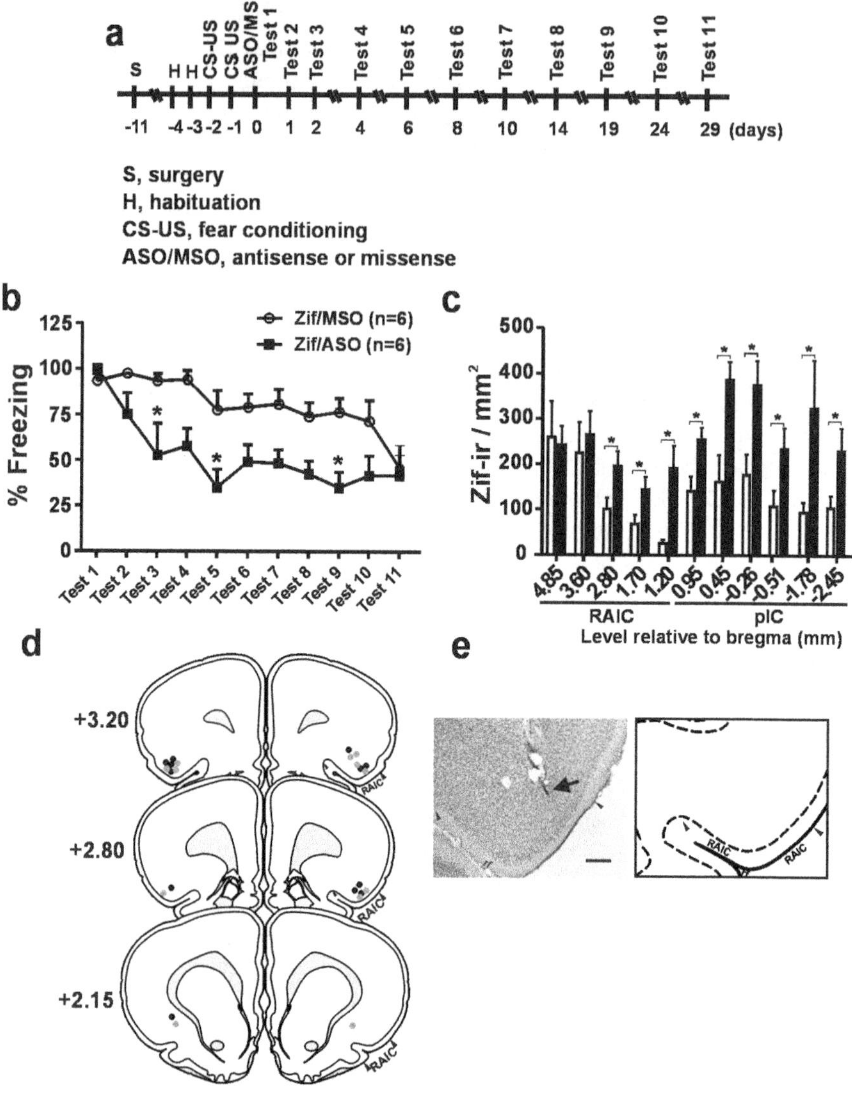

Fig. 4.2 Interference with Zif268 expression in the RAIC at the time of fear memory reactivation decreases fear responses for nearly a week. (**a**) Time line of the experiments. The surgery to bilaterally implant injection cannulae in the RAIC was performed on day −11. The rats were conditioned to a tone (CS) paired with a mild foot shock on days −2 and −1. On day 0 the rats received either antisense (ASO) or missense (MSO) Zif268 oligonucleotides and were tested 90 min after, and on subsequent days. (**b**) The infusion of Zif268 antisense oligonucleotides into the RAIC decreased freezing, a marker of fear expression, for nearly 3 weeks. (**c**) Zif268 immunoreactivity in the IC in the caudal RAIC and in pIC. (**d**) Schematic drawings based on Swanson's atlas to show the location of the tip of the injection needles into the RAIC. (**e**) A Nissl stained section through the RAIC showing the tip of an injection needle, indicated by an *arrow* (*left side*), and a schematic outline of the section. The *arrowheads* indicate the boundaries of the RAIC. Scale bar, 200 μm

Addiction to drugs of abuse is a chronic, recurring disease where the subjects keep using the drug in spite of the negative consequences. Relapse into drug use is a common trait of addicted individuals (Koob and Volkow 2010) and is due to an irresistible craving triggered by stress (Erb et al. 2000), by cues associated with previous drug use (See 2002; O'Brien et al. 1998) and/or by a single exposure to the drug (Weiss et al. 2001; Everitt and Wolf 2002). Drug addiction is the result of prolonged use of an addictive drug, owing to an individual predisposition and to the addictive potential of a particular drug (Koob and Le Moal 2006). The road to addiction has been described as a spiraling process that involves profound plastic changes in a number of brain structures (Everitt and Robbins 2005; Koob and Le Moal 2005) such as the orbitofrontal cortex involved in association functions (London et al. 2000); the ventral and dorsal striatum involved in habit formation (Everitt and Robbins 2013), and the mesencephalic dopamine projection to those large forebrain regions involved in the reinforcement of behaviors (Aberman et al. 1998).

It has been postulated that the emotional changes that appear during withdrawal and that lead to the shared symptoms that characterize chronic drug users are represented in the IC (Contreras et al. 2007, 2012). In addition to the specific signs and symptoms of withdrawal for each different drug, irritability, sadness and anxiety are present in addicted people and perhaps in other addicted mammals. The place where drug was consumed or other contextual cues (for example in animal experiments when the drug is administered) may become associated with the symptoms that preceded intake and that cause an intense desire to consume.

In support of these ideas, experiments done in our laboratory in a place preference task performed on amphetamine-experienced rats reveal that the reversible inactivation of the pIC with lidocaine (Na^+ channel blocker) infused into this primary interoceptive cortex changed the rats preference for a compartment not associated with drug. We speculated that possibly the inactivation of insula disrupted the interoceptive feelings associated with drug clues, which prevented the urge to seek amphetamine (Contreras et al. 2007). In agreement with this idea, interoceptive insula inactivation also blunted the signs of malaise induced by acute lithium administration (Contreras et al. 2007). The changes that occur in the IC during the addictive process are still largely unknown, but the evidence shows that neural plasticity is also induced in different areas of IC, suggesting a more extended involvement of the IC in addiction.

In further support of a role of the interoceptive insula in drug addiction, we showed that the local inhibitor of protein synthesis in either the pIC or the RAIC causes a long-lasting, but not permanent, loss the preference for the place where the rats used to receive amphetamine. We interpreted these results as indicative that an amnesic intervention in the IC produces a loss of context-induced craving that lasts for a long time. This amnesic effect lasts longer (24 days) when the RAIC received the amnesic treatment. The amnesia lasted 15 days when the pIC was affected. The memory that associated drug effect with context needed to be reactivated for the protein synthesis inhibitor to be effective (Contreras et al. 2012). Moreover, these rats that became reversible amnesic had decreased expression of zif268 IC, a protein involved in drug memory reconsolidation (Lee et al. 2005).

In addition to its role in the learning process of amphetamine-conditioned rats, the IC is also important in learning related to opiate addiction (Li et al. 2013) and in alcohol self-administration (Pushparaj and Le Foll 2015). Therefore, the plastic changes that occur in IC during the progression to addiction are just beginning to be addressed.

Concluding Remarks

Learning and memory are primary examples of plasticity in the insular cortex of adult rats. Here we addressed emotion learning using conditioned fear as an example; drug addiction as a condition that involves plastic changes in the IC; and conditioned taste aversion and neophobia as the most studied plasticity functions of the IC. We emphasized that these plastic functions of the IC share common underlying mechanisms with similar functions in other forebrain structures. We also made the point that the specific learning or memory capabilities of the IC can only be fully understood if one considers the IC as a complex cortex with several functional and structural subdivisions The internal organization of the IC as well as the interconnections with other forebrain and hindbrain regions are, we believe, better understood when the IC is considered a sensory cortex, with primary interoceptive and gustatory cortices, and a yet undetermined number of higher order insular cortices.

References

Aberman JE, Ward SJ, Salamone JD (1998) Effects of dopamine antagonists and accumbens dopamine depletions on time-constrained progressive-ratio performance. Pharmacol Biochem Behav 61:341–348

Acuña-Goycolea C, Fuentealba P, Torrealba F (2000) Anatomical substrate for separate processing of ascending and descending visceral information in the nucleus of the solitary tract of the rat. Brain Res 883:229–232

Allen G, Saper CB, Hurley KM, Cechetto DF (1991) Organization of visceral and limbic connections in the insular cortex of the rat. J Comp Neurol 311(1):1–16

Alves FH, Gomes FV, Reis DG, Crestani CC, Correa FM, Guimaraes FS, Resstel LB (2013) Involvement of the insular cortex in the consolidation and expression of contextual fear conditioning. Eur J Neurosci 38:2300–2307

Andrew D, Craig AD (2001) Spinothalamic lamina I neurons selectively sensitive to histamine: a central neural pathway for itch. Nat Neurosci 4:72–77

Berman DE, Hazvi S, Rosenblum K, Seger R, Dudai Y (1998) Specific and differential activation of mitogen-activated protein kinase cascades by unfamiliar taste in the insular cortex of the behaving rat. J Neurosci 18:10037–10044

Bermúdez-Rattoni F, McGaugh JL (1991) Insular cortex and amygdala lesions differentially affect acquisition on inhibitory avoidance and conditioned taste aversion. Brain Res 549:165–170

Bermúdez-Rattoni F, Introini-Collison I, Coleman-Mesches K, McGaugh JL (1997) Insular cortex and amygdala lesions induced after aversive training impair retention: effects of degree of training. Neurobiol Learn Mem 67:57–63

Bermúdez-Rattoni F, Ramírez-Lugo L, Gutiérrez R, Miranda MI (2004) Molecular signals into the insular cortex and amygdala during aversive gustatory memory formation. Cell Mol Neurobiol 24:25–36

Blessing WW (1997) The lower brainstem and bodily homeostasis. Oxford University Press, Oxford

Bozon B, Davis S, Laroche S (2003) A requirement for the immediate early gene zif268 in reconsolidation of recognition memory after retrieval. Neuron 40:695–701

Campeau S, Davis M (1995) Involvement of subcortical and cortical afferent to the lateral nucleus of the amygdala in fear conditioning measured with fear-potentiated startle in rats trained concurrently with auditory and visual conditioned stimuli. J Neurosci 15(3):2312–2327

Casanova JP, Madrid C, Contreras M, Rodríguez M, Vasquez M, Torrealba F (2016) A role for the interoceptive insular cortex in the consolidation of learned fear. Behav Brain Res 296:70–77

Cechetto DF, Saper CB (1987) Evidence for a viscerotopic sensory representation in the cortex and thalamus in the rat. J Comp Neurol 262:27–45

Cechetto DF, Saper CB (1990) Role of the cerebral cortex in autonomic function. In: Loewy AD, Spyer KM (eds) Central regulation of autonomic functions. Oxford University Press, Oxford, pp 208–223

Christianson JP, Benison AM, Jennings J, Sandsmark EK, Amat J, Kaufman RD, Baratta MV, Paul ED, Campeau S, Watkins LR, Barth DS, Maier SF (2008) The sensory insular cortex mediates the stress-buffering effects of safety signals but not behavioral control. J Neurosci 28:13703–13711

Christianson JP, Jennings JH, Ragole T, Flyer JG, Benison AM, Barth DS, Watkins LR, Maier SF (2011) Safety signals mitigate the consequences of uncontrollable stress via a circuit involving the sensory insular cortex and bed nucleus of the stria terminalis. Biol Psychiatry 70:458–464

Claps A, Torrealba F (1988) The carotid body connections: a WGA-HRP study in the cat. Brain Res 455(1):123–133

Condé F, Audinat E, Maire-Lepoivre E, Crépel F (1995) Afferent connections of the medial frontal cortex of the rat. A study using retrograde transport of fluorescent dyes. I. Thalamic afferents. Brain Res Bull 24:341–354

Contreras M, Ceric F, Torrealba F (2007) Inactivation of the interoceptive insula disrupts drug craving and malaise induced by lithium. Science 318:655–658

Contreras M, Billeke P, Vicencio S, Madrid C, Perdomo G, Gonzalez M, Torrealba F (2012) A role for the insular cortex in long-term memory for context-evoked drug craving in rats. Neuropsychopharmacology 37:2101–2108

Craig AD (2002) How do you feel? Interoception: the sense of the physiological condition of the body. Nat Rev Neurosci 3:655–666

Craig AD (2004) Interoception: the sense of the physiological condition of the body. Curr Opin Neurobiol 13:500–505

Dunn LT, Everitt BJ (1988) Double dissociations of the effects of amygdala and insular cortex lesions on conditioned taste aversion, passive avoidance, and neophobia in the rat using the excitotoxin ibotenic acid. Behav Neurosci 102:3–23

Erb S, Hitchcott PK, Rajabi H, Mueller D, Shaham Y, Stewart J (2000) Alpha-2 adrenergic receptor agonists block stress-induced reinstatement of cocaine seeking. Neuropsychopharmacology 23(2):138–150

Everitt BJ, Robbins TW (2005) Neural systems of reinforcement for drug addiction: from actions to habits to compulsion. Nat Neurosci 8:1481–1489

Everitt BJ, Robbins TW (2013) From the ventral to the dorsal striatum: devolving views of their roles in drug addiction. Neurosci Biobehav Rev 79:1946–1954

Everitt BJ, Wolf ME (2002) Psychomotor stimulant addiction: a neural systems perspective. J Neurosci 22:3312–3320

Figueroa-Guzmán Y, Reilly S (2008) NMDA receptors in the basolateral amygdala and gustatory neophobia. Brain Res 1210:200–203

Figueroa-Guzmán Y, Kuo JS, Reilly S (2006) NMDA receptor antagonist MK-801 infused into the insular cortex prevents the attenuation of gustatory neophobia in rats. Brain Res 1114:183–186

Gabbott PL, Warner TA, Jays PR, Bacon SJ (2003) Areal and synaptic interconnectivity of prelimbic (area 32), infralimbic (area 25) and insular cortices in the rat. Brain Res 993:59–71

Garcia J, Hankins WG, Rusiniak KW (1974) Behavioral regulation of the milieu interne in man and rat. Science 185:824–831

Gauriau C, Bernard JF (2002) Pain pathways and parabrachial circuits in the rat. Exp Physiol 87:251–258

Gauriau C, Bernard JF (2004) A comparative reappraisal of projections from the superficial laminae of the dorsal horn in the rat: the forebrain. J Comp Neurol 468:24–56

Guldin W, Markowitsch HJ (1983) Cortical and thalamic afferent connections of the insular and adjacent cortex of the rat. J Comp Neurol 215(2):135–153

Gutiérrez R, Téllez LA, Bermúdez-Rattoni F (2003) Blockade of cortical muscarinic but not NMDA receptors prevents a novel taste from becoming familiar. Eur J Neurosci 17:1556–1562

Hall J, Thomas KL, Everitt BJ (2001) Cellular imaging of zif268 expression in the hippocampus and amygdala during contextual and cued fear memory retrieval: selective activation of hippocampal CA1 neurons during the recall of contextual memories. J Neurosci 21:2186–2193

Hoover WB, Vertes RP (2007) Anatomical analysis of afferent projections to the medial prefrontal cortex in the rat. Brain Struct Funct 212:149–179

Iggo A (1986) Afferent C-fibres and visceral sensation. In: Cervero F, Morrison JFB (eds) Visceral sensation. Elsevier, Amsterdam, pp 29–36

Jasmin L, Burkey AR, Granato A, Ohara PT (2004) Rostral agranular insular cortex and pain areas of the central nervous system: a tract-tracing study in the rat. J Comp Neurol 468:425–440

Kalia M, Richter D (1988) Rapidly adapting pulmonary receptor afferents: II. Fine structure and synaptic organization of central terminal processes in the nucleus of the tractus solitarius. J Comp Neurol 274:574–594

Kiefer SW, Rusiniak KW, Garcia J (1982) Flavor-illness aversions: gustatory neocortex ablations disrupt taste but not taste-potentiated odor cues. J Comp Physiol Psychol 96:540–548

Koob GF, Le Moal M (2005) Plasticity of reward neurocircuitry and the 'dark side' of drug addiction. Nat Neurosci 8:1442–1444

Koob GF, Le Moal M (2006) Neurobiology of addiction. Academic, London

Koob GF, Volkow ND (2010) Neurocircuitry of addiction. Neuropsychopharmacology 35:217–238

Krettek JE, Price JL (1977) The cortical projections of the mediodorsal nucleus and adjacent thalamic nuclei in the rat. J Comp Neurol 171(2):157–191

Lacroix L, Spinelli S, Heidbreder CA, Feldon J (2000) Differential role of the medial and lateral prefrontal cortices in fear and anxiety. Behav Neurosci 114:1119–1130

LeDoux JE (1996) The emotional brain. Simon and Schuster, New York

LeDoux JE (2000) Emotion circuits in the brain. Annu Rev Neurosci 23:155–184

Lee JL, Everitt BJ, Thomas KL (2004) Independent cellular processes for hippocampal memory consolidation and reconsolidation. Science 304:839–843

Lee JL, Di Ciano P, Thomas KL, Everitt BJ (2005) Disrupting reconsolidation of drug memories reduces cocaine seeking behavior. Neuron 47:795–801

Lee JL, Milton AL, Everitt BJ (2006) Cue-induced cocaine seeking and relapse are reduced by disruption of drug memory reconsolidation. J Neurosci 26:5881–5887

Li CL, Zhu N, Meng XL, Li YH, Sui N (2013) Effects of inactivating the agranular or granular insular cortex on the acquisition of the morphine-induced conditioned place preference and naloxone-precipitated conditioned place aversion in rats. J Psychopharmacol 27(9):837–844

Lin JY, Reilly S (2012) Amygdala-gustatory insular cortex connections and taste neophobia. Behav Brain Res 235:182–188

Lin JY, Roman C, Arthurs J, Reilly S (2012) Taste neophobia and c-Fos expression in the rat brain. Brain Res 1448:82–88

Lin JY, Arthurs J, Reilly S (2015) Gustatory insular cortex, aversive taste memory and taste neophobia. Neurobiol Learn Mem 119:77–84

London ED, Ernst M, Grant S, Bonson K, Weinstein A (2000) Orbitofrontal cortex and human drug abuse: functional imaging. Cereb Cortex 10:334–342

Mcdonald AJ, Mascagni F, Guo L (1996) Projections of the medial and lateral prefrontal cortices to the amygdala: a Phaseolus vulgaris leucoagglutinin study in the rat. Neuroscience 71:55–75

Mesulam M, Mufson EJ (1982) Insula of the old world monkey I. Architectonics in the insulo-orbito-temporal component of the paralimbic brain. J Comp Neuro 212:1–22

Morgan MA, LeDoux JE (1999) Contribution of ventrolateral prefrontal cortex to the acquisition and extinction of conditioned fear in rats. Neurobiol Learn Mem 72(3):244–251

Nerad L, Ramirez-Amaya V, Ormsby CE, Bermudez-Rattoni F (1996) Differential effects of anterior and posterior insular cortex lesions on the acquisition of conditioned taste aversion and spatial learning. Neurobiol Learn Mem 66:44–50

O'Brien CP, Childress AR, Ehrman R, Robbins SJ (1998) Conditioning factors in drug abuse: can they explain compulsion? J Psychopharmacol 12:15–22

Öngur D, Price JL (2000) The organization of networks within the orbital and medial prefrontal cortex of rats; monkeys and humans. Cerb Cortex 10(3):206–219

Pushparaj A, Le Foll B (2015) Involvement of the caudal granular insular cortex in alcohol self-administration in rats. Behav Brain Res 293:203–207

Ray J, Price JL (1992) The organization of the thalamocortical connections of the mediodorsal thalamic nucleus in the rat, related to the ventral forebrain-prefrontal cortex topography. J Comp Neurol 323(2):167–197

Reilly S, Bornovalova MA (2005) Conditioned taste aversion and amygdala lesions in the rat: a critical review. Neurosci Biobehav Rev 29:1067–1088

Reynolds SM, Zahm DS (2005) Specificity in the projections of prefrontal and insular cortex to ventral striatopallidum and the extended amygdala. J Neurosci 25:11757–11767

Romanski L, LeDoux JE (1992) Bilateral destruction of neocortical and perirhinal projection targets of the acoustic thalamus does not disrupt auditory fear conditioning. Neurosci Lett 142(2):228–232

Rosen J, Hitchcock JM, Miserendino MJ, Falls WA, Campeau S, Davis M (1992) Lesions of the perirhinal cortex but not of the frontal, medial prefrontal, visual, or insular cortex block fear-potentiated startle using a visual conditioned stimulus. J Neurosci 12(12):4624–4633

Rosenblum K, Meiri N, Dudai Y (1993) Taste memory: the role of protein synthesis in gustatory cortex. Behav Neural Biol 59:49–56

Rosenblum K, Berman DE, Hazvi S, Lamprecht R, Dudai Y (1997) NMDA receptor and the tyrosine phosphorylation of its 2B subunit in taste learning in the rat insular cortex. J Neurosci 17:5129–5135

Santini E, Ge H, Ren K, Peña de Ortiz S, Quirk GJ (2004) Consolidation of fear extinction requires protein synthesis in the medial prefrontal cortex. J Neurosci 24:5704–5710

Saper CB (1982) Convergence of autonomic and limbic in the insular cortex of the rat. J Comp Neurol 210(2):163–173

Saper CB (1995) The rat nervous system, 2nd edn. Academic, San Diego

Schafe GE, Bernstein IL (1998) Forebrain contribution to the induction of a brainstem correlate of conditioned taste aversion. II. Insular (gustatory) cortex. Brain Res 800:40–47

See RE (2002) Neural substrates of conditioned-cued relapse to drug-seeking behavior. Pharmacol Biochem Behav 71:517–529

Sesack SR, Deutch AY, Roth RH, Bunney BS (1989) Topographical organization of the efferent projections of the medial prefrontal cortex in the rat: an anterograde tract-tracing study with Phaseolus vulgaris leucoagglutinin. J Comp Neurol 290(2):213–242

Shi CJ, Cassell MD (1998a) Cortical, thalamic, and amygdaloid connections of the anterior and posterior insular cortices. J Comp Neurol 399:440–468

Shi CJ, Cassell MD (1998b) Cascade projections from somatosensory cortex to the rat basolateral amygdala via the parietal insular cortex. J Comp Neurol 399:469–491

Stehberg J, Acuña-Goycolea C, Ceric F, Torrealba F (2001) The visceral sector of the thalamic reticular nucleus in the rat. Neuroscience 106:745–755
Torrealba F, Claps A (1988) The carotid sinus connections: a WGA- HRP study in the cat. Brain Res 455:134–143
Vertes RP (2004) Differential projections of the infralimbic and prelimbic cortex in the rat. Synapse 51(1):32–58
Weiss F, Ciccocioppo R, Parsons LH, Katner S, Liu X, Zorrilla EP, Valdez GR, Ben-Shahar O, Angeletti S, Richter RR (2001) Compulsive drug-seeking behavior and relapse. Neuroadaptation, stress, and conditioning factors. Ann N Y Acad Sci 937:1–26
Widdicombe JG (1998) Afferent receptors in the airways and cough. Respir Physiol 114(1):5–15

Chapter 5
Learning as a Functional State of the Brain: Studies in Wild-Type and Transgenic Animals

José M. Delgado-García and Agnès Gruart

Abstract Contemporary neuroscientists are paying increasing attention to subcellular, molecular, and electrophysiological mechanisms underlying learning and memory processes. Recent studies have examined the development of transgenic mice affected at different stages of the learning process, or have emulated in animals various human pathological conditions involving cognition and motor learning. However, a parallel effort is needed to develop stimulating and recording techniques suitable for use in behaving mice in order to understand activity-dependent synaptic changes taking place during the very moment of the learning process. The in vivo models should incorporate information collected from different molecular and in vitro approaches. Long-term potentiation (LTP) has been proposed as the neural mechanism underlying synaptic plasticity, and NMDA receptors have been proposed as the molecular substrate of LTP. It now seems necessary to study the relationship of both LTP and NMDA receptors to functional changes in synaptic efficiency taking place during actual learning in selected cerebral cortical structures. Here, we review data collected in our laboratory during the past 10 years on the involvement of different hippocampal synapses in the acquisition of the classically conditioned eyelid responses in behaving mice. Overall the results indicate a specific contribution of each cortical synapse to the acquisition and storage of new motor and cognitive abilities. Available data show that LTP, evoked by high-frequency stimulation of Schaffer collaterals, disturbs both the acquisition of conditioned eyelid responses and the physiological changes that occur at hippocampal synapses during learning. Moreover, the administration of NMDA-receptor antagonists is able not only to prevent LTP induction in vivo, but also to hinder both the formation of conditioned eyelid responses and functional changes in the strength of the CA3-CA1 synapse. Nevertheless, many other neurotransmitter receptors, intracellular mediators, and transcription factors are also involved in learning and memory processes. In summary, it can be proposed that learning and memory in behaving mammals are the result of the activation of

J.M. Delgado-García (✉) • A. Gruart
Division of Neurosciences, Pablo de Olavide University,
Ctra. de Utrera, Km. 1, Seville 41013, Spain
e-mail: jmdelgar@upo.es; http://www.divisiondeneurociencias.es

© Springer International Publishing AG 2017
R. von Bernhardi et al. (eds.), *The Plastic Brain*, Advances in Experimental Medicine and Biology 1015, DOI 10.1007/978-3-319-62817-2_5

complex and distributed functional states involving many different cerebral cortical synapses, with the participation also of various neurotransmitter systems.

Keywords Hippocampal synapses • Classical conditioning • Field excitatory postsynaptic potentials • Long-term potentiation • Mice • NMDA receptors

Abbreviations

AMPA	α-amino-3-hydroxy-5-methyl-4-isoxazolepropionic acid
CGP 39551	(*E*)-(±)-2-amino-4-methyl-5-phosphono-3-pentenoic acid ethyl ester
CR	Conditioned response
CREB	cAMP response element-binding protein
CS	Conditioned stimulus
fEPSP	Field excitatory postsynaptic potentials
HFS	High-frequency stimulation
LTD	Long-term depression
LTP	Long-term potentiation
NBQX	2,3-dioxo-6-nitro-1,2,3,4-tetrahydrobenzo[f]quinoxaline-7-sulfonamide disodium salt
NMDA	N-methyl-aspartate
trkB	Tropomyosin receptor kinase B
trkC	Tropomyosin receptor kinase C
US	Unconditioned stimulus

> *Conduct may be founded on the hard rock or the wet marshes,*
> *but after a certain point I don't care what is founded on*
>
> (*The Great Gatsby*, F. Scott Fitzgerald, 1925)

Introduction

One of the most-basic assumptions of contemporary neuroscience is that newly acquired information and abilities are registered and stored in the form of functional and ultrastructural changes in synaptic strength (Ramón y Cajal 1909–1911; Konorski 1948; Hebb 1949; Marr 1971). There are many excellent studies on the subcellular and molecular events underlying activity-dependent changes in synaptic efficiency and on the electrophysiological (in vitro) processes conceivably related to learning and memory phenomena generated in vivo (Bliss and Collingridge 1993; Malenka and Nicoll 1999; Kandel 2001; Lynch 2004; Neves et al. 2008; Wang and

Morris 2010). But even now, not much information is available regarding synaptic functional events taking place in multiple synaptic sites during actual learning in alert behaving animals. This experimental limitation has been an important drawback to our understanding of functional neural states underlying the acquisition of knowledge and new motor abilities (Delgado-García and Gruart 2002).

At the same time, it is generally assumed that long-term potentiation (LTP) is the most-plausible neural mechanism supporting associative learning (Bliss and Gardner-Medwin 1973; Bliss and Lømo 1973; McNaughton et al. 1978). LTP is usually evoked (both in vitro and in vivo) by high-frequency stimulation (HFS) of selected afferent pathways, resulting in a long-lasting enhancement of synaptic efficacy. As mentioned in Chap. 1, and although there are important exceptions, the necessary and sufficient condition for inducing LTP is the activation of N-methyl-D-aspartate (NMDA) receptors (Collingridge et al. 1983a, b; Harris et al. 1984; Bliss and Collingridge 1993; Malenka and Nicoll 1999). Thus, it can be assumed that the experimental blockage of NMDA channels in behaving animals should be able to prevent expression of LTP, as well as the acquisition of associative learning and the associated changes in synaptic efficiency and strength (Hebb 1949; Marr 1971).

From an experimental point of view, the hippocampus and related cortical structures appear to be an excellent model for the study of the plastic changes taking place at the synaptic level during the acquisition and storage of new memories. Indeed, the hippocampus has been implicated in a wide variety of learning and memory tasks, such as object recognition (Clarke et al. 2010), spatial orientation (Moser et al. 2008), and operant conditioning (Jurado-Parras et al. 2013). One of the most-used experimental models for the study of neural processes underlying associative learning is the classical conditioning of eyelid responses (Hoehler and Thompson 1980; Berger et al. 1983; McEchron and Disterhoft 1997; Múnera et al. 2001). Typically for eyeblink conditioning, a conditioned stimulus (CS) such as a tone precedes an unconditioned stimulus (US) such as a puff of air presented to the eye, causing an eyeblink, which is repeated until the CS alone elicits an eyeblink $\approx$ 90% of the time. In delay conditioning, the CS terminates when the US does, whereas in trace eyeblink conditioning the CS ends before the US occurs. In this regard, it has been proposed that hippocampal lesions impair the acquisition, but not the retention, of trace eyeblink conditioning, whereas they do not alter delay conditioning (Thompson 1988, 2005; Moyer et al. 1990), a fact supported by unitary recordings (McEchron et al. 2003). Although the cerebellum seems to participate in both types of conditioning, many other cerebral structures are also involved in this model of associative learning (Caro-Martin et al. 2015; Ammann et al. 2016). Classical eyelid conditioning studies have usually been carried out in species such as cat and rabbit, but the availability of transgenic and knock-out mice has prompted researchers to extend learning and memory studies to those small mammals. With regard to mice, it was shown years ago that they are capable of acquiring classically conditioned eyelid responses using either delay or trace paradigms (Takatsuki et al. 2003; Domínguez-del-Toro et al. 2004).

In general terms, it is reasonable to assume that functional changes evoked by learning should be detectable at synaptic sites relevant to the acquisition process. For example, it has been reported that inferior olive synaptic contacts on cerebellar interpositus neurons potentiate the evoked synaptic field potentials recorded there during the acquisition of a classically conditioned eyelid response elicited using a delay paradigm (Gruart et al. 1997). Weisz et al. (1984) reported years ago a significant modification of the synaptic activation of dentate granule cells by perforant pathway axons during the acquisition of conditioned nictitating membrane responses. Recently, it has been reported that field excitatory postsynaptic potentials (fEPSPs) evoked in the hippocampal CA1 area by single pulses presented to the ipsilateral Schaffer collateral-commissural pathway are modulated in slope by the acquisition and extinction of Pavlovian conditioning of eyelid responses in conscious mice (Gruart et al. 2006) and by other types of associative learning tasks (Whitlock et al. 2006).

Nevertheless, the availability of in vivo models of learning should be used with the simultaneous and multiple recording of the largest possible number of involved synapses, distributed across cortical and subcortical sites. In our opinion this is a necessary step towards the understanding of neural functional states underlying learning and memory processes. This chapter will address this key issue, presenting recent experimental studies from our laboratory (Carretero-Guillén et al. 2015; Gruart et al. 2014). But first, we will address three preliminary questions: (i) Is it possible to study changes in synaptic strength during the acquisition of new learning? (ii) Are activity-dependent changes in synaptic strength related to LTP? And (iii) Does the process depend upon NMDA receptors? The chapter is based on three previous reviews from our laboratory (Delgado-Garcia and Gruart 2006; Gruart and Delgado-García 2007; Gruart et al. 2013), with the addition of recently data on the separate activity of different cortical and subcortical synaptic sites during the acquisition and storage of new memories (Carretero-Guillén et al. 2015; Jurado-Parras et al. 2013; Gruart et al. 2014).

A Feasible Experimental Approach to the Study of Synaptic Events Taking Place in Selected Hippocampal Sites During the Acquisition of New Motor Abilities

In a now classic report, Gruart et al. (2006) attempted to determine whether the acquisition of associative learning modifies the synaptic strength of the hippocampal CA3-CA1 synapse during learning (Fig. 5.1). As a learning task, the classical conditioning of eyelid responses involving a trace paradigm was used—a training process involving the hippocampal circuit as well as the cerebellum (Hoehler and Thompson 1980; Berger et al. 1983; McEchron and Disterhoft 1997; Múnera et al. 2001; McEchron et al. 2003). For this, mice were presented with a brief tone as a CS followed 500 ms after its end by an electrical shock presented to the trigeminal

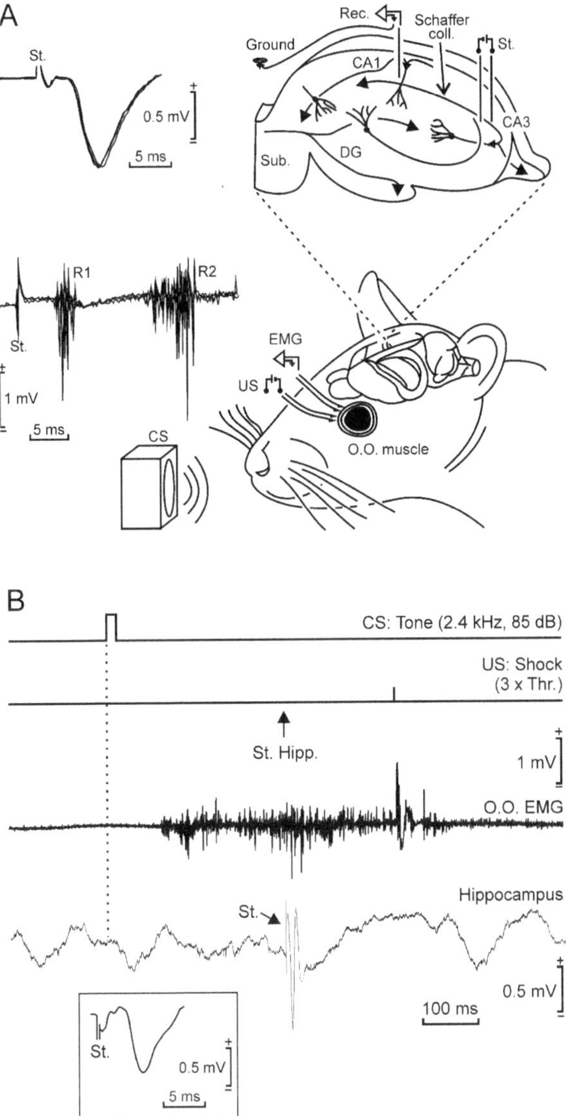

Fig. 5.1 Experimental design for in vivo studies of activity-dependent changes in synaptic strength during classical eyeblink conditioning in behaving mice. (**a**) Electrodes aimed to record the electromyographic (EMG) activity of the orbicularis oculi (O.O.) muscle were implanted in the upper eyelid. Stimulating electrodes were implanted on the ipsilateral supraorbital nerve; those electrodes were used for US presentations. A brief tone delivered from a loudspeaker located in front of the animal's head was used as a CS. Animals were also implanted (*top diagram*) with bipolar stimulating electrodes (St.) in the Schaffer collateral pathway, and with a recording electrode (Rec.) in the stratum radiatum of the ipsilateral CA1 area. Superimposed recordings at the top left illustrate the fEPSP recorded in the CA1 area following electrical stimulation (St.) of Schaffer collaterals. The records illustrated at the *bottom left* correspond to reflex blinks evoked in the O.O. muscle by the electrical stimulation of the trigeminal nerve. *R1* and *R2* indicate the two successive activation of the O.O. muscle during the eyeblink reflex (see Kugelberg 1952). (**b**) From *top* to *bottom* are illustrated the trace conditioning paradigm, the EMG activity of the O.O. muscle, and the activity recorded in the hippocampal CA1 area. The inset illustrates the recorded fEPSP on a faster time base. The maximum slope of the fast downward component of the recorded fEPSP was measured and stored. *Abbreviations*: *DG* dentate gyrus, *Sub.* subiculum (Taken with permission and modified from Gruart et al. 2006)

nerve as an US. Eyelid responses were determined by the electromyographic activity of the orbicularis oculi muscle ipsilateral to the side of US presentation. In order to follow synaptic events taking place at the hippocampal CA3-CA1 synapse during the acquisition process, Gruart et al. (2006) recorded the fEPSPs evoked at the apical dendrites of CA1 pyramidal cells by electrical stimulation of the ipsilateral Schaffer collateral pathway. As illustrated in Fig. 5.2, the slopes of fEPSPs evoked a few milliseconds after CS presentations increased steadily during conditioning sessions, but not during habituation or pseudoconditioning sessions. It is important to clarify that the electrical stimulation did not affect the training, i.e., that the acquisition process was the same with and without this stimulation. Interestingly, fEPSP slopes decreased proportionally to the percentage of CRs evoked during extinction sessions. In accordance with these results, Gruart et al. (2006) proposed that the CA3-CA1 synapse undergoes a slow modulation (i.e., potentiation and decrease) in synaptic strength (Hebb 1949) across the different conditioning situations in parallel with the acquisition and/or extinction of eyelid CRs. Accordingly, in answer to our first question, concerning feasibility of measuring changes in synaptic strength related to conditioning, it is possible to follow in alert, behaving mice activity-dependent changes in synaptic strength evoked in selected cortical synapses by associative learning tasks.

Relationships Between Experimentally Evoked Long-Term Potentiation and Functional (in vivo) Modifications of Synaptic Strength

It has already been reported that place representation in hippocampal networks can be modified experimentally by LTP (Dragoi et al. 2003), and that LTP saturation of hippocampal circuits disrupts spatial learning (Barnes et al. 1994). Apparently, hippocampal CA1 kindling also has similar disrupting effects on spatial memory performance in behaving rats (Leung and Shen 2006). Thus, it can be predicted that the experimental induction of LTP, restricted to relevant synapses, will disturb the physiological synaptic changes taking place during the different stages (acquisition, extinction, retrieval, etc.) of the learning process. This prediction also explains the relationships between experimentally evoked LTP and the rather small potentiation recorded at selected hippocampal synapses during associative learning (Weisz et al. 1984; Gruart et al. 2006). In their experimental approach to this interesting question, Gruart et al. (2006) were able to evoke LTP in behaving mice prior to selected conditioning sessions in order to determine its effects on classical conditioning of eyelid responses. In Fig. 5.3 are illustrated the ability of LTP induction to disturb both the acquisition of the associative learning and the profile of evoked fEPSPs. LTP was equally effective in blocking CRs during recall and reconditioning tasks (Gruart et al. 2006). Accordingly, it can be proposed that the functional changes in synaptic strength taking place in the CA3-CA1 synapse during associative learning

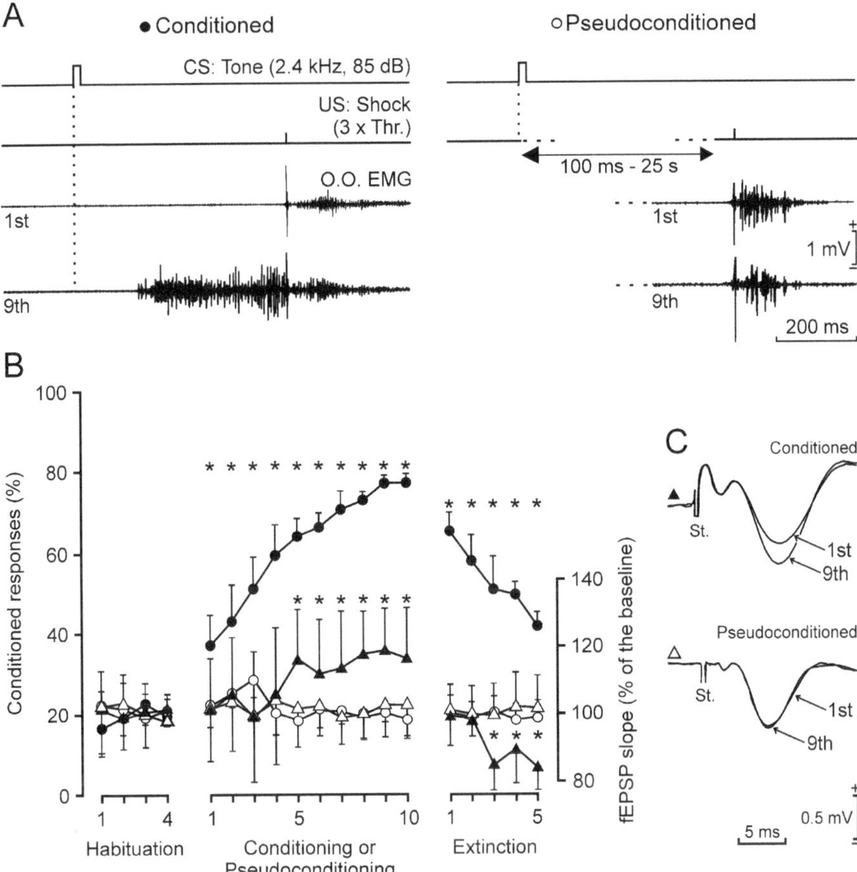

Fig. 5.2 The slope of fEPSPs evoked at the CA3-CA1 synapse increases in parallel with the acquisition of classically conditioned eyelid responses. (**a**) A representation of the experimental paradigm, illustrating CS and US and representative examples of EMG recordings from the orbicularis oculi (O.O.) muscle obtained from the 1st and the 9th conditioning sessions collected from a conditioned (left set of records) and a pseudoconditioned (right) mouse. For pseudoconditioning, CS and US were unpaired to prevent any putative association between them. (**b**) The *graphs* show the percentage (%) of CRs during successive sessions for conditioned (*black circles*) and pseudoconditioned (*white circles*) groups. The fEPSP slope is also indicated for conditioned (*black triangles*) and pseudoconditioned (*white triangles*) groups, expressed as the % change with respect to mean values collected during the four habituation sessions. Mean % values are followed by ± SD. (**c**) Representative fEPSPs recorded in the CA1 area following a single pulse presented to the ipsilateral Schaffer collaterals 300 ms after CS presentation, in a conditioned (*black triangle*) and in a pseudoconditioned (*white triangle*) animal, during the 1st and 9th conditioning sessions. Calibration as indicated (Taken with permission and modified from Gruart et al. 2006)

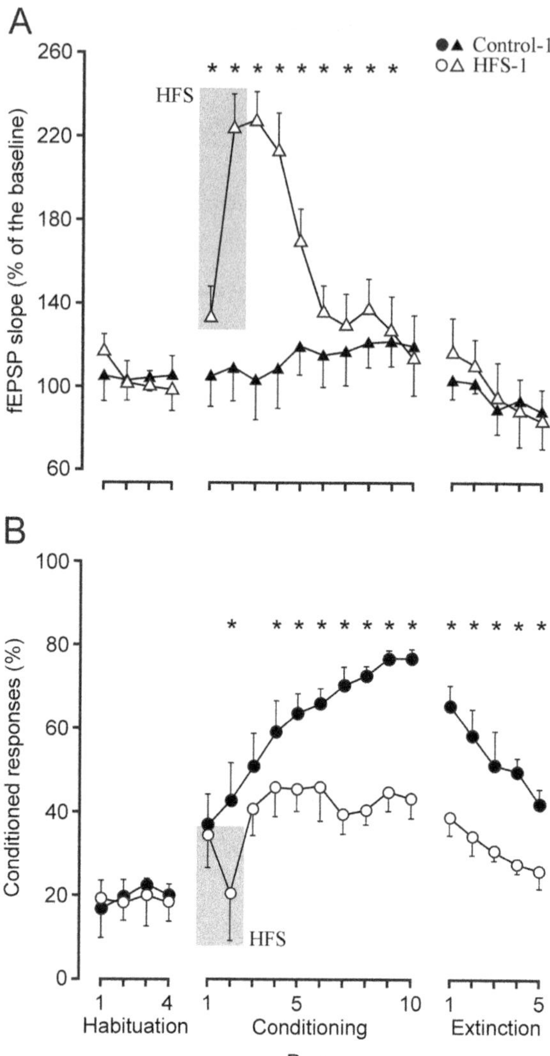

Fig. 5.3 LTP prevents the acquisition of classically conditioned eyelid responses. (**a, b**) fEPSP slope (**a**, *triangles*) and percentage of eyelid CRs (**b**, *circles*) in controls (*black*) and in mice that received a high-frequency stimulation (HFS) protocol 10 min before the first two conditioning sessions. The HFS consisted of five trains (200 Hz, 100 ms) of pulses (100 μs, square, biphasic) at a rate of 1/s. This protocol was presented six times in total. As a result, the fEPSP slope for the HFS group was significantly above baseline values during the first 9 days of conditioning (**a**). The acquisition and extinction curves presented by the HFS group were also significantly different from those of controls (**b**). Values are expressed as mean ± SD. *, $P < 0.001$ (Taken with permission and modified from Gruart et al. 2006)

are similar to (although more physiologically induced than) those evoked by the experimental induction of LTP.

The similitudes and differences between learning-dependent changes in synaptic activities and those evoked by experimentally triggered LTP deserve a further comment. It is well known that the generation of classically conditioned eyelid responses requires a considerable number (<300) of paired CS-US presentations, as recorded in mice, rabbits, and cats (Woody 1986; Thompson 1988, 2005; Gruart et al. 1995, 2000a, b; Takatsuki et al. 2003; Domínguez-del-Toro et al. 2004). Moreover, eyelid CRs present a characteristic ramp-like profile and a long latency (>50 ms) from CS onset, as well as a quantum-by-quantum increase in amplitude and duration (Domingo et al. 1997). All these procedural and kinetic characteristics suggest that the neural processes underlying the generation of CRs are not directly related to LTP (or to long-term depression, LTD) mechanisms by which an immediate acquisition of the evoked synaptic response is obtained (Ito 1989; Bliss and Collingridge 1993). Thus, the neural response expected from associative learning is not a sharp, sustained increase in fEPSP profiles or in neuronal discharge rates, but a distributed and limited increase in the number of neurons recruited to respond to an initially irrelevant sensory stimulus represented by the CS (Woody 1986). In this regard, we should pay attention to reliable proposals suggesting a gradual, feed-forward generation of learned movements, involving many pre-motor centers and/or circuits (Houk et al. 1996; Delgado-García and Gruart 2002). Obviously, experimental in vitro and in vivo procedures to evoke LTP involve a strong activation of involved synaptic contacts (Bliss and Gardner-Medwin 1973; Bliss and Lømo 1973), but LTP has the property of associability, suggesting that a weak input can still be potentiated if activated (within a given time window) by a strong stimulus evoked through a different but convergent input (McNaughton et al. 1978; Bliss and Collingridge 1993; Levy and Steward 1983). It is feasible, therefore, to suggest that the potentiation process observed at the CA1-CA3 synapse during classical conditioning of eyelid responses is a physiological resemblance of the LTP mechanism evoked experimentally both in vitro and in vivo using cruder procedures. As recently shown, the experimental induction of LTP at different stages of conditioning is capable of introducing a noticeable disturbance in the acquisition (or extinction) process (Fig. 5.3; Gruart et al. 2006). Thus, the answer to our second question, whether activity-dependent changes are related to LTP, is that although there are important functional differences between experimentally evoked LTP and activity-dependent changes in synaptic strength, these two neural mechanisms have similar properties. A more detailed description of the functional similitudes and differences between these two neural phenomena has been presented elsewhere (Madroñal et al. 2007, 2009).

Comparative Roles of NMDA Channels in Associative Learning, Synaptic Plasticity, and Long-Term Potentiation

In the preceding section, we have shown that activity-dependent synaptic changes in functional efficiency and LTP are related phenomena. In this regard, it seems interesting to determine the role of NMDA receptors in these two functionally related neural phenomena. As indicated above, NMDA receptors are intimately related to LTP induction (Collingridge et al. 1983a, b; Bliss and Collingridge 1993; Harris et al. 1984; Malenka and Nicoll 1999). Moreover, it was proposed some time ago that hippocampal NMDA receptors are involved in the acquisition of eyelid CRs (Kishimoto et al. 2001; Sanders and Fanselow 2003; Mokin and Keifer 2005).

Two of the conditions needed for evoking LTP at the CA3-CA1 synapse are postsynaptic depolarization of CA1 pyramidal cells and the simultaneous activation of NMDA receptors located on those neurons (Bliss and Collingridge 1993; Malenka and Nicoll 1999). In fact, there is already enough experimental evidence showing that hippocampal NMDA receptors are involved in the acquisition of classically conditioned eyelid responses (Servatius and Shors 1996; Kishimoto et al. 2001; Sanders and Fanselow 2003; Mokin and Keifer 2005). For example, it has been shown that the administration of (E)-($\pm$)-2-amino-4-methyl-5-phosphono-3-pentenoic acid ethyl ester [CGP 39551, a competitive antagonist of the NMDA receptor, and frequently used for in vivo studies (Maren et al. 1992; Servatius and Shors 1996; D'Hooge et al. 1999)] prevents the acquisition of a classically conditioned eyelid response (Gruart et al. 2006). Moreover, CGP 39551 administration to behaving mice blocks the fEPSP potentiation of the hippocampal CA3-CA1 synapse observed in controls across conditioning (Fig. 5.4b). As already reported in anesthetized rats for the perforant path-dentate gyrus synapse (Maren et al. 1992), CGP 39551 also prevented the induction of LTP in the hippocampal CA1 area following high frequency stimulation (HFS) of Schaffer collaterals (Fig. 5.4a). It is important to note that CGP 39551 had no effect on the monosynaptic fEPSP evoked in the CA1 area by single pulses presented to the ipsilateral Schaffer collaterals in the absence of conditioning (CS-US) stimuli. In contrast, the fEPSPs evoked in CA1 pyramidal cells are attenuated by NBQX (2,3-dioxo-6-nitro-1,2,3,4-tetrahydrobenzo[f]quinoxaline-7-sulfonamide disodium salt), a selective and competitive α-amino-3-hydroxy-5-methyl-4-isoxazolepropionic acid (AMPA)-receptor antagonist also used for in vivo studies (Parada et al. 1992; Namba et al. 1994; Gruart et al. 2006). The effects of NBQX indicate that CA3-CA1 synapses are activated in normal conditions by the opening of AMPA channels.

As an answer to our third question, and in accordance with the available information collected from both in vitro and in vivo studies, there is substantial evidence regarding the intrinsic relationships between LTP, activity-dependent synaptic plasticity, NMDA-receptor activation, and associative learning in mammals.

Learning is Not the Result of the Activation of a Single Neurotransmitter Receptor

Since the 1960s, the species of choice for classical conditioning of nictitating membrane (or eyelid) responses has been the rabbit (Gormezano et al. 1983; Thompson 1988, 2005). As already mentioned, the availability of genetically manipulated mice (transgenic, knock-in, knock-out, etc.) has prompted behavioral and electrophysiological researchers to use these mutated animals as an interesting experimental tool for the study of neural processes underlying this type of associative learning task (Gruart and Delgado-García 2007). An important result of this series of studies is that, besides the significant contribution of NMDA receptors to learning and memory processes, many other neurotransmitters, neurotransmitter receptors, intracellular

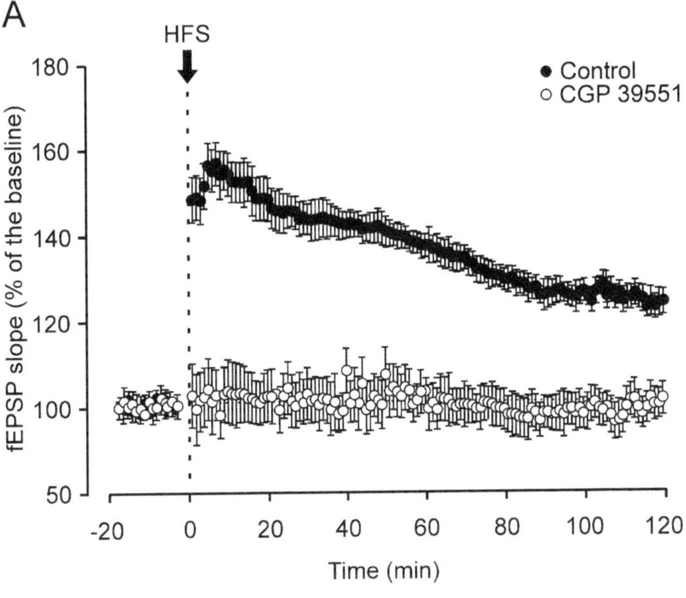

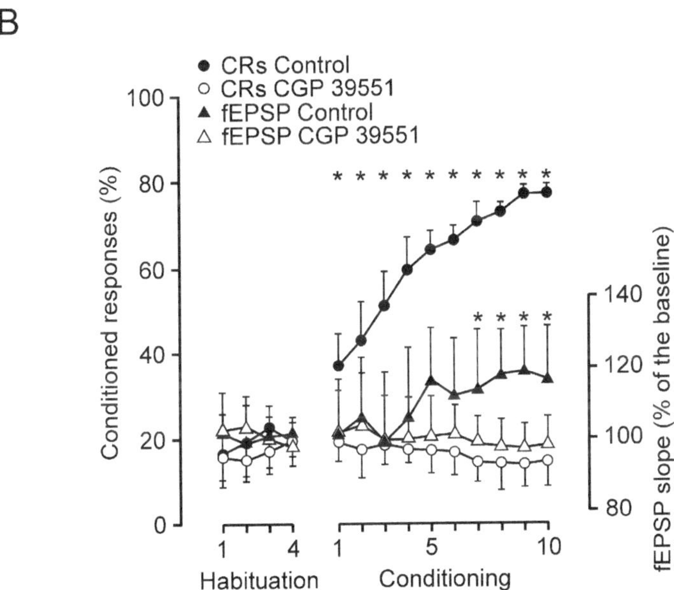

Fig. 5.4 NMDA antagonists are able to block both LTP and classical conditioning in alert behaving mice. (**a**) LTP was easily evoked in controls using an HFS protocol, but the i.p. administration of 6.5 mg/kg CGP 39551 (an NMDA antagonist) 1 h before HFS prevented LTP induction. (**b**) Learning curves (*black circles* for controls and *white circles* for CGP-39551-injected animals) and fEPSPs evoked in the hippocampal CA1 area (*black triangles* for controls and *white triangles* for CGP-39551-injected animals) by the electrical stimulation of the ipsilateral Schaffer collaterals. The CGP 39551 group was injected 30 min before each conditioning session (6.5 mg/kg, i.p.). Note that animals from the CGP 39551 group were unable to acquire the expected CRs. Furthermore, fEPSPs evoked in the CGP 39551 group did not change across conditioning. Values are expressed as mean ± SD (Taken with permission and modified from Gruart et al. 2006)

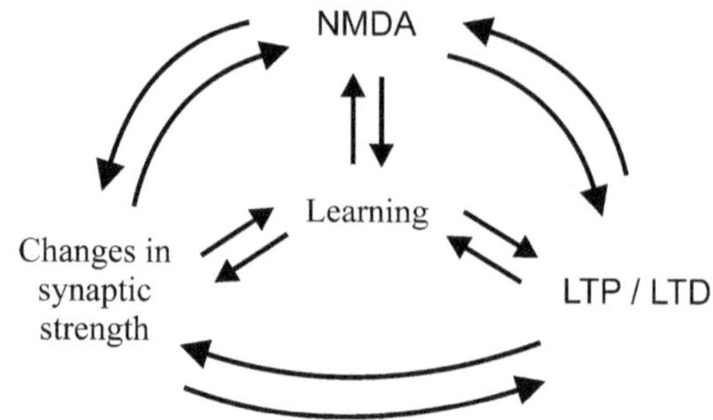

Fig. 5.5 A representation of the respective interactions between NMDA receptors, LTP (or LTD) processes, and changes in synaptic strength in relation with associative learning. See text for details (Taken with permission from Gruart and Delgado-García 2007)

enzymes and mediator factors, transcription factors, etc., have also been implicated in those neural phenomena. We are including below a short description of the studies carried out in our laboratory during the past 10 years with regard to the involvement of many different molecular components in the acquisition of new memories.

Besides describing the role of NMDA receptors in alert behaving mice during classical eyeblink conditioning, we have reported that the targeted disruption of the mGLUR1 gene also modifies the learning capabilities of these genetically manipulated mice (Gil-Sanz et al. 2008). Other neurotransmitter receptors present in hippocampal circuits, such as adenosine A_{2A} (Fontinha et al. 2009), endogenous cannabinoid CB1 (Madroñal et al. 2012), and dopamine Drd1a (Ortiz et al. 2010), are also involved in the generation of conditioned eyelid responses. In addition, neurotrophin receptors such as the tropomyosin receptor kinase B (TrkB; Gruart et al. 2007) and the tropomyosin receptor kinase C (TrkC; Sahún et al. 2007) and their corresponding intraneuronal cascades, as well as development-related proteins such as reelin (Pujadas et al. 2010), seem to play a role in the generation of memory traces. Finally, the overexpression of the transcription factor CREB (VP16-CREB mice; Gruart et al. 2012) (see Chap. 2), the lack of glycogen synthase in the central nervous system (GYS1$^{Nestin-KO}$ mice; Duran et al. 2013), the inhibition of the protein kinase Mζ (Madroñal et al. 2010), and a deficit in DNA polymerase μ (Polμ$^{-/-}$ mice; Lucas et al. 2013) can modify the proper acquisition of a classical eyeblink conditioning task by alert behaving mice. Interestingly, this deficit in DNA polymerase μ evoked a surprising improvement in the learning capabilities of these mice!

This brief description of studies carried out in our laboratory during recent years on learning capabilities of different types of genetically manipulated mice is intended only to point out the molecular complexity underlying the acquisition, storage, and retrieval of new motor and cognitive abilities. A more complete picture of these in vivo studies in genetically manipulated mice has recently been offered elsewhere (Gruart et al. 2013). In conclusion, Fig. 5.5 should be understood as an oversimplification of molecular events taking place in synaptic contacts related to the acquisition and storage of memory traces.

Learning as a Functional State of Cortical and Subcortical Structures

A frequent error of restricting the study of synaptic plastic events during motor learning to a single synaptic contact is the acceptance that other related synapses will follow a similar (activation or depression) functional pattern. We have addressed this important matter in a recent study (Gruart et al. 2014) in which we followed the activity-dependent changes in synaptic strength of nine different hippocampal synapses (corresponding to the intrinsic hippocampal circuitry and to its main inputs and outputs) during the acquisition of a trace eyeblink conditioning in behaving mice. The timing and degree of synaptic changes across the acquisition process were determined with the help of analytical tools developed in our laboratory. The time course of changes in synaptic strength indicated that the synaptic contacts were not modified in anatomical sequence (Fig. 5.6). Furthermore, we explored the functional relevance of the extrinsic and intrinsic afferents to CA3 and CA1 pyramidal neurons, and evaluated the distinct input patterns to the intrinsic hippocampal circuit. Collected results confirmed that the acquisition of a classical eyeblink conditioning is a multi-synaptic process in which the contribution of each synaptic cortical contact is different in strength, and takes place at different moments across learning. Thus, the precise and timed activation of multiple hippocampal synaptic contacts during classical eyeblink conditioning evokes a specific, dynamic map of functional synaptic states in that circuit (Fig. 5.6). These results strongly support our early proposal of learning being considered the result of the activation of complex and disseminated functional cortical states with the participation of large populations of neuronal networks (Delgado-García and Gruart 2002).

Another interesting aspect of learning is the role of context in this type of experimental study with behaving animal models. We have addressed this issue in a recent study (Carretero-Guillén et al. 2015). It shows that both context and pseudoconditioning training evoke early, lasting changes in synaptic strength in perforant pathway synapses to dentate gyrus (PP-DG) and hippocampal CA3 (PP-CA3) and CA1 (PP-CA1) areas. Pseudoconditioning also evoked early, non-lasting changes in strength within the intrinsic hippocampal circuit (CA3-CA1 and CA3-cCA1 synapses). In contrast, during both trace and delay training sessions, changes in synaptic strength were mostly noticed within the intrinsic hippocampal circuit (DG-CA3, CA3-CA1; CA3-cCA1). It should be noticed that, although the hippocampus is not required for the acquisition of a delay conditioning, hippocampal synapses are also modified during this type of conditioning paradigm. In addition, the response of hippocampal synapses to afferent impulses seems to be modulated differentially by both context and cues during associative learning in behaving rabbits.

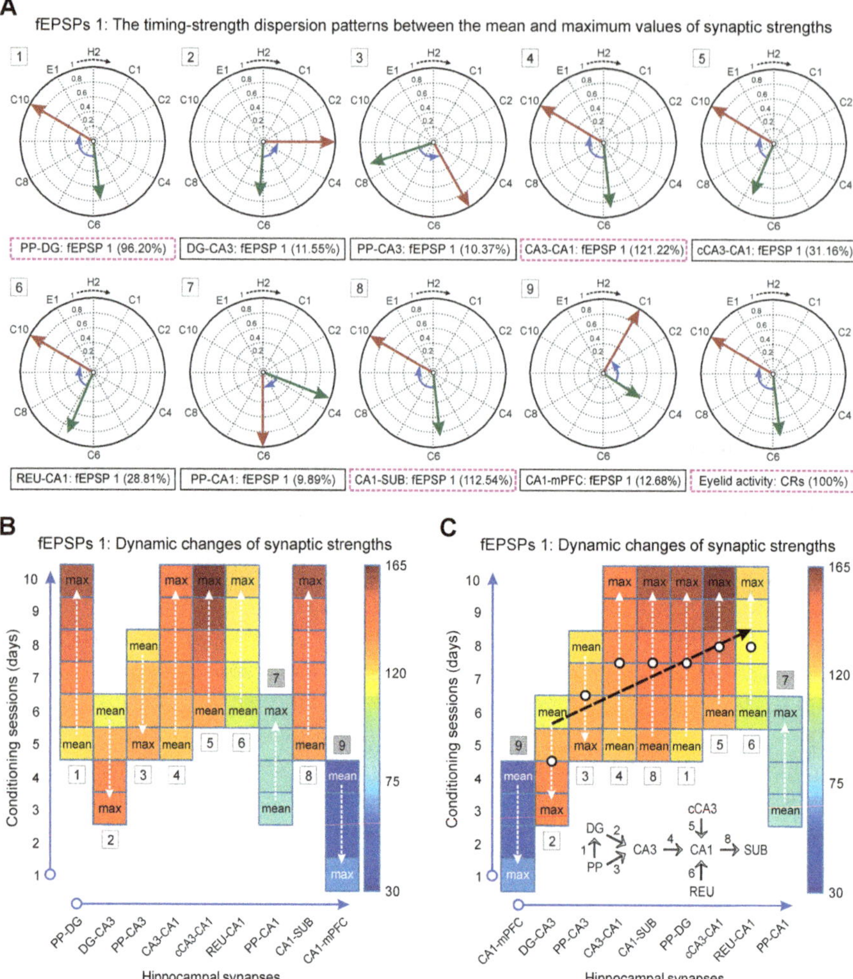

Fig. 5.6 Graphs of timing and strength and functional organization of the nine selected hippocampal synapses (including the main inputs and outputs) show changes across learning. The nine selected synapses were perforant pathway (PP)-dentate gyrus (DG), DG-hippocampal CA3, PP-CA3, CA3-CA1, contralateral CA3 (cCA3)-CA1, thalamic reuniens nucleus (REU)-CA1, PP-CA1, CA1-subiculum (SUB), and CA1-medial prefrontal cortex (mPFC). (**a**) Timing-strength dispersion patterns between the mean (*green arrows*) and maximum (*red arrows*) values of synaptic strengths reached across conditioning sessions. The normalized fEPSP slope determined the strength/magnitude of the vector, while the training session (habituation, H2; conditioning, C1–C10; or extinction, E1) determined the timing/orientation of the vector. Three synapses (PP-DG, CA3-CA1, and CA1-SUB) showed a dispersion pattern similar to that of eyelid activity (% CRs), whilst other synapses (cCA3–CA1, REU–CA1, and PP–CA1) presented smaller timing-strength dispersion patterns. Finally, the synapses DG–CA3, PP–CA3, and CA1–mPFC showed smaller values of the dispersion indices but with angular displacements in opposite directions (see *blue bent arrow* inside each circumference). (**b, c**) *Color map* representations of synaptic strengths (fEPSP slopes, as % of baseline) between mean and maximum (max) values (*white dashed arrows*) for all the synapses. In **b** is illustrated the preliminary distribution of the nine synapses according to anatomical criteria and connectivity, while in **c** is illustrated the functional distribution of synapses according to the timing-strength dispersion index for each synapse, the relative dispersion index between the circular patterns, and the trend (see *black dashed arrow*) of the synaptic evolution index across synapses (Taken with permission from Gruart et al. 2014)

Concluding Remarks

It has already been proposed (Delgado-García and Gruart 2002) that learning is a precise functional state of the brain, and that we should take a dynamic approach to the study of neural and synaptic activities in ensembles of sensorimotor and cognition-related circuits during actual learning in alert behaving animals. In this regard, it can be suggested that each environmental and social situation demanding a behavioral response will evoke a corresponding differential state of synaptic weights in hippocampal circuits. Obviously, additional neural, synaptic, and motor information can be collected experimentally and added to the better determination of ongoing functional states. However, it is important to point out that our information with regard to brain functioning during a given learning situation is greatly constrained by the difficulty of recording a large enough number of kinetic (i.e., firing and synaptic activities of neuronal elements) and kinematic (i.e., biomechanical characteristics of evoked motor responses) parameters in simultaneity with the newly acquired ability (Delgado-García and Gruart 2002). To resolve these constraints, it seems necessary to record enough different neural kinetic data at the same time as collecting data from enough kinematic variables. For example, in a recent study we were able to collect up to 24 kinetic variables (related to neural firing activities in the facial and cerebellar interpositus nuclei) together with 36 kinematic variables (related to eyelid biomechanics and to the electrical activity of the orbicularis oculi muscle) from alert behaving cats during classical eyeblink conditioning (Sánchez-Campusano et al. 2007). Results collected by our group in recent years allow a dynamic interpretation of the hippocampal role in learning and memory processes underlying the acquisition of new motor and cognitive abilities, as opposed to an excessively localizationist view of hippocampal functions (McHugh et al. 2007). Indeed, the hippocampus and related cortical structures will have an almost infinite repertoire of functional states corresponding to the enormous possibilities of sensory stimulations and the different needs of behavioral responses.

Acknowledgements This study was supported by Spanish MINECO (BFU2011-29089 and BFU2011-29286) and Junta de Andalucía (BIO122 and CVI7222) grants to A.G. and J.M. D.-G. We thank Mr. Roger Churchill for his help in editing the manuscript.

References

Ammann C, Márquez-Ruiz J, Gómez-Climent MA, Delgado-García JM, Gruart A (2016) The motor cortex is involved in the generation of classically conditioned eyelid responses in behaving rabbits. J Neurosci 36:6988–7001

Barnes CA, Jung MW, McNaughton BL, Korol DL, Andreasson K, Worley PF (1994) LTP saturation and spatial learning disruption: effects of task variables and saturation levels. J Neurosci 14:5793–5806

Berger TW, Rinaldi P, Weisz DJ, Thompson RF (1983) Single-unit analysis of different hippocampal cell types during classical conditioning of rabbit nictitating membrane response. J Neurophysiol 50:1197–1219

Bliss TVP, Collingridge GL (1993) A synaptic model of memory: long-term potentiation in the hippocampus. Nature 361:31–39

Bliss TVP, Gardner-Medwin AR (1973) Long-lasting potentiation of synaptic transmission in the dentate area of the unanaesthetized rabbit following stimulation of the perforant path. J Physiol 232:357–374

Bliss TVP, Lømo T (1973) Long-lasting potentiation of synaptic transmission in the dentate area of the anaesthetized rabbit following stimulation of the perforant path. J Physiol 232:331–356

Caro-Martín CR, Leal-Campanario R, Sánchez-Campusano R, Delgado-García JM, Gruart A (2015) A variable oscillator underlies the measurement of time intervals in the rostral medial prefrontal cortex during classical eyeblink conditioning in rabbits. J Neurosci 35:14809–14821

Carretero-Guillén A, Pacheco-Calderón R, Delgado-García JM, Gruart A (2015) Involvement of hippocampal inputs and intrinsic circuit in the acquisition of context and cues during classical conditioning in behaving rabbits. Cereb Cortex 25:1278–1289. doi:10.1093/cercor/bht321

Clarke JR, Cammarota M, Gruart A, Izquierdo I, Delgado-García JM (2010) Plastic modifications induced by object recognition memory processing. Proc Natl Acad Sci U S A 107:2652–2657

Collingridge GL, Kehl SJ, McLennan H (1983a) The antagonism of amino acid-induced excitations of rat hippocampal CA1 neurones in vitro. J Physiol 334:19–31

Collingridge GL, Kehl SJ, McLennan H (1983b) Excitatory amino acids in synaptic transmission in the Schaffer collateral-commissural pathway of the rat hippocampus. J Physiol 334:33–46

D'Hooge R, Raes A, Van de Vijver G, Van Bogaert PP, De Deyn PP (1999) Effects of competitive NMDA receptor antagonists on excitatory amino acid-evoked currents in mouse spinal cord neurones. Fundam Clin Pharmacol 13:67–74

Delgado-García JM, Gruart A (2002) The role of interpositus nucleus in eyelid conditioned responses. Cerebellum 1:289–308

Delgado-García JM, Gruart A (2006) Building new motor responses: eyelid conditioning revisited. Trends Neurosci 29:330–338

Domingo JA, Gruart A, Delgado-García JM (1997) Quantal organization of reflex and conditioned eyelid responses. J Neurophysiol 78:2518–2530

Domínguez-del-Toro E, Rodríguez-Moreno A, Porras-García E, Sánchez-Campusano R, Blanchard V, Lavilla M, Böhme GA, Benavides J, Delgado-García JM (2004) An in vitro and in vivo study of early deficits in associative learning in transgenic mice that over-express a mutant form of human APP associated with Alzheimer's disease. Eur J Neurosci 20:1945–1952

Dragoi G, Harris KD, Buzsáki G (2003) Place representation within hippocampal networks is modified by long-term potentiation. Neuron 39:843–853

Duran J, Saez I, Gruart A, Guinovart J, Delgado-García JM (2013) Impairment in long-term memory formation and learning-dependent synaptic plasticity in mice lacking glycogen synthase in the brain. J Cereb Blood Flow Metab 33:550–556

Gil-Sanz C, Delgado-García JM, Fairén A, Gruart A (2008) Involvement of the mGluR1 receptor in hippocampal synaptic plasticity and associative learning in behaving mice. Cereb Cortex 18:1653–1663

Gormezano I, Kehoe EJ, Marshall BS (1983) Twenty years of classical conditioning research with the rabbit. Prog Psychobiol Physiol Psychol 10:197–275

Gruart A, Delgado-García JM (2007) Activity-dependent changes of the hippocampal CA3-CA1 synapse during the acquisition of associative learning in conscious mice. Genes Brain Behav Suppl 1:24–31

Gruart A, Blázquez P, Delgado-García JM (1995) Kinematics of unconditioned and conditioned eyelid movements in the alert cat. J Neurophysiol 74:226–248

Gruart A, Pastor AM, Armengol JA, Delgado-García JM (1997) Involvement of cerebellar cortex and nuclei in the genesis and control of unconditioned and conditioned eyelid motor responses. Prog Brain Res 114:511–528

Gruart A, Guillazo-Blanch G, Fernández-Mas R, Jiménez-Díaz L, Delgado-García JM (2000a) Cerebellar posterior interpositus nucleus as an enhancer of classically conditioned eyelid responses in alert cats. J Neurophysiol 84:2680–2690

Gruart A, Schreurs BG, Domínguez-del-Toro E, Delgado-García JM (2000b) Kinetic and frequency-domain properties of reflex and conditioned eyelid responses in the rabbit. J Neurophysiol 83:836–852

Gruart A, Muñoz MD, Delgado-Garcia JM (2006) Involvement of the CA3-CA1 synapse in the acquisition of associative learning in behaving mice. J Neurosci 26:1077–1087

Gruart A, Sciarreta C, Valenzuela-Harrington M, Delgado-García JM, Minichiello L (2007) Mutation at the TrkB PLCγ-docking site affects hippocampal LTP and associative learning in conscious mice. Learn Mem 14:54–62

Gruart A, Benito E, Delgado-García JM, Barco A (2012) Enhanced cAMP response element-binding protein activity increases neuronal excitability, hippocampal long-term potentiation, and classical eyeblink conditioning in alert behaving mice. J Neurosci 32:17431–17441

Gruart A, Madroñal N, Jurado-Parras MT, Delgado-García JM (2013) Synaptic plasticity studies and their applicability in mouse models of neurodegenerative diseases. Translat Neurosci 4:134–143

Gruart A, Sánchez-Campusano R, Fernández-Guizán A, Delgado-García JM (2014) A differential and timed contribution of identified hippocampal synapses to associative learning in mice. Cereb Cortex 25(9):2542–2555. doi:10.1093/cercor/bhu054

Harris EW, Ganong AH, Cotman CW (1984) Long-term potentiation in the hippocampus involves activation of N-methyl-D-aspartate receptors. Brain Res 323:132–137

Hebb DO (1949) The organization of behavior. Wiley, New York

Hoehler FK, Thompson RF (1980) Effect of the interstimulus (CS-UCS) interval on hippocampal unit activity during classical conditioning of the nictitating membrane response of the rabbit (*Oryctolagus cuniculus*). J Comp Physiol Psychol 94:201–215

Houk JC, Buckingham JT, Barto AG (1996) Models of cerebellum and motor learning. Behav Brain Sci 19:368–383

Ito M (1989) Long-term depression. Annu Rev Neurosci 12:85–102

Jurado-Parras MT, Sánchez-Campusano R, Castellanos NP, del-Pozo F, Gruart A, Delgado-García JM (2013) Differential contribution of hippocampal circuits to appetitive and consummatory behaviors during operant conditioning of behaving mice. J Neurosci 33:2293–2304

Kandel ER (2001) The molecular biology of memory storage: a dialogue between genes and synapses. Science 294:1030–1038

Kishimoto Y, Kawahara S, Mori H, Mishima M, Kirino Y (2001) Long-trace interval eyeblink conditioning is impaired in mutant mice lacking the NMDA receptor subunit ε1. Eur J Neurosci 13:1221–1227

Konorski J (1948) Conditioned reflexes and neuron organization. Cambridge University Press, Cambridge, MA

Kugelberg E (1952) Facial reflexes. Brain 75:385–396

Leung LS, Shen B (2006) Hippocampal CA1 kindling but not long-term potentiation disrupts spatial memory performance. Learn Mem 13:18–26

Levy WB, Steward O (1983) Temporal contiguity requirements for long-term associative potentiation/depression in the hippocampus. Neuroscience 8:791–797

Lucas D, Delgado-García JM, Escudero B, Albo C, Aza A, Acín-Pérez R, Torres Y, Moreno P, Enríquez JA, Samper E, Blanco L, Fairén A, Bernard A, Gruart A (2013) Increased learning and brain long-term potentiation in aged mice lacking DNA polymerase μ. PLoS One 8(1):e53243. doi:10.1371/journal.pone.0053243

Lynch MA (2004) Long-term potentiation and memory. Physiol Rev 84:87–136

Madroñal N, Delgado-García JM, Gruart A (2007) Differential effects of long-term potentiation evoked at the CA3-CA1 synapse before, during, and after the acquisition of classical eyeblink conditioning in behaving mice. J Neurosci 27:12139–12146

Madroñal N, Gruart A, Delgado-García JM (2009) Differing presynaptic contributions to LTP and associative learning in behaving mice. Front Behav Neurosci 3:7. doi:10.3389/neuro.08.007.2009

Madroñal N, Gruart A, Sacktor TC, Delgado-García JM (2010) PKMζ inhibition reverses learning-induced increases in hippocampal synaptic strength and memory during trace eyeblink conditioning. PLoS One 5(4):e10400. doi:10.1371/journal.pone.0010400

Madroñal N, Gruart A, Valverde O, Espadas I, Moratalla R, Delgado-García JM (2012) Involvement of cannabinoid CB1 receptor in associative learning and in hippocampal CA3-CA1 synaptic plasticity. Cereb Cortex 22:550–566

Malenka R, Nicoll RA (1999) Long-term potentiation – a decade of progress? Science 285:1870–1874

Maren S, Baudry M, Thompson RF (1992) Effects of the novel NMDA receptor antagonist, CGP 39551, on field potentials and the induction and expression of LTP in the dentate gyrus in vivo. Synapse 11:221–228

Marr D (1971) Simple memory: a theory of archicortex. Philos Trans R Soc Lond (Biol) 262:23–81

McEchron MD, Disterhoft JF (1997) Sequence of single neuron changes in CA1 hippocampus of rabbits during acquisition of trace eyeblink conditioned responses. J Neurophysiol 78:1030–1044

McEchron MD, Tseng W, Disterhoft JF (2003) Single neurons in CA1 hippocampus encode trace interval duration during trace heart rate (fear) conditioning in rabbit. J Neurosci 23:1535–1547

McHugh TJ, Jones MW, Quinn JJ, Balthasar N, Coppari R, Elmquist JK, Lowell BB, Fanselow MS, Wilson MA, Tonegawa S (2007) Dentate gyrus NMDA receptors mediate rapid pattern separation in the hippocampal network. Science 317:94–99

McNaughton BL, Douglas RM, Goddard GV (1978) Synaptic enhancement in fascia dentata: cooperativity among coactive afferents. Brain Res 157:277–293

Mokin M, Keifer J (2005) Expression of the immediate-early gene-encoded protein Egr-1 (zif268) during in vitro classical conditioning. Learn Mem 12:144–149

Moser EI, Kropff E, Moser MB (2008) Place cells, grid cells, and the brain's spatial representation system. Rev Neurosci 31:69–89

Moyer JR Jr, Deyo RA, Disterhoft JF (1990) Hippocampectomy disrupts trace eye-blink conditioning in rabbits. Behav Neurosci 104:243–252

Múnera A, Gruart A, Muñoz MD, Fernández-Mas R, Delgado-García JM (2001) Discharge properties of identified CA1 and CA3 hippocampus neurons during unconditioned and conditioned eyelid responses in cats. J Neurophysiol 86:2571–2582

Namba T, Morimoto K, Sato K, Yamada N, Kuroda S (1994) Antiepileptogenic and anticonvulsant effects of NBQX, a selective AMPA receptor antagonist, in the rat kindling model of epilepsy. Brain Res 638:36–44

Neves G, Cooke SF, Bliss TV (2008) Synaptic plasticity, memory and the hippocampus: a neural network approach to causality. Nat Rev Neurosci 9:65–75

Ortiz O, Delgado-García JM, Espadas I, Bahí A, Trullas R, Dreyer JL, Gruart A, Moratalla R (2010) Associative learning and CA3-CA1 synaptic plasticity are impaired in D1R null, Drd1a-/- mice and in hippocampal siRNA silenced Drd1a mice. J Neurosci 30:12288–12300

Parada J, Czuczwar SJ, Turski WA (1992) NBQX does not affect learning and memory tasks in mice: a comparison with D-CPPene and ifenprodil. Brain Res Cogn Brain Res 1:67–71

Pujadas L, Gruart A, Delgado L, Teixeira CM, Rossi D, de Lecea L, Martínez A, Delgado-García JM, Soriano E (2010) Reelin regulates postnatal neurogenesis and enhances spine hypertrophy and long-term potentiation. J Neurosci 30:4636–4649

Ramón y Cajal S (1909-1911) Histologie du Système Nerveux de l'Homme et des Vertébrés. Maloine, Paris

Sahún I, Delgado-García JM, Amador-Arjona A, Giralt A, Alberch J, Dierssen M, Gruart A (2007) Dissociation between CA3-CA1 synaptic plasticity and associative learning in TgNTRK3 transgenic mice. J Neurosci 27:2253–2260

Sánchez-Campusano R, Gruart A, Delgado-García JM (2007) The cerebellar interpositus nucleus and the dynamic control of learned motor responses. J Neurosci 27:6620–6632

Sanders MJ, Fanselow MS (2003) Pre-training prevents context fear conditioning deficits produced by hippocampal NMDA receptor blockade. Neurobiol Learn Mem 80:123–129

Servatius RJ, Shors TJ (1996) Early acquisition, but not retention, of classically conditioned eyeblink responses is N-methyl-D-aspartate (NMDA) receptor dependent. Behav Neurosci 110:1040–1048

Takatsuki K, Kawahara S, Kotani S, Fukunaga S, Mori H, Mishina M, Kirino Y (2003) The hippocampus plays an important role in eyeblink conditioning with a short trace interval in glutamate receptor subunit δ2 mutant mice. J Neurosci 23:17–22

Thompson RF (1988) The neural basis of basic associative learning of discrete behavioral responses. Trends Neurosci 11:152–155

Thompson RF (2005) In search of memory traces. Annu Rev Psychol 56:1–23

Wang SH, Morris RG (2010) Hippocampal-neocortical interactions in memory formation, consolidation, and reconsolidation. Annu Rev Psychol 61:49–79

Weisz DJ, Clark GA, Thompson RF (1984) Increased responsivity of dentate granule cells during nictitating membrane response conditioning in rabbit. Behav Brain Res 12:145–154

Whitlock JR, Heynen AJ, Shuler MG, Bear MF (2006) Learning induces long-term potentiation in the hippocampus. Science 313:1093–1097

Woody CD (1986) Understanding the cellular basis of memory and learning. Annu Rev Psychol 37:433–493

Part II
Neural Plasticity in Early Postnatal Development

Chapter 6
Bidirectional Effects of Mother-Young Contact on the Maternal and Neonatal Brains

Gabriela González-Mariscal and Angel I. Melo

Abstract Adaptive plasticity occurs intensely during the early postnatal period through processes like proliferation, migration, differentiation, synaptogenesis, myelination and apoptosis. Exposure to particular stimuli during this critical period has long-lasting effects on cognition, stress reactivity and behavior. Maternal care is the main source of social, sensory and chemical stimulation to the young and is, therefore, critical to "fine-tune" the offspring's neural development. Mothers providing a low quantity or quality of stimulation produce offspring that will exhibit reduced cognitive performance, impaired social affiliation and increased agonistic behaviors. Transgenerational transmission of such traits occurs epigenetically, i.e., through mechanisms like DNA methylation and post-translational modification of nucleosomal histones, processes that silence or increase gene expression without affecting the DNA sequence. Reciprocally, providing maternal care profoundly affects the behavior, learning, memory and fine neuroanatomy of the adult female. Such effects are in many cases permanent and sometimes they involve the hormones of pregnancy and lactation. The above evidence supports the idea that the mother-young dyad exerts profound and permanent effects on the brains of both adult and developing organisms, respectively. Effects on the latter can be explained by the neural developmental processes taking place during the early postnatal period. In contrast, little is known about the mechanisms mediating the plasticity of the adult maternal brain. The bidirectional effects that mother and young exert on each other's brains exemplify a remarkable plasticity of this organ for organizing itself and provide an immense source of variability for adaptation and evolution in mammals.

Keywords Brain plasticity • Cognition • Critical periods • Dendritic arborization • DNA methylation • Epigenetic mechanisms • Lactation • Maternal behavior • Memory • Stress reactivity

G. González-Mariscal (✉) • A.I. Melo
Centro de Investigación en Reproducción Animal, CINVESTAV-Universidad Autónoma de Tlaxcala, Apdo Postal 62, Tlaxcala, Tlax 90000, Mexico
e-mail: gabygmm@gmail.com

© Springer International Publishing AG 2017
R. von Bernhardi et al. (eds.), *The Plastic Brain*, Advances in Experimental Medicine and Biology 1015, DOI 10.1007/978-3-319-62817-2_6

Abbreviations

5-HIAA	5-Hydroxyindoleacetic acid
5-HT	5-Hydroxytryptamine
ACTH	Adrenocorticotrophin hormone
ADP-ribosyl	Adenosin diphosphate-ribosyl
AMG	Amygdala
AMPA	α-amino-3-hydroxy-5-methyl-4-isoxazolepropionic acid
AR	Artificial rearing
AVP	Arginine vasopressin
BDNF	Brain-derived neurotrophic factor
bFGF	Basic Fibroblast growth factor
BNST	Bed nucleus of the stria terminalis
CA1	Field CA1 of hippocampus
CA3	Field CA3 of hippocampus
CBZ	Benzodiazepines
CeA	Central nucleus of the amygdala
CD38	Transmembrane glycoprotein
CNS	Central nervous system
CORT	Corticosterone
CREB	cAMP response element-binding protein
CRF	Corticotropin-releasing factor
CRF *ir*	Corticotrophin-releasing factor -like immunoreactivity
CSF	Cerebrospinal fluid
DA	Dopamine
DNA	Deoxyribonucleid acid
DRN	Dorsal Raphe nucleus
EPM	Elevated plus maze
FC	Frontal cortex
FS	Forced swimming test
GABA	γ-aminobutyric acid
GAP-43	Growth associated protein 43
GC	Glucocorticoid
GFAP	Glial fibrillary acidic protein
LC/PBN	Locus coeruleus/parabrachial nucleus
LG-ABN	Licking/grooming arched-back nursing
LH	Luteinizing hormone
MAPK	Mitogen-activated protein kinase
mRNA	Ribonucleic acid messenger
MB	Maternal behavior
MD	Maternal deprivation
MeA	Medial amygdale
mPFC	Medial Prefrontal cortex
MPOA	Medial preoptic area

MR   Mother rearing
MS   Maternal Separation
NAcc   Nucleus accumbens
NCAM   Neural cell adhesion molecule
Neu-N   Neural structural protein
NMDA   N-Methyl-D-aspartic acid
NTS   Nucleus tractus solitarius
OT   Oxytocin
PFC   Prefrontal cortex
P   Posnatal day
PRL   Prolactin
PVN   Paraventricular nucleus
S-100β Astrocyte marker protein
SERT   Serotonin transporter
SHR   Spontaneous hypertensive rats
SON   Supraoptic nucleus
UV   Ultrasonic vocalization
WKY   Wistar Kyoto rat

Effects of Mothering on Offspring Development

An organism has the ability to adapt to an environment according to its genotype
and the environment, in turn, can modify gene expression (Godfrey et al. 2007;
Portha et al. 2014). The maximal expression of adaptive plasticity occurs during the
early postnatal period, when the nervous system is particularly sensitive to environmental insults and is undergoing temporally and regionally specific processes of
proliferation, migration, differentiation, synaptogenesis, myelination and apoptosis
(Rice and Barone 2000). The exposure to physical and/or biological stimuli during
this critical or sensitive period has long-lasting effects on cognitive, emotional, neuroendocrine, and behavioral outcomes in mammals (Fleming et al. 2002; Maestripieri
and Megna 2000; Teicher et al. 2012; González-Mariscal and Kinsley 2009;
González-Mariscal and Melo 2013). The mother, by providing maternal care, is the
main biological source of social, sensory (visual, auditory, olfactory and tactile),
and chemical (through milk) stimulation (Hofer 1984; see Melo 2015) which is
essential not only to allow survival of the young but also to "fine-tune" the offspring's neural development. Thus, a "good" mother that provides her newborn with
the optimal stimuli produces offspring that are resilient to disturbed environments
and less likely to show emotional alterations in adulthood. Conversely, a mother that
provides her young with a low quantity or quality of stimulation produces offspring
that will exhibit reduced cognitive performance, impaired affiliative behaviors, and
increased agonistic behaviors. These effects, in turn, can be transmitted to the next
generation (transgenerational transmission) in an epi-genetic fashion, i.e., through

Table 6.1 Effects of early life experience

	Positive		Negative	
	High LG-ABN	Handling	MS	AR
Emotionality				
Anxiety	Low	Low	High	High
Depression	Low	Low	High	High
Fear	Low	Low	High	High
Stress responses	Low	Low	High	High
Social interaction	Efficient	Efficient	Impairment	Impairment
Exploration	High	High	Low	High
Locomotion	Efficient	High	High	High
Cognition				
Learn and memory	Efficient	Efficient	Impairment	Impairment
Attention	Efficient	Efficient	Deficit	Deficit
Behaviors				
Maternal	Efficient	Efficient	Deficient	Deficient
Play-fight	Low	Low	High	–
Aggression	Low	Low	High	High
Vulnerability to drugs abuse	Low	Low	High	–
Neurochemical systems				
Serotonin	High	Low	Low/high	–
Dopamine	High	Low	High	High

sensory, hormonal and social stimuli from the mother. The mechanisms involved in this form of transmission include DNA methylation and post-translational modification of nucleosomal histones, processes that silence or increase gene expression without affecting the DNA sequence (Feng et al. 2007; Fagiolini et al. 2009; Razin 1998; Champagne 2008; Champagne et al. 2006).

There are two main experimental approaches to study the influence of maternal care on offspring development: (1) increasing the amount of licking/grooming arched-back nursing (high LG-ABN) or, conversely, decreasing the amount of LG-ABN, or (2) eliminating this stimulation altogether and using artificial rearing (AR).

1. Mothers that, due to natural variations in behavior, provide high LG-ABN to their offspring, and dams that are experimentally induced to exhibit more pup licking ("handling" procedure) positively affect the developing brain of their offspring (see Table 6.1). The handling procedure consists of removing the pups from the nest for 15 min/day across postnatal days 1–14. When the pups are returned to the nest, the mother licks them more than she would have if pups had remained in the nest.

Behavioral changes Many physiological and behavioral changes are induced in the young by high LG-ABN mothers and by the "handling" procedure. As adults such offspring are less emotional, adapt better to new environments, become "good" mothers, have more efficient cognitive responses (attention, learning) and are less aggressive. These offspring also show more exploratory behavior in a novel environment, spend less time burying a shock-probe, and display fewer defensive responses in the resident-intruder test. They display reduced anxiety, depression and stress responses, as reflected in a moderate HPA axis reactivity and behavioral responses to a stressful situation and lower levels of fear and startle responses (Menard and Hakvoort 2007; Caldji et al. 2000; Macri et al. 2004; Cirulli et al. 2003). Moreover, these rats perform better in spatial learning and memory tests (Liu et al. 2000). Finally, females (either mothers or non-lactating "sensitized" virgins) reared by high LG-ABN dams display maternal behavior with a short latency and express high levels of maternal licking to foster pups, compared to females reared by low LG-ABN dams (Pryce et al. 2001; Francis et al. 1999).

Physiological changes Offspring reared by high LG-ABN dams show: reduced plasma ACTH and corticosterone responses to restraint stress, enhanced glucocorticoids (GC) negative feedback sensitivity, increased transcription of the*N-r3cl*gene for GC receptors (Francis et al. 1999), and increased of GC receptor mRNA expression in the hippocampus. These animals also show: decreased Corticotropin Releasing Factor (CRF) mRNA expression in the hypothalamus, decreased central GABA/benzodiazepine receptor levels, increased benzodiazepine receptors in the amygdala and increased dopamine release in the Nucleus Accumbens (NAcc) (Liu et al. 1997, 2000; Caldji et al. 1998, 2000; Francis et al. 1999; Sapolsky et al. 1985; Champagne et al. 2004). The enhanced maternal behavior (MB) displayed by these animals may be explained by higher levels of estrogen receptor alpha in the medial preoptic area (MPOA) (Champagne et al. 2006) and high levels of oxytocin receptors in the MPOA, amygdala, lateral septum, and bed nucleus of *stria terminalis* (Francis et al. 2000). The physiological responses to stress of "handled" animals are similar to those reported in high LG-ABN rats (Macri et al. 2004; Plotsky and Meaney 1993).

Neuroanatomical changes High LG-ABN offspring show increased hippocampal neuronal survival and synaptogenesis (Liu et al. 1997), increased dendritic branch length and high spine density of pyramidal neurons in the CA1, as well increased hippocampal neuronal survival (Liu et al. 1997). Furthermore, these rats displayed high Brain-derived neurotrophic factor (BDNF) and more complex granule cell dendritic trees (van Hasselt et al. 2012). In addition, low methylation across the exon 1_7 GR promoter sequence in the hippocampus has been observed (Weaver et al. 2004) in such offspring.

2. The partial or total maternal separation (MS) paradigms reduce or eliminate, respectively, the sensory cues the offspring receive from their mother. In the latter case, litters are reared artificially (Hall 1998; Gonzalez et al. 2001; Lévy et al. 2003; Melo et al. 2006, 2009). The first paradigm consists of removing pups, as a litter, from the nest for 1–6 h daily across postnatal days (P) 1–14 (see Cirulli et al. 2003). In contrast, the artificial rearing (AR) paradigm consists of removing the pups from the nest on P 3–4 and rearing them in complete isolation through an AR system (see Table 6.2).

Table 6.2 Effects of maternal separation paradigm

References	Methods	Effects
Caldji et al. (2000)	MS: 3 h/day, 1–14 PND Pups separated as litter	↓ GABA$_A$ receptors in LC and NTS ↓ CBZ receptors in central and lateral nucleus of the AMG, FC and NTS ↓ Levels of the mRNA of α_2 subunit of the GABA$_A$ receptor
Brake et al. (2004)	MS: 3 h/day, 1–14 PND Pups separated as litter	↑ Motor activity ↑ DA in NAcc > Sensitivity to amphetamine-locomotor action under repeated stress ↓ D3 DA receptor-binding ↓ mRNA levels in NAcc (Shell) < Density of DA transporter at NAcc (core) and striatum > Dose-dependent sensitive to cocaine to middle stressor
Francis et al. (2002)	MS: 3 h/day, 1–14 PND Pups separated as litter	↑ CORT 20 min after restrain stress < Time into the inner area of an open field box ↑ Latency to eat into a novel environment ↓ Level of GC receptors mRNA at dentate gyrus and CA1 and CA3 of the hippocampus ↑ CRF mRNA in PVN Environment enrichment from PND 22 to 70 reverse almost all the effects of MS
Plotsky et al. (2005)	MS: 3 h/day, 2–14 PND Pups separated as litter	↑ ACTH and CORT, 10–15′ post-air puff ↑ CRF *ir* in lumbar CSF ↑ CRF mRNA in PVN, CeA, BNST and LC/PBN ↑ CRF *ir* content in same areas ↑ Total CRF receptors in PVN, LC/PBN ↑ CFR *ir* binding in PVN, LC/PBN
Rhees et al. (2001)	MS: 6 h/day, 2–10 PND Pups separated as litter	♂ ↑ Latency to mount ↑ Latency to intromission ↓ % of males that display ejaculation
Rees et al. (2006)	MS: 5 h/day, 2–14 PND Pups separated as litter	♀ Juveniles > CORT at 5′ than MD ♀ Adults →↓ proportion at inner grid crossing in open field
	MD: 5 h/day, 2–14 PND Pups separated from mother and littermate	♀ Juveniles → < CORT stress response at 5′ than MS ♀ Adults → < ACTH basal than MS at 5′ > CORT basal than MS and control group
Boccia and Pedersen (2001)	MS: 3 h/day, 2–14 PND Pups separated as litter	Low nursing, pup licking and nest building ↑ Maternal aggression (> latency to attacks) Low proportion time in open arm in EPM at 6 postpartum day

(continued)

Table 6.2 (continued)

References	Methods	Effects
Lovic et al. (2001)	MD: 5 h/day, 1–17 PND pups separated from mother and littermate	♀ Juveniles → < Time licking < Time nest-building ♀Adults → ↓ maternal licking ↓ maternal crouching over pups ♂
Sterley et al. (2011)	MS: 3 h/day, 2–14 PND Pups separated as litter (WKY rats vs SHR)	↓ Open arm entries in WKY ↑ Activity in SHR in EPM ↓ Activity in WKY in EPM ♀ SHR — ↑ basal plasma CORT ↑ active in EPM and FS
Huot et al. (2001)	MS: 3 h/day, 2–14 PND Pups separated as litter	↑ ACTH and CORT by air puff startle < Less time in open arm at EPM ↓ Dranking water-sucrose solution ↑ Dranking ethanol-sucrose solution ↑ Correlation: ethanol ingest and CORT response Negative correlation between % time in the open arm and ethanol ingest Chronic 5-HT reuptake inhibitor reduced the amount at ethanol ingest, and the stress response
Uchida et al. (2010)	MS: 3 h/day, 2–14 PND Pups separated as litter	↑ Plasma CORT at 30′ and 60′ post restraint stress ↓ Sucrose preference after repeated restraint stress ↓ Latency to immobility ↑ Immobility time in FS test
Plotsky and Meaney (1993)	MS: 3 h/day, 2–14 PND Pups separated as litter	↑ Hypothalamic CRF mRNA Restraint stress — ↑ plasma CORT ↓ CRF
Bravo et al. (2014)	MS: 3 h/day, 2–12 PND Pups separated as litter	↑ 5-HT$_{1A}$ receptor mRNA in the AMG ↓ 5-HT$_{1A}$ receptor mRNA in the DRN ↓ 5-HT transporter (SERT) mRNA in the DRN
Kalinichev et al. (2002)	MS: 3 h/day, 2–14 PND Pups separated as litter	♂ ↑ CORT to mild stress (2.5–5 times) ♂ and ♀ < exploration in EPM ♂ ↑ Startle amplitudes (35%) ♂ High UV than ♀
Veenema et al. (2006)	MS: 3 h/day, 1–14 PND Pups separated as litter	↑ Immobility ↑ ACTH response to acute stress ↑ Aggression: Lateral threat, offensive upright ↑ AVP mRNA and AVP *ir* in the PVN and SON ↓ 5-HT *ir* in anterior hypothalamus and SON = negative correlation with aggression

(continued)

Table 6.2 (continued)

References	Methods	Effects
Veenema and Neumann (2009)	MS: 3 h/day 1–14 PND Pups separated as litter	♂ Juvenile ↑ Offensive play-fight time ↑ Nape attack, offensive pulling behavior and biting ↓ Frequency of submissive play behavior ↑ Plasma CORT ↓ Vasopressin mRNA in PVN and BNST
Eiland and McEwen (2012)	MS: 3 h/day, 2–14 PND Pups separated as litter	Impaired spatial memory ↑ CORT after chronic stress ↑ Anxiety in EPM
Arborelius and Eklund (2007)	MS: 3 h/twice/day 1–13 PND Pups separated as litter	Aged ♀ ↑ 5-HT and 5-HIAA in DRN ↑ 5-HIAA in NAcc
Pascual and Zamora-León (2007)	MS: 3 h/day 6–21 PND	↓ Basal dendritic length in layer II/III pyramidal neurons ↓ Number of entries and time into the open arms of EPM Exposure to enrichment environment at 23–35 PND did not reverses the MS effect
Monroy et al. (2010)	MS: 2 h/day, 1–12 PND Pups separated as litter	↓ Dendritic length → PFC (35 and 60 PND); → CA1 hippocampus (at PND 60); → NAcc (at PND 35 and 60) ↓ Dendritic density → PFC (at PND 60); → CA 1 Ventral hippocampus (at PND 60); → NAcc (at PND 35 and 60)
Xue et al. (2013)	MD: 4 h/day, 1–21 PND Pups separated alone	♂ Juvenile and adults: ↑ 5-HT and 5-HIAA in NAcc ↑ 5-HT in PFC juvenile and adult ↓ Ratio between 5-HIAA and 5-HT in PFC

Behavioral changes Adult rats that were separated from the mother early in life show increased locomotor activity, stress response, anxiety and depression-like behavior (Uchida et al. 2010; Francis et al. 2002; Kalinichev et al. 2002; Franklin et al. 2010; Kember et al. 2012; Lambás-Señas et al. 2009); impaired passive-coping and learning and memory (Eiland and McEwen 2012; Sterley et al. 2011); reduced male copulatory behavior (Rhees et al. 2001); deficient maternal behavior (Lovic et al. 2001; Boccia and Pedersen 2001); and increased play-fighting and aggressive behavior (Veenema et al. 2006; Veenema and Neumann 2009).

Furthermore, these animals are more vulnerable to drug abuse (Brake et al. 2004) and ethanol intake (Huot et al. 2001; see Nylander and Roman 2013).

AR female rats, tested as juveniles or as adults, spend less time crouching over foster pups and display lower levels of pup licking (Gonzalez et al. 2001; Melo et al. 2006, 2009; Gonzalez and Fleming 2002) with respect to their maternally-reared (MR) counterparts. They also present deficits in maternal memory (Lévy et al. 2003), exhibit a reduced 'fear' in a plus maze task (Burton et al. 2006), show an exaggerated response to a novel object (Melo et al. 2009), engage in poor social interactions, and cross a higher proportion of central squares in the open field test (Gonzalez et al. 2001). Male AR rats exhibit deficits in *ex copula* penile reflexes: longer latencies of penile reflexes and reduction in the length of dendrites from the spinal nucleus of the bulbocavernosus (Lenz et al. 2008). Moreover, AR non-pregnant virgins or lactating females show high levels of aggressive behaviors (Melo et al. 2006, 2009). AR adult females also display low levels of attention in pre-pulse inhibition and attention set-shifting tasks (Lovic and Fleming 2004, Lovic et al. 2006; Burton et al. 2006). Finally, in comparison to mother reared (MR) male rats, the electrophysiological properties and myelination of the sural nerve (sensory nerve) of AR animals are abnormal (Segura et al. 2014).

Physiological changes Partial MS paradigms lead to increased and prolonged stress-induced concentrations of plasma corticosterone and ACTH (Plotsky et al. 2005; Rees et al. 2006; Huot et al. 2001); reduced expression of GC in the hippocampus and hypothalamus and elevations in CRF mRNA levels in the hypothalamus (Plotsky et al. 2005; Plotsky and Meaney 1993; Francis et al. 2002); elevation in DNA methylation of the GC promoter, which causes a reduced expression of GC receptors in the hippocampus (See Meaney and Szyf 2005; Zhou 2012), though other authors have not found this effect (Daniels et al. 2009). AR rats, in turn, show high levels of corticosterone in plasma after exposure to a novel environment, compared with maternally-reared animals (Belay et al. 2011).

Partial MS also provokes high levels of 5-Hydroxytryptamine (5-HT) and 5-Hydroxyindoleacetic acid (5-HIAA) in the NAcc (Arborelius and Eklund 2007), but 5-HIAA and 5-HT is low in the prefrontal cortex (PFC) (Xue et al. 2013) and decreased expression of 5-HT $_{1A}$ receptor mRNA in hippocampus, medial prefrontal cortex (mPFC), and serotonin transporter mRNA in the dorsal raphe nucleus (DRN; Bravo et al. 2014). Such rats also show high levels of DA in the NAcc and a low density in NAcc (core) of the striatal dopamine transporter (Brake et al. 2004). AR rats show increased basal extracellular DA levels in the NAcc and reduced pup-related DA elevations (Afonso et al. 2008); increases in amphetamine-induced locomotor activity above and beyond what is typically observed in MR rats (Lovic et al. 2006).

Neuroanatomical changes Rats that were MS have reduced levels of dendritic spine density in Layer II/III of pyramidal neurons in the anterior cingulate prefrontal cortex (Pascual and Zamora-León 2007); reduced total dendritic length and dendritic spine density in neurons of the PFC, the CA1 ventral hippocampus and the NAcc at a postpubertal age (Monroy et al. 2010).

Brains of AR rats show low levels of synaptophysin and BDNF proteins in the mPFC and NAcc (Burton et al. 2007), and N-CAM (cell-adhesion molecule), GAP-43 (axon elongation), as well as high levels of Neu-N (a neuronal marker) and S-100β (an astrocyte marker) in several brain sites (e.g., mPFC, amygdala, NAcc) (Chatterjee et al. 2007). Moreover, in comparison to MR mothers, AR mothers show no increments in cell survival in the Bed nucleus of the stria terminalis (BNST) and the NAcc, but they show a significant reduction in cell survival in the amygdala (Akbari et al. 2007). AR juvenile females also show significant reductions in *c-fos* protein immunoreactivity in the MPOA, parietal and pyriform cortices after pup interactions (Gonzalez and Fleming 2002).

By providing food (Suchecki et al. 1993), tactile and social stimuli (Suchecki et al. 1993; van Oers et al. 1998; Kuhn et al. 1990), or passive contact with a female, the effects of MS on stress responses are largely prevented (Levine et al. 1988). Similarly, most of the effects provoked by AR were partially or completely prevented by: (a) providing pups with daily additional stroking stimulation with a paintbrush 5–8 times per day, a procedure that is designed to simulate the mother's licking (Fleming et al. 2002; Gonzalez et al. 2001; Lévy et al. 2003), or (b) by providing peer-derived social stimulation during AR (Melo et al. 2006, 2009; Segura et al. 2014).

Most of the effects observed in animal models have been observed in analogous human situations. Thus, orphaned children or those that received poor quality/quantity of maternal care (from depressed or teenage mothers, for example) show low IQ, anxiety, and attention-deficit and hyperactivity disorder, together with high basal levels of GC (Nicolson 2004; Lupien et al. 2011); enlarged amygdala volumes (Tottenham et al. 2010; Lupien et al. 2011); important reductions in gray and white matter volume, and reduced total brain volume (Nelson et al. 2013; Mehta et al. 2009).

Childhood maltreatment and sexual abuse are extreme forms of early adverse childhood experience (Teicher et al. 2012; Weiss et al. 1999). Adults that underwent such experiences in childhood show: reduced hippocampal volume (Bremner et al. 1997; Weiss et al. 1999); sleep disorders, nervousness, anxiety, depression, deficits in learning and memory (De Bellis et al. 2002; Sapolsky 1996); alterations in neurotransmitter activity, stress response, fewer neural connections, CNS instability, and aberrant cortical development (Teicher et al. 1997; Ito et al. 1993; Kaufman et al. 2000). Conversely, tactile stimulation to premature or low birth-weight infants increased their weight and improved their performance in development tasks (Field et al. 1986; Kuhn et al. 1991).

Effects of Motherhood on the mother's Brain

Abundant experimental evidence has revealed that mothering exerts profound effects on the behavior and abilities of an adult female that extend beyond the display of maternal behavior itself. These effects have been studied mainly along three lines: (a) neuroanatomy; (b) learning and memory; (c) neuroendocrinology. Such work has also investigated if the observed effects are transitory or permanent, whether the hormones of pregnancy or lactation are involved, and the possibility of a correlation between the behavioral effects observed and specific changes in a neural substrate or neurotransmitter. In the following sections we will review the evidence, obtained mainly in rodents, supporting the plasticity of the maternal brain along the lines referred to above.

(a) *Neuroanatomical studies.* Hippocampal morphology has consistently revealed plastic changes following mothering. Primiparous rats show dendritic remodelling in the CA1 and CA3 regions, compared to nulliparous, and multiparity leads to enhanced spine density in the CA1 region (Pawluski and Galea 2006). In the mPFC, the dentate gyrus, and the CA1 hippocampal region late-lactating dams show a larger number of dendritic spines than nulliparous, an effect not observed in the orbitofrontal cortex (Leuner and Gould 2010). Moreover, in the caudate nucleus primiparous dams show *decreased* dendritic complexity relative to nulliparous, while in the NAcc shell opposite effects occur (Shams et al. 2012). Furthermore, multiparous dams show an increased *percentage* of surviving neurons relative to *new* neurons, in the dentate gyrus (Pawluski and Galea 2007) and also an increased estrogen-induced cell proliferation in middle age (Barha and Galea 2011). In the medial amygdala (MeA) multiparity exerts heterogeneous effects on dendritic spine density depending on the subnucleus. Thus, while multiparous dams show a larger number of dendritic spines than nulliparous in the anterodorsal MeA, the reverse is seen in the posterodorsal MeA, while no effects are evident in the posteroventral MeA (Rasia-Filho et al. 2004).

   Glial fibrillary acidic protein (GFAP), the main protein component of the astrocytic cytoskeleton, has been reliably used as an indicator of plasticity in various learning paradigms. The number of GFAP-immunoreactive (IR) cells following a brief contact with pups is larger in the MPOA of multiparous than primiparous rats, although opposite effects occur in the MeA and habenula (Featherstone et al. 2000). The hormones of pregnancy and the stimuli received during mothering interact to increase the number of GFAP-IR and astrocytic basic fibroblast growth factor-2 (bFGF)-IR cells in the cingulate cortex (area 2) and the MPOA (Salmaso et al. 2005). Interestingly, these increases are already evident at 3 h postpartum in the cingulate cortex and persist for at least 24 days in lactating dams, even if direct contact with the litter is interrupted on day 16 (Salmaso and Woodside 2008). Moreover, GFAP- and bFGF-IR cells are equally raised in virgins given estradiol benzoate plus progesterone and allowed 3 h of experience with foster pups as in lactating dams, relative to the corresponding groups not exposed to a litter (Salmaso et al. 2009).

(b) *Learning and memory.* Primiparous rats make fewer errors than nulliparous when tested in an eight-arm radial maze to assess reference and working memory. These effects, triggered by a single maternal experience, last for several weeks post-weaning (Pawluski et al. 2006b). Similar results are seen in newly parturient dams, exposed to the hormones of pregnancy but with pups removed

24 h post-partum (Pawluski et al. 2006a), indicating a contribution of endocrine and somatosensory stimuli to the effects observed. In the dry-land maze, which tests spatial memory, multiparous rats take less time to find baited food than virgins and they also make more correct choices than nulliparous in the radial maze (Kinsley et al. 1999). Moreover, multiparous dams also learn sooner a new location of baited food (in the dry-land maze) than primiparous or nulliparous animals, and these differences persist for up to 24 months (Gatewood et al. 2005). In another test of spatial memory, the Morris water maze, primiparous dams outperform nulliparous and these differences persist for at least 16 months. By contrast, no differences between these groups are seen in exploration of a novel object, in the open field test, or the elevated plus maze, indicating a specific effect of motherhood on spatial information. However, when tested at different ages, the differences between nulliparous, primiparous, and multiparous dams are not uniform. For instance, the latency to find baited food in a dry-land maze is largest in nulliparous and lowest in primiparous rats at 9 months, but the latter outperform multiparous and primiparous at 17 months. Moreover, multiparous dams retrieve more "fruit loops" in a food-competition task, but spend more time in contact with a novel stimulus than the other two groups (Love et al. 2005). Clearly, complex interactions between pregnancy, lactation, and aging occur to influence long-lasting modifications in particular cognitive abilities.

These behavioral observations are consistent with the fact that application of short electrical pulses (1 s, 100 Hz), NMDA or AMPA agonists induces a larger long-term potentiation response in hippocampal slices from multiparous dams than from nulliparous (Lemaire et al. 2006). Other studies, however, have found that multiparous dams show reduced fear conditioning, relative to primiparous or nulliparous females (Kinsley et al. 2007), and display reduced fear and anxiety (Wartella et al. 2003). Interestingly, the reduced anxiety responses observed in the elevated plus maze in primiparous rats, relative to nulliparous, are reversed in middle age (Byrnes and Bridges 2006).

In a maternal memory test (measuring latencies to retrieve, crouch, lick, and nurse pups on postpartum day 10) electrolytic lesions to the NAcc shell provoked deficits only in dams allowed 1 h of pup exposure but not in those allowed 24 h of interaction with the young (Li and Fleming 2003). Similarly, in rabbits, mother-young separation at parturition abolishes maternal behavior 24 h later in 70% of primiparous does while multiparous mothers are not affected (González-Mariscal et al. 1998).

(c) *Neuroendocrinology studies.* Motherhood has also been found to impact the responsiveness to specific hormones, peptides, and neurotransmitters beyond the period of lactation. For instance, in middle-aged rats, estrogen injections (estradiol-alpha, estradiol-beta, or estrone) increase cell proliferation in the dentate gyrus of the hippocampus *only* in multiparous females, but not in nulliparous or ovariectomized ones (Barha and Galea 2011).

Regarding corticosteroids, the concentration of corticosterone in blood on postpartum day 1 is larger in primiparous than multiparous dams while the levels of corticosterone binding globulin are lower in the former. Consequently, the concentration of *free* corticosterone is elevated in primiparous rats during their first lactation, relative to multiparous animals (Pawluski et al. 2009).

Maternal experience also alters the concentration of oxytocin (OT) in the hypothalamus and plasma. In both wild type mice and knock-out females lacking the transmembrane glycoprotein CD38, which promotes OT secretion through activation of ADP-ribosyl cyclase, primiparous animals have higher concentrations of OT in the hypothalamus and plasma than do nulliparous. In knock-out maternal experience also leads to a faster retrieval of foster pups than that shown by naïve animals (Lopatina et al. 2011). In hippocampal slices from virgin mice a one-train tetanus provokes long-term potentiation and cAMP response element-binding protein (CREB) phosphorylation only after the application of OT. In multiparous females these effects are seen without the application of OT. Moreover, intraventricular injections of OT to virgins improve their performance in the dry-land maze, while injections of an OT antagonist to multiparous mice prevent the behavioral improvement, long-term potentiation, and CREB phosphorylation (Tomizawa et al. 2003) (for further discussion on CREB see Chap. 2).

The sensitivity to opiates is also modified by motherhood: injections of morphine in early to middle lactation abolish the display of maternal behavior in 87% of primiparous but only in 37% of multiparous dams. Additionally, morphine exerts a smaller analgesic effect in multiparous rats, as they show *shorter* tail-flick latencies than primiparous females (Kinsley and Bridges 1988). These findings agree with the observation that multiparous dams show a greater sensitivity to naloxone (an opiate antagonist) regarding the facilitation of LH release on lactation day 10 (Hopwood et al. 1998). Multiparity is also associated with a higher density of opiate receptors in the MPOA in early lactation, relative to primiparous dams (Bridges and Hammer 1992). This observation agrees with the fact that the maternal behavior of multiparous dams is *less inhibited* by infusions of beta-endorphin into the MPOA than primiparous rats (Mann and Bridges 1992).

The sensitivity to prolactin (PRL) is increased in primiparous dams relative to nulliparous, as the former show *lower* concentrations of plasma PRL following injections of *ovine* PRL that provoke a negative feedback. Moreover, the expression of the mRNA for the long form of the PRL receptor is higher in primiparous dams than in nulliparous in the MPOA and arcuate nucleus and injections of ovine PRL further enhance this difference (Anderson et al. 2006). Furthermore, the facilitation of PRL release by haloperidol (a dopamine antagonist), is reduced in multiparous dams, relative to nulliparous. This effect is abolished if parturient dams are not allowed to nurse the litter, indicating that lactation is essential to modify the responsiveness of the dopaminergic system (Bridges et al. 1997).

Future Directions, Unsolved Questions

The above evidence expands the view that the interaction between mother and young is a dyad whose complexity exerts profound and permanent effects on the brains of an adult and a developing organism, respectively. Effects on the latter can be explained by the neural development processes that are still taking place during the early postnatal period. The mechanisms involved are epigenetic, i.e., they include processes that regulate gene expression but do not modify the DNA sequence. Yet, as they occur *during* brain development, such changes are permanent and can be transmitted to the next generation through the MB that the female offspring will display towards their own litters as adults. In contrast, the mechanisms mediating the plasticity of the adult maternal brain are not clear. If the "mothering state" is a *window of opportunity, a critical period* during which (semi) permanent changes in the brain can be brought about, is it because of the hormonal "cocktail" the brain is exposed to across pregnancy and lactation? Is the first maternal experience the most important one? Answers to these questions are not unequivocal as there is a marked heterogeneity in the direction of the observed changes, i.e., heightened or diminished responses in mothers vs. virgins across brain regions. This variability complicates the analysis of the mechanisms involved and the generation of a hypothesis that can be used for predictions. Indeed, little is known about the subcellular mechanisms underlying the *onset* and *permanence* of mothering-induced plasticity, although obvious candidates worth investigating are the pathways that mediate learning and memory (e.g., Mitogen-activated protein kinase, MAPK; see Chap. 2). Yet, regardless of the mechanisms involved it is clear that mothering-induced brain plasticity includes changes in "extra-maternal" abilities and also on MB itself. In summary, the bidirectional effects that mother and young exert on each other's brain exemplify not only a remarkable plasticity of this organ for organizing itself but also an immense source of variability for allowing adaptation and evolution in mammalian species.

Acknowledgments This work was supported by grants # 128625 and 156413 (to GGM and AIM, respectively) from CONACYT (National Council of Science and Technology, Mexico). Partial support was received from grant # 181334 to Grupo de Investigación G3, CONACYT. The authors thank Mayra Flores-Jiménez for her help in the preparation of the tables.

References

Afonso VM, Grella SL, Chatterjee D, Fleming AS (2008) Previous maternal experience affects accumbal dopaminergic responses to pup-stimuli. Brain Res 1198:115–123

Akbari EM, Chatterjee D, Lévy F, Fleming AS (2007) Experience-dependent cell survival in the maternal rat brain. Behav Neurosci 121:1001–1011

Anderson GM, Grattan DR, van den Ancker W, Bridges RS (2006) Reproductive experience increases prolactin responsiveness in the medial preoptic area and arcuate nucleus of female rats. Endocrinology 147:4688–4694

Arborelius L, Eklund MB (2007) Both long and brief maternal separation produces persistent changes in tissue levels of brain monoamines in middle-aged female rats. Neuroscience 145:738–750

Barha CK, Galea LAM (2011) Motherhood alters the cellular response to estrogens in the hippocampus later in life. Neurobiol Aging 32:2091–2095

Belay H, Burton CL, Lovic V, Meaney MJ, Sokolowski M, Fleming AS (2011) Early adversity and serotonin transporter genotype interact with hippocampal glucocorticoid receptor mRNA expression, corticosterone, and behavior in adult male rat. Behav Neurosci 125:150–160

Boccia ML, Pedersen CA (2001) Brief vs. long maternal separations in infancy: contrasting relationships with adult maternal behavior and lactation levels of aggression and anxiety. Psychoneuroendocrinology 26:657–672

Brake WG, Zhang TY, Diorio J, Meaney MJ, Gratton A (2004) Influence of early postnatal rearing conditions on mesocorticolimbic dopamine and behavioural responses to psychostimulants and stressors in adult rats. Eur J Neurosci 19:1863–1874

Bravo JA, Dinan TG, Cryan JF (2014) Early-life stress induce persistent alterations in 5-HT$_{1A}$ receptor and serotonin transporter mRNA expression in the adult rat brain. Front Mol Neurosci 7:1–9

Bremner JD, Randall P, Vermetten E, Staib L, Bronen RA, Mazure C, Capelli S, McCarthy G, Innis RB, Charney DS (1997) Magnetic resonance imaging-based measurement of hippocampal volume in posttraumatic stress disorder related to childhood physical and sexual abuse – a preliminary report. Biol Psychiatry 41:23–32

Bridges RS, Hammer RP Jr (1992) Parity-associated alterations of medial preoptic opiate receptors in female rats. Brain Res 578:269–274

Bridges TRS, Henriquez BM, Sturgis JD, Mann PE (1997) Reproductive experience reduces haloperidol-induced prolactin secretion in female rats. Neuroendocrinology 66:321–326

Burton C, Lovic V, Fleming AS (2006) Early adversity alters attention and locomotion in adults Sprague-Dawley rats. Behav Neurosci 120:665–675

Burton CL, Chatterjee D, Chatterjee-Chakraborty M, Lovic V, Grella SL, Steiner M, Fleming AS (2007) Prenatal restraint stress and mother less rearing disrupts expression of plasticity markers and stress-induced corticosterone release in adult female Sprague-Dawley rats. Brain Res 1158:28–38

Byrnes EM, Bridges RS (2006) Reproductive experience alters anxiety-like behavior in the female rat. Horm Behav 50:70–76

Caldji C, Tannenbaum B, Sharma S, Francis D, Plotsky PM, Meaney MJ (1998) Maternal care during infancy regulates the development of neural systems mediating the expression of fearfulness in the rat. Proc Natl Acad Sci U S A 95:5335–5340

Caldji C, Francis D, Sharma S, Plotsky PM, Meaney MJ (2000) The effects of early rearing environment on the development of GABA$_A$ and central benzodiazepine receptor levels and novelty-induced fearfulness in the rat. Neuropsychopharmacology 22:219–229

Champagne FA (2008) Epigenetic mechanisms and the transgenerational effects of maternal care. Front Neuroendocrinol 29:386–397

Champagne FA, Chretien P, Stevenson CW, Zhang TY, Gratton A, Meaney MJ (2004) Variations in nucleus accumbens dopamine associated with individual differences in maternal behavior in the rat. J Neurosci 24:4113–4123

Champagne FA, Weaver ICG, Diorio J, Dymov S, Szyf M, Meaney MJ (2006) Maternal care associated with methylation of the estrogen receptor-alpha1b promoter and estrogen receptor-alpha expression in the medial preoptic area of female offspring. Endocrinology 147:2909–2915

Chatterjee D, Chatterjee-Chakraborty M, Rees S, Cauchi J, de Medeiros CB, Fleming AS (2007) Maternal isolation alters the expression of neural proteins during development: 'Stroking' stimulation reverses these effects. Brain Res 1158:11–27

Cirulli F, Berry A, Alleva E (2003) Early disruption of the mother-infant relationship: effects on brain plasticity and implications for psychopathology. Neurosci Biobehav Rev 27:73–82

Daniels WMU, Fairbairn LR, van Tilburg G, McEvoy CRE, Zigmond MJ, Russell VA, Stein DJ (2009) Maternal separation alters nerve growth factor and corticosterone levels of the exon 17 glucocorticoid receptor promoter region. Metab Brain Dis 24:615–627

De Bellis MD, Keshavan MS, Shifflett H, Iyengar S, Beers SR, Hall J, Moritz G (2002) Brain structures in pediatric maltreatment-related post-traumatic stress disorder: a sociodemographically matchet study. Biol Psychiatry 52:1066–1078

Eiland L, McEwen BS (2012) Early life stress followed by subsequent adult chronic stress potentiates anxiety and blunts hippocampal structural remodeling. Hipp 22:82–91

Fagiolini M, Jensen CL, Champagne FA (2009) Epigenetic influences on brain development and plasticity. Curr Opin Neurobiol 19:207–212

Featherstone RE, Fleming AS, Ivy GO (2000) Plasticity in the maternal circuit: effects of experience and partum condition on brain astrocyte number in female rats. Behav Neurosci 114:158–172

Feng J, Fouse S, Fan G (2007) Epigenetic regulation of neural gene expression and neuronal function. Pediatr Res 61(5 Pt 2):58R–63R

Field TM, Schanberg SM, Scafidi F, Bauer CR, Vega-Lahr N, Garcia R, Nystrom J, Kuhn CM (1986) Tactile/kinesthetic stimulation effects on preterm neonates. Pediatrics 77:654–658

Fleming AS, Kraemer GW, Gonzalez A, Lovic V, Ress A, Melo AI (2002) Mothering begets mothering: the transmission of behavior and its neurobiology across generation. Pharmacol Biochem Behav 73:61–75

Francis D, Diorio J, Liu D, Meaney MJ (1999) Nongenomic transmission across generations of maternal behavior and stress responses in the rat. Science 286:1155–1158

Francis DD, Champagne FC, Meaney MJ (2000) Variations in maternal behaviour are associated with differences in oxytocin receptor levels in the rat. J Neuroendocrinol 12:1145–1148

Francis DD, Diorio J, Plotsky PM, Meaney MJ (2002) Environmental enrichment reverses the effects of maternal separation on stress reactivity. J Neurosci 22:7840–7843

Franklin TB, Russing H, Weiss IC, Graff J, Linder N, Michalon A, Vizi S, Mansury IM (2010) Epigenetic transmission of the impact of early stress across generations. Biol Psychiatry 68:408–415

Gatewood JD, Morgan MD, Eaton M, McNamara IM, Stevens LF, Macbeth AH, Meyer EAA, Lomes LM, Kozub FJ, Lambert KG, Kinsley CH (2005) Motherhood mitigates aging-related decrements in learning and memory and positively affects brain aging in the rat. Brain Res Bull 66:91–98

Godfrey KM, Lillycrop KA, Burdge GC, Gluckman PD, Hanson MA (2007) Epigenetic mechanisms and the mismatch concept of the developmental origins of health and disease. Pediatr Res 61:5R–10R

Gonzalez A, Fleming AS (2002) Artificial rearing causes changes in maternal behavior and *c-fos* expression in juvenile female rats. Behav Neurosci 116:999–1013

Gonzalez A, Lovic V, Ward GR, Wainwright PE, Fleming AS (2001) Intergenerational effects of complete maternal deprivation and replacement stimulation on maternal behavior and emotionality in female rats. Dev Psychobiol 38:11–32

González-Mariscal G, Kinsley CH (2009) From indifference to ardor: the onset, maintenance, and meaning of the maternal brain. In: Pfaff D, Arnold AP, Etgen AM, Fahrbach SE, Rubin RT (eds) Hormones, brain and behavior, 2nd edn. Academic, San Diego, pp 109–136

González-Mariscal G, Melo AI (2013) Parental behavior. In: Pfaff DW (ed) Neuroscience in the 21st century. Springer Science Business Media, New York, pp 2069–2100

González-Mariscal G, Melo AI, Chirino R, Jiménez P, Beyer C, Rosenblatt JS (1998) Importance of mother/young contact at parturition and across lactation for the expression of maternal behavior in rabbits. Dev Psychobiol 32:101–111

Hall FS (1998) Social deprivation of neonatal, adolescent, and adults has distinct neurochemical and behavioral consequences. Crit Rev Neurobiol 2:129–162

Hofer MA (1984) Relationships as regulators: a psychobiologic perspective on bereavement. Psychosom Med 46:183–197

Hopwood RM, Mann PE, Bridges RS (1998) Reproductive experience and opioid regulation of luteinizing hormone release in female rats. J Reprod Fertil 114:259–265

Huot RL, Thrivikraman KV, Meaney MJ, Plotsky PM (2001) Development of adult ethanol preference and anxiety as a consequence of neonatal maternal separation in Long Evans rats and reversal with antidepressant treatment. Psychopharmacology 158:366–373

Ito Y, Teicher MH, Glod CA, Harper D, Magnus E, Gelbard HA (1993) Increased prevalence of electrophysiological abnormalities in children with psychological, physical, and sexual abuse. J Neuropsychiatr Clin Neurosci 5:401–408

Kalinichev M, Easterling KW, Plotsky PM, Holtzman SG (2002) Long-lasting changes in stress-induced corticosterone response and anxiety-like behaviors as a consequence of neonatal maternal separation in Long-Evans rats. Pharmacol Biochem Behav 73:131–140

Kaufman J, Plotsky PM, Nemeroff CB, Charney DS (2000) Effects of early adverse experiences on brain structure and function: clinical implications. Biol Psychiatry 48:778–790

Kember RL, Dempster EL, Lee TH, Schalkwyk LC, Mill J, Fernandes C (2012) Maternal separation is associated with strain-specific responses to stress and epigenetic alterations to Nr3c1, Avp, and Nr4a1 in mouse. Brain Behav 2:455–467

Kinsley CH, Bridges RS (1988) Parity-associated reductions in behavioral sensitivity to opiates. Biol Reprod 39:270–278

Kinsley CH, Madonia L, Gifford GW, Tureski K, Griffin GR, Lowry C, Williams J, Collins J, McLearie H, Lambert KG (1999) Motherhood improves learning and memory. Nature 402:137–138

Kinsley CH, Bardi M, Karelina K, Rima B, Christon L, Friedenberg J, Griffin G (2007) Motherhood induces and maintains behavioral and neural plasticity across the lifespan in the rat. Arch Sex Behav 37:43–56

Kuhn CM, Pauk J, Schanberg SM (1990) Endocrine responses to mother-infant separation in developing rats. Dev Psychobiol 23:395–410

Kuhn CM, Schanberg SM, Field T, Symanski R, Zimmerman E, Scafidi F, Roberts J (1991) Tactile-kinesthetic stimulation effects on sympathetic and adrenocortical function in preterm infants. J Pediatr 119:434–440

Lambás-Señas L, Mnie-Filali O, Certin V, Faure C, Lemoine L, Zimmer L, Haddjeri N (2009) Functional correlates for 5-HT (1A) receptors in maternally deprived rats displaying anxiety and depression-like behaviors. Prog Neuro-Psychopharmacol Biol Psychiatry 33:262–268

Lemaire V, Billard JM, Dutar P, George O, Piazza PV, Epelbaum J, Le Moal M, Mayo W (2006) Motherhood-induced memory improvement persists across lifespan in rats but is abolished by a gestational stress. Eur J Neurosci 23:3368–3374

Lenz KM, Graham MD, Parada M, Fleming AS, Sengelaub DR, Monks DA (2008) Tactile stimulation during artificial rearing influences adult function and morphology in a sexually dimorphic neuromuscular system. Dev Neurobiol 68:542–557

Leuner B, Gould E (2010) Dendritic growth in the medial prefrontal cortex and cognitive flexibility are enhanced during the postpartum period. J Neurosci 30:13499–13503

Levine S, Stanton ME, Gutierrez YR (1988) Maternal modulation of pituitary-adrenal activity during ontogeny. Adv Exp Med Biol 245:295–310

Lévy F, Melo AI, Gale BG Jr, Madden M, Fleming AS (2003) Complete maternal deprivation affects social, but not spatial, learning in adult rats. Dev Psychobiol 43:177–191

Li M, Fleming AS (2003) Differential involvement of nucleus accumbens shell and core subregions in maternal memory in postpartum female rats. Behav Neurosci 117:426–445

Liu D, Diorio J, Tannenbaum B, Caldji C, Francis D, Freedman A (1997) Maternal care, hippocampal glucocorticoid receptors and hypothalamic-pituitary-adrenal responses to stress. Science 227:1659–1662

Liu D, Diorio J, Day JC, Francis DD, Meaney MJ (2000) Maternal care, hippocampal synaptogenesis and cognitive development in rats. Nat Neurosci 3:799–806

Lopatina O, Inzhutova A, Pichugina YA, Okamoto H, Salmina AB, Higashida H (2011) Reproductive experience affects parental retrieval behaviour associated with increased plasma oxytocin levels in wild-type and CD38-knockout mice. J Neuroendocrinol 23:1125–1133

Love G, Torrey N, McNamara L, Morgan M, Banks M, Hester NW, Glasper ER, De Vries AC, Kinsley CH, Lambert KG (2005) Maternal experience produces long-lasting behavioral modifications in the rat. Behav Neurosci 119:1084–1096

Lovic V, Fleming AS (2004) Artificially-reared female rats show reduced prepulse inhibition and deficits in the attentional set-shifting task-reversal of effects with maternal-like licking stimulation. Behav Brain Res 148:209–219

Lovic V, Gonzalez A, Fleming AS (2001) Maternally separated rats show dificits in maternal care in adulthood. Dev Psychobiol 39:19–33

Lovic V, Fleming AS, Fletcher PJ (2006) Early life tactile stimulation changes adult rat responsiveness to amphetamine. Pharmacol Biochem Behav 84:497–503

Lupien SJ, Parent S, Evans AC, Tremblay RE, Zelazo PD, Corbo V, Pruessner JC, Séguin JR (2011) Larger amygdala but no change in hippocampal volume in 10-year-old children exposed to maternal depressive symptomatology since birth. Proc Natl Acad Sci U S A 108:14324–14329

Macri S, Mason GJ, Würbel H (2004) Dissociation in the effects of neonatal maternal separations on maternal care and the offspring's HPA and fear responses in rats. Eur J Neurosci 20:1017–1024

Maestripieri D, Megna NL (2000) Hormones and behavior in rhesus macaque abusive and nonabusive mothers 1. Social interactions during late pregnancy and early lactation. Physiol Behav 71:35–42

Mann PE, Bridges RA (1992) Neural and endocrine sensitivities to opioids decline as a function of multiparity in the rat. Brain Res 580:241–248

Meaney MJ, Szyf M (2005) Environment programming of stress responses through DNA methylation: life at the interface between a dynamic environment and a fixed genome. Dial Clin Neurosci 7:103–123

Mehta MA, Golembo NI, Nosarti C, Colvert E, Mota A, Williams SC, Rutter M, Sonuga-Barke EJ (2009) Amygdala, hippocampal and corpus callosum size following severe early institutional deprivation: the English and Romanian adoptees study pilot. J Child Psychol Psychiatry 50:943–995

Melo AI (2015) Role of sensory, social, and hormonal signals from the mother on the development of offspring. In: Antonelli MC (ed) Perinatal programming of neurodevelopment. Springer Science+Business Media, New York, pp 219–248

Melo AI, Lovic V, Gonzalez A, Madden M, Sinopoli K, Fleming AS (2006) Maternal and littermate deprivation disrupt maternal behavior and social-learning of food preference in adulthood: tactile stimulation, nest odor, and social rearing prevent these effects. Dev Psychobiol 48:209–219

Melo AI, Hernández-Curiel M, Hoffman KL (2009) Maternal and peer contact during the postnatal period participate in the normal development of maternal aggression, maternal behavior, and the behavioral response to novelty. Behav Brain Res 201:14–21

Menard JL, Hakvoort RM (2007) Variations of maternal care alter offspring levels of behavioural defensiveness in adulthood: evidence for a threshold model. Behav Brain Res 176:302–313

Monroy E, Hernández-Torres E, Flores G (2010) Maternal separation disrupts dendritic morphology of neurons in prefrontal cortex, hippocampus, and nucleus accumbens in male rat offspring. J Chem Neuroanat 40:93–101

Nelson CA 3rd, Fox NA, Zeanah CH Jr (2013) Anguish of the abandoned child. Sci Am 308:62–67

Nicolson NA (2004) Childhood parental loss and cortisol levels in adult men. Psychoneuroendocrinology 29:1012–1018

Nylander I, Roman E (2013) Is the rodent maternal separation model a valid and effective model for studies on the early-life impact on ethanol comsumption? Psychopharmacology 229:555–569

Pascual R, Zamora-León SP (2007) Effects of neonatal maternal deprivation and postweaning environmental complexity on dendritic morphology of prefrontal pyramidal neurons in the rat. Acta Neurobiol Exp (Wars) 67:471–479

Pawluski JL, Galea LAM (2006) Hippocampal morphology is differentially affected by reproductive experience in the mother. J Neurobiol 66:71–81

Pawluski JL, Galea LAM (2007) Reproductive experience alters hippocampal neurogenesis during the postpartum period in the dam. Neurocience 149:53–67

Pawluski JL, Brandy L, Vanderbyl BL, Ragan K, Galea LAM (2006a) First reproductive experience persistently affects spatial reference and working memory in the mother and these effects are not due to pregnancy or 'mothering' alone. Behav Brain Res 175:157–165

Pawluski JL, Walker SK, Galea LAM (2006b) Reproductive experience differentially affects spatial reference and working memory performance in the mother. Horm Behav 49:143–149

Pawluski JL, Charlier TD, Lieblich SE, Hammond GL, Galea LAM (2009) Reproductive experience alters corticosterone and CBG levels in the rat dam. Physiol Behav 96:108–114

Plotsky PM, Meaney MJ (1993) Early, postnatal experience alters hypothalamic corticotrophin-releasing factor (CRF) mRNA, median eminence CRF content and stress-induced release in adult rats. Mol Brain Res 18:195–200

Plotsky PM, Thrivikraman KV, Nemeroff CB, Caldji C, Sharma S, Meaney MJ (2005) Long-term consequences of neonatal rearing on central corticotropin-releasing factor systems in adult male rat offspring. Neuropsychopharmacology 30:2192–2204

Portha B, Fournier A, Kioon MD, Mezger V, Movassat J (2014) Early environmental factors, alteration of epigenetic marks and metabolic disease susceptibility. Biochimie 97:1–15

Pryce CR, Bettschen D, Feldon J (2001) Comparison of the effects of early handling and early deprivation on maternal care in the rat. Dev Psychobiol 38:239–251

Rasia-Filho AA, Fabian C, Rigoti KM, Achaval M (2004) Influence of sex, estrous cycle and motherhood on dendritic spine density in the rat medial amygdale revealed by the Golgi method. Neuroscience 126:839–847

Razin A (1998) CpG methylation, chromatin structure and gene silencing-a three-way connection. EMBO J 17:4905–4908

Rees SL, Steiner M, Fleming AS (2006) Early deprivation, but not maternal separation, attenuates rise in corticosterone levels after exposure to a novel environment in both juvenile and adult female rats. Behav Brain Res 175:383–391

Rhees RW, Lephart ED, Eliason D (2001) Effects of maternal separation during early postnatal development on male sexual behavior and female reproductive function. Behav Brain Res 123:1–10

Rice D, Barone S Jr (2000) Critical periods of vulnerability for the developing nervous system: evidence from human and animal. Environ Health Perspect 108:511–533

Salmaso N, Woodside B (2008) Fluctuations in astrocytic basic fibroblast growth factor in the cingulate cortex of cycling, ovariectomized and postpartum animals. Neuroscience 154:932–939

Salmaso N, Popeski N, Peronace LA, Woodside B (2005) Differential effects of reproductive and hormonal state on basic fibroblast growth factor and glial fibrillary acid protein immunoreactivity in the hypothalamus and cingulate cortex of female rats. Neuroscience 134:1431–1440

Salmaso N, Nadeau J, Woodside B (2009) Steroid hormones and maternal experience interact to induce glial plasticity in the cingulate cortex. Eur J Neurosci 29:786–794

Sapolsky RM (1996) Stress, glucocorticoids, and damage to the nervous system: the current state of confusion. Stress 1:1–19

Sapolsky RM, Meaney MJ, McEwen BS (1985) The development of the glucocorticoid receptor system in the rat limbic brain. III. Negative-feedback regulation. Brain Res 350:169–173

Segura B, Melo AI, Fleming AS, Mendoza-Garrido ME, Gonzalez del Pliego M, Aguirre-Benitez EL, Hernández-Falcón J, Jiménez-Estrada I (2014) Early social isolation provoques electophysiological and structural changes in cutaneous sensory nerves of adult male rats. Dev Neurobiol 74:1184–1193

Shams S, Pawluski JL, Chatterjee-Chakraborty M, Oatley H, Mastroianni A, Fleming AS (2012) Dendritic morphology in the striatum and hypothalamus differentially exhibits experience-dependent changes in response to maternal care and early social isolation. Behav Brain Res 233:79–89

Sterley TL, Howells FM, Russell A (2011) Effects of early life trauma are dependent on genetic predisposition: a rat study. Behav Brain Funct 7:11

Suchecki D, Rosenfeld P, Levine S (1993) Maternal regulation of the hypothalamic-pituitary-adrenal axis in the infant rat: the roles of feeding and stroking. Brain Res Dev Brain Res 75:185–192

Teicher MH, Ito Y, Glod CA, Andersen SL, Dumont N, Ackerman E (1997) Preliminary evidence for abnormal cortical development in physically and sexually abused children using EEG coherence and MRI. Ann N Y Acad Sci 821:160–175

Teicher MH, Anderson CM, Polcari A (2012) Children maltreatment is associated with reduced volume in the hippocampal subfields CA3, dentate gyrus, and subiculum. Proc Natl Acad Sci USA 109(9):E563–E572

Tomizawa K, Iga N, Lu YF, Morikawi A, Matsushita M, Li ST, Miyamoto O, Itano T, Matsui H (2003) Oxytocin improves long-lasting spatial memory during motherhood through MAP kinase cascade. Nat Neurosci 6:384–390

Tottenham N, Hare TA, Quinn BT, McCarry TW, Nurse M, Gilhooly T, Millner A, Galvan A, Davidson MC, Eigsti IM, Thomas KM, Freed PJ, Booma ES, Gunnar MR, Altemus M, Aronson J, Casey BJ (2010) Prolonged institutional rearing is associated with atypically large amygdala volume and difficulties in emotion regulation. Dev Sci 13:46–61

Uchida A, Hara K, Kobayashi A, Funato H, Hobara T, Otsuki K, Yamagata H, McEwen BS, Watanabe Y (2010) Early life stress enhances behavioral vulnerability to stress through the activation of REST4-mediating gene transcription in the medial prefrontal cortex of rodents. J Neurosci 30:15007–15018

Van Hasselt FN, Cornelisse S, Zhang TY, Meaney MJ, Velzing EH, Krugers HJ, Joëls M (2012) Adult hippocampal glucocorticoid receptor expression and dentate synaptic plasticity correlate with maternal care received by individuals early in life. Hippocampus 22:255–266

Van Oers HJ, de Kloet ER, Whelan T, Levine S (1998) Maternal deprivation effect on the infant's neural stress markers is reversed by tactile stimulation and feeding but not by suppressing corticosterone. J Neurosci 18:10171–10179

Veenema AH, Neumann ID (2009) Maternal separation enhances offensive play-fighting, basal corticosterone and hypothalamic vasopressin mRNA expression in juvenile male rats. Psychoneuroendocrinology 34:463–467

Veenema AH, Blume A, Niederle D, Buwalda B, Neumann ID (2006) Effects of early life stress on adult male aggression and hypothalamic vasopressin and serotonin. Eur J Neurosci 24:1711–1720

Wartella J, Amory E, Madonia-Lomas L, Macbeth AH, McNamara I, Stevens L (2003) Single or multiple reproductive experiences attenuate neurobehavioral stress and fear responses in the female rat. Physiol Behav 79:373–381

Weaver IC, Cervoni N, Champagne FA, D'Alessio AC, Sharma S, Seckl JR, Dymov S, Szyf M, Meaney MJ (2004) Epigenetic programming by maternal behavior. Nat Neurosci 7:847–854

Weiss EL, Longhurst JG, Mazure CM (1999) Childhood sexual abuse as a risk factor for depression in women: psychosocial and neurobiological correlates. Am J Psychiatry 156:816–828

Xue X, Shao S, Li M, Shao F, Wang W (2013) Maternal separation induces alterations of serotonergic system in different aged rats. Brain Res Bull 95:15–20

Zhou FC (2012) DNA methylation program during development. Front Biol 7:485–494

Chapter 7
Prenatal Stress and Neurodevelopmental Plasticity: Relevance to Psychopathology

María Eugenia Pallarés and Marta C. Antonelli

Abstract Prenatal development constitutes a critical time for shaping adult behaviour and may set the stage for vulnerability to disease later in life. A wealth of information from humans as well as from animal research has revealed that exposure to hostile conditions during gestation may result in a series of coordinated biological responses aimed at enhancing the probability of survival, but could also increase the susceptibility to mental illness. Prenatal stress has been linked to abnormal cognitive, behavioural and psychosocial outcomes both in animals and in humans, but the underlying molecular and physiological mechanisms remain largely unknown. In this chapter, we shall review experimental data from studies reported for rats, since more information is available for them than for other species. The major focus of the present chapter is to update and discuss data on behavioural, functional and morphological effects of prenatal stress in rats that may have counterparts in prospective and/or retrospective studies of gestational stress in humans. This work contributes to understanding the role of neuronal plasticity in the long-term effects of developmental adversity on brain function and its implications for vulnerability to disease.

Keywords Gestational stress • Developmental programming • Glucocorticoids • Neurotrophins • Hypothalamic-pituitary-adrenal (HPA) axis

Abbreviations

11-beta-HSD2	11-beta-hydroxysteroid dehydrogenase type II enzyme
BDNF	Brain derived neurotrophic factor
FGF-2	Fibroblast grow factor type II
GCs	Glucocorticoids
GR	Glucocorticoid receptor type II
HPA	Hypothalamic-pituitary-adrenal axis

M.E. Pallarés • M.C. Antonelli (✉)
Instituto de Biología Celular y Neurociencia "Prof. Eduardo De Robertis",
Facultad de Medicina, Universidad de Buenos Aires, Buenos Aires, Argentina
e-mail: mca@fmed.uba.ar

© Springer International Publishing AG 2017
R. von Bernhardi et al. (eds.), *The Plastic Brain*, Advances in Experimental
Medicine and Biology 1015, DOI 10.1007/978-3-319-62817-2_7

Introduction

As discussed in Chap. 6, preclinical and clinical studies have shown that adverse events early in life induce profound and persistent effects on brain structure and function, which may lead to cognitive deficits and increased incidence of psychopathology later in life (Fumagalli et al. 2007). The concept of '*developmental programming*' was put forward in an attempt to explain the association between environmental challenges during pregnancy, altered foetal growth and development, and later pathophysiology. During *programming*, non-genetic factors such as stress or exposure to excess glucocorticoids, the main hormonal mediator of stress, are transmitted to the foetus and act on specific tissues during sensitive windows of development to change developmental trajectories and thus their organisation and function (Maccari et al. 2003; Harris and Seckl 2011). Since different cells and tissues are sensitive to various factors at different times, the effects of adversity on an animal's biology will be tissue, time and challenge specific (Sandman et al. 2011; Harris and Seckl 2011; Connors et al. 2008). It can be speculated that prenatal plasticity of physiological systems is biologically adaptive, allowing environmental factors acting on the mother or the foetus or both to alter the functions of an organ or tissue system to prepare the unborn animal optimally for the environmental conditions *ex-utero* (Del Giudice 2012; Maccari et al. 2003). Nevertheless, in extreme conditions like gestational stress, offspring display short and long-term physiological and behavioural abnormalities that increase the risk of disease. In humans, current information available from retrospective and prospective studies has reported that foetuses exposed to maternal stress at various times during gestation are at greater subsequent risk for cardiovascular and metabolic disorders than species with shorter lifespans. Such stress can result from a wide range of natural or man-made disasters or pressures, including earthquakes, floods, freezing storms, war, terrorist acts, chronic interpersonal tensions, or adverse conditions in the home or workplace (Weinstock 2008). Prenatal stress during pregnancy is associated with intrauterine growth restrictions and an increased risk of premature birth (Rondo et al. 2003), as well as with emotional and cognitive deficits in early life. Higher incidence of developing autism, hyperactivity and attention-deficit hyperactivity disorder were found in prenatally stressed children (Weinstock 2008; Cottrell and Seckl 2009). In adults, prenatal stress is linked to depression, anxiety and schizophrenia (Charil et al. 2010; Fumagalli et al. 2007; Weinstock 2008). Studies performed on mothers that self-reported elevated levels of anxiety during pregnancy indicated that their infants were more irritable and showed a higher incidence of sleeping and feeding problems than those of non-anxious mothers (Huizink et al. 2002).

By studying the effects of prenatal stress in experimental animals, one can control the timing, intensity and duration of stress exposure and evaluate the interaction of the mother with her offspring in a controlled environment (Fumagalli et al. 2007; Weinstock 2008). Most animal studies have been performed in rodents, for which the most comprehensive behavioural, morphological and histological information is available. The studies show the deleterious effects of prenatal stress on the offspring's neuronal development and brain morphology, as well as changes in cerebral asymmetry that persist into adulthood (Fride and Weinstock 1988; Barros et al. 2006a; Adrover et al. 2007; Baier et al. 2012; Pallares et al. 2013b; Sandman et al. 2011; Zuena et al. 2008).

Foetal development can also be compromised by factors other than prenatal stress, including maternal malnutrition, mental illness, drugs, alcohol, viral infection, or environmental toxins (Valenzuela et al. 2012; Sandman et al. 2011; Connors et al. 2008; Barker 1998). Nevertheless, in this chapter, we will concentrate only on some effects of gestational stress in rats on behaviour, particularly in relation to the possible underlying morphological and neurochemical changes in brain regions that are involved in human disease vulnerability later in life.

Prenatal Stress Effects on Offspring Brain Plasticity and Related Behavioural Dysfunctions in Rats

Experimental Paradigms

Animal models of prenatal stress were developed to mimic etiological features of human diseases. Most studies with those models have exposed pregnant dams to a stressful situation and later explored short- and long-term behavioural and molecular changes in the newborn. Numerous protocols are described in the literature, which vary in the type of stressor applied, daily frequency, length of application, and week of gestation chosen. The types of stressors used in rodents have been summarized by Baier et al. (2012) and range from suspension, crowding, repeated tail shocks, restraint, immobilization, and saline injections to unpredictable stress with noise and flashing lights. According to Archer and Blackman (1971), the intensity of the response of the mother is more important than the intensity of the stimulus.

In our laboratory, we employ a prenatal restraint stress model in which pregnant rats are randomly assigned either to a control group that stays undisturbed throughout the pregnancy or to the stress group. "Stressed dams" are placed individually in a transparent plastic restrainer fitted closely to body size for 45 min three times a day (9.00, 12.00 and 16.00 h) during the last week of gestation (i.e. from gestational day 14 until delivery), which approximately coincides with the second trimester of gestation in humans and is considered critical for vulnerability to psychiatric disorders (Bayer et al. 1993). The restraint stress protocol is a preferred means of stressing animals because it is painless, straightforward, does not involve bodily harm and is inexpensive. Moreover, the physiological changes associated with restraint result from the distress and aversive nature of having to remain immobile, and the changes mainly manifest as increases in adrenocorticotropic hormone and corticosterone (Buynitsky and Mostofsky 2009; Ward and Weisz 1984). In our hands, the intensity and duration of the stress applied are sufficient to induce alterations in a variety of assessed parameters, including behavioural and physiological dysfunction related to glutamatergic and dopaminergic system development in the offspring (Berger et al. 2002; Barros et al. 2006a, b; Adrover et al. 2007; Katunar et al. 2010; Pallares et al. 2013b). We have also observed that prenatal stress induces long-term effects on the male offspring reproductive system and spermatogenesis development, particularly by inducing a long term imbalance of circulating sexual hormone levels (Pallares et al. 2013a).

Prenatal Stress and Behavioural Dysfunctions in Rats

The effect of prenatal stress on baseline motor activity in rodents has been regarded as a main parameter of altered behaviour in order to evaluate coping and responsiveness of specific neurotransmitter pathways related to locomotion. However, the results that have been reported are controversial, since increased, decreased and unchanged basal motor activity have all been reported in prenatal stress animals (Pallares et al. 2007; Koenig et al. 2005; Fumagalli et al. 2007). On the other hand, amphetamine injection increases locomotion in prenatally stressed rats (Lehmann et al. 2000; Koenig et al. 2005). Schizophrenic patients respond abnormally to psychostimulants much as do prenatally stressed rats, suggesting that prenatal exposure to stress might cause an enhanced responsiveness of the dopaminergic system, which represents a core feature of schizophrenia (Fumagalli et al. 2007). In support of this observation, our group demonstrated that basal and amphetamine-stimulated dopamine output in the Nucleus Accumbens of adolescent and adult prenatal stress rats is higher than in controls (Silvagni et al. 2008). In addition, schizophrenic patients have deficits in filtering or discriminating relevant from irrelevant information (i.e. signal to noise ratio) and the pathology is characterized by deficits in operative measures of filter ability, such as prepulse inhibition. It has been demonstrated that exposure to chronic stress during gestation disrupts prepulse inhibition (Koenig et al. 2005).

Despite the similarities between schizophrenia and long-lasting behavioural effects resulting from prenatal stress, this paradigm has also been proposed as a model of major depression. The symptoms of clinical depression in humans involve changes in mood that cannot be assessed in animal models. Nevertheless, behavioural patterns related to human depression, such as loss of active coping, social withdrawal and anhedonia (inability to feel pleasure), can be measured under appropriate conditions. Prenatal stress rats and mice show increased duration of immobility in the forced swim Porsolt test when compared to rats with normal gestation (Morley-Fletcher et al. 2003; Alonso et al. 1991). Furthermore, prenatal stress reduces exploratory behaviour in adult animals, presumably due to decreased motivation (Vallee et al. 1997), and the animals are more likely to develop anhedonia, as indicated by a decrease in sucrose solution consumed when performing the sucrose preference test (Sun et al. 2013).

Evidence from experimental research also demonstrates that prenatal stress affects emotions of the offspring by increasing anxiety-like behaviour. In 1957,

Thompson was the first to demonstrate that trauma in females during pregnancy increased emotionality in their offspring (Thompson 1957). Anxiety state in experimental animals can be measured in tests that evaluate fear-like reactions, which can be elicited in unfamiliar open spaces like the "open field" or in the "elevated plus maze." Increased anxiety in both sexes of offspring of mothers subjected to unpredictable noise throughout gestation were reported by Fride and Weinstock (1988) and Vallee et al. (1997). In our hands, prenatally stressed adult rats spend more time in the closed arms of the elevated plus maze than control arms, suggesting anxiety (Barros et al. 2006b).

Cognitive impairment is also related to gestational stress. Chronic maternal restraint stress slows the acquisition of spatial learning (Lemaire et al. 2000; Son et al. 2006; Yang et al. 2006) and reduces long-term potentiation, a primary physiological model of memory (Son et al. 2006) in offspring of rats and mice (see Chap. 5). The principal region for the regulation of cognition is the hippocampus, which also has the highest density of glucocorticoid receptors. Hippocampal integrity is essential for the development of physiological cognitive processes, including learning and memory. The cytoarchitecture of the rat hippocampus is altered by maternal gestational stress; prenatal stress decreases the hippocampal volume of the progeny by reducing the number of granule cells within the hippocampal dentate gyrus (Lemaire et al. 2000) and neurogenesis in this region (Lemaire et al. 2006). By mimicking the effects of prenatal stress through intramuscular injection of synthetic glucocorticoids during pregnancy, Uno et al. (1994) have demonstrated a dose-dependent degenerative change and reduction of the offspring hippocampal neurons. Moreover, hippocampus is smaller in schizophrenic and depressed subjects, so gestational stress might be etiologically relevant for the morphological changes that contribute to cognitive deterioration in psychiatric patients (Fumagalli et al. 2007). Furthermore, prenatal stress reduces the number of hippocampal synapses in the CA3 area (Hayashi et al. 1998) and, as we reported in our studies, prenatal stress induces a long term reduction in dendritic arborisation of the hippocampal CA1 region (Barros et al. 2006a; Pallares et al. 2013b).

Other brain areas that play a key role in controlling cognitive processes, such as the amygdala and the cerebral cortex, are also reported to be affected by prenatal stress. The amygdala is involved in mood regulation and in the mediation of emotional processes of fear and anxiety. Prenatal stress diminishes the size of most of the amygdalar nuclei, with exception of the lateral amygdalar nuclei, for which the effects of prenatal stress on size are age-specific: during early postnatal development a decreased volume and reduced number of neurons and glial cells are observed in prenatally stressed rats, whereas postnatally prenatal stress increases the number of neurons and glial cells in the same nuclei (Salm et al. 2004; Kraszpulski et al. 2006; Kawamura et al. 2006). The cerebral cortex is responsible for a variety of higher order brain processes such as memory, planning, problem solving and voluntary muscle movement. We found that prenatal stress alters subtypes of dopamine and glutamate receptors in different cortical regions in males offspring (Berger et al. 2002) and that prenatal stress induces a long term reduction of dendritic arborisations (Barros et al. 2006a; Pallares et al. 2013a). Furthermore, dendritic spine densities on both the apical and basal dendrites of pyramidal neurons are reduced by approximately 20% in the dorsal anterior cingulate cortex and orbitofrontal cortex for both male and female prenatally stressed rats (Murmu et al. 2006).

Possible Mechanisms of Prenatal Stress-Related Brain Programming

The mechanism by which the pregnant mother translates stress to the developing brain is still a matter of debate. Nevertheless, in this chapter we present several hypotheses that have been proposed (Fig. 7.1).

Maternal Stress Affects by Altering Blood Flow to the Placenta

Maternal stress has also been shown to constrict the placental arteries. The placenta has a high number of adrenergic receptors that in presence of hormones reduce foetal blood flow and the supply of essential nutrients and oxygen from the uterus. Thereby, the morphology and development of the placenta is affected as well as the foetus' oxygenation and nutrition, inducing, for example, low birth weight among other effects (Weinstock 2005; Charil et al. 2010). Moreover, fluctuations in the oxygen availability induce oxidative stress in the placenta, increasing the number of oxygen free radicals and the release of cytokines that could also contribute to the development of neural disorders by oxidative damage (Charil et al. 2010).

Deregulation of Neurotrophins

Neurotrophins are small polypeptides that mediate the enduring effects of perinatal adversities on brain function by exerting a complex array of actions on various cellular phenotypes. In the central nervous system, they regulate survival and maturation of developing neurons including axonal growth and synaptic plasticity. Moreover, they also modulate neurotransmitter function (Fumagalli et al. 2007). Therefore, neurotrophins may be a route through which developmental manipulation, such as stress, can alter cellular resilience and modify brain structure. In addition, interfering with neurotrophin receptor function during development determines behavioural abnormalities such as dopaminergic hyper-responsivity and disrupted prepulse inhibition of acoustic startle, both of which are characteristic of schizophrenia (Rajakumar et al. 2004). Brain derived neurotrophic factor (BDNF) regulates synaptic transmission across a broad temporal spectrum ranging from short-term modulation, which occurs in the order of seconds to minutes, to

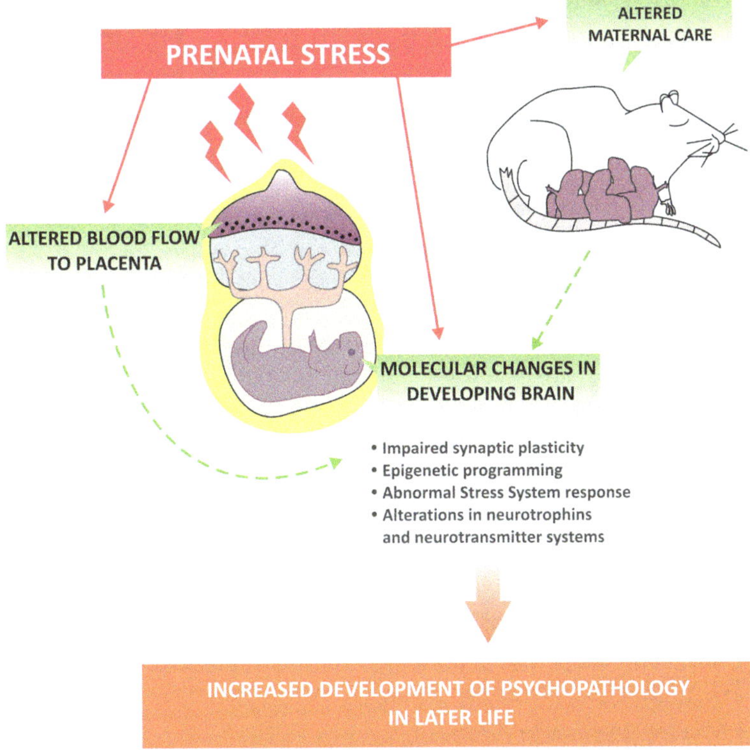

Fig. 7.1 Possible mechanism linking prenatal stress with plastic changes in the developing brain. The scheme depicts three of the possible links between prenatal stress and changes in the developing brain of the offspring, As explained in the text one of the mechanism proposed suggests that the increase of glucocorticoids and catecholamines in the stressed pregnant dam binds to their specific receptors in the placenta, reducing foetal blood flow, decreasing the supply of essential nutrients and oxygen to the foetus concomitantly with oxidative damage. A second mechanism directly relates the increase of circulating hormones in the mother to induce a dysfunctional HPA axis in the offspring, triggering alterations in the capability to cope with a stressful event. A third mechanism is related to changes in maternal behaviour that produces a long-term deregulation of the functional activity of the offspring HPA axis and alterations of DNA methylation

prolonged effects that persist for many hours, such as long-term potentiation (Fumagalli et al. 2007). Several studies have demonstrated that prenatal stress may reduce the biosynthesis of the BDNF in the infant (Van den Hove et al. 2006) and adult (Zuena et al. 2008; Monteleone et al. 2014) hippocampus of prenatally stressed male rats. Moreover, the expression of the fibroblast growth factor (FGF-2), which plays a relevant role as a neuroprotective molecule during development and adulthood, but is also involved in several psychiatric disorders, was found to be reduced in prefrontal cortex of prenatal stress animals (Fumagalli et al. 2007). The reduced expression of both trophic factors could occur as a result of perinatal stressors and may account for the vulnerability of specific neuronal systems.

Excess Glucocorticoids and Dysfunctional Stress Response in the Offspring

One potential mechanism whereby prenatal stress influences foetal development is by modifying the programming of the **hypothalamic-pituitary-adrenal (HPA)** axis, a major system controlling the organism's response to stress and regulating certain circadian activity. Behavioural alterations induced by a dysfunctional HPA axis (i.e. alterations in the individual capability to cope with a stressful event) show similarities with psychiatric disorders including major depression and schizophrenia (Fumagalli et al. 2007). Maternal stress can affect the offspring's stress response later in life (Levine 1967). Moreover, prolonged restraint stress exposure in pregnant rats during the last week of gestation reprograms their foetal HPA axis (Weinstock 2005, 2008): prenatally stressed rats secrete higher amounts of total- and free basal corticosterone at the end of the light period (Koehl et al. 1997) and exhibit prolonged elevation in plasma glucocorticoid levels following acute exposure to restraint stress (Vallee et al. 1997). That both prenatal stress and prenatal synthetic glucocorticoid exposure during the last week of gestation in rats permanently diminish corticosterone receptors in the hippocampus and hypothalamus could explain the reported imbalance in stress hormone levels during resting or even after a stressful episode of exposure. Reduction of central glucocorticoid receptors leads to an attenuation of the HPA axis feedback loop sensitivity (Henry et al. 1994; Maccari et al. 1995; Welberg et al. 2000).

During pregnancy, glucocorticoids (GCs) are naturally elevated. GCs are essential for foetal growth and induction of certain substances, such as pulmonary surfactant. GCs are also involved in normal brain development, where they exert a wide spectrum of effects in most regions of the developing brain, ranging from subcellular re-organization to neuron-neuron and neuron-glial interactions. Since sustained elevation of these hormones or their removal from the foetal brain is detrimental to normal processes, it is not surprising that most studies agree that glucocorticoids are the main agent conveying the effects of maternal stress to developing foetuses (Mastorci et al. 2009; Charil et al. 2010; Matthews 2001). In rats, maturation of the HPA axis starts early in development and extends to the early postnatal period. The glucocorticoid type II receptor (GR) mRNA can be detected in the hippocampus, hypothalamus and pituitary from day 13 of gestation and increases rapidly after birth (Cintra et al. 1993). The long-term effects of prenatal stress might be the consequence of excessive exposure of the foetus to maternal corticosterone. Such effects were shown to be prevented if the adrenal glands were surgical removed from the dams (Barbazanges et al. 1996). Additionally, inhibition of the enzyme 11beta-hydroxysteroid dehydrogenase type 2 (11beta-HSD2), which rapidly inactivates glucocorticoids when passing through placenta and other foetal tissues, induces permanent alterations of the HPA axis and increases anxiety-like behaviour (Welberg et al. 2000) suggesting that foetal overexposure to endogenous glucocorticoids may represent a common link between the prenatal environment and disorders linked to adult HPA axis dysfunction.

On the other hand, persistent effects of prenatal stress on HPA axis activity can be simulated by experimental models of *maternal deprivation* from postnatal day 2 to 14 (Liu et al. 2000), or it can be reversed by *early adoption* (Maccari et al. 1995) and *neonatal handling* (Meaney 2001), indicating that maternal behaviour may be crucial for the long-term regulation of functional activity of their offspring's HPA axis. Indeed, as discussed in Chap. 6, in recent years it has been demonstrated that the mother's behaviour produces stable alterations of DNA methylation and chromatin structure in the offspring (Meaney and Szyf 2005). Offspring in high mother-pup interactions have a reduced number of methylation of CpG dinucleotides in the GR promoter sequence than do offspring from lower maternal care mothers. That hypomethylation might be responsible for increased transcription of the GR gene, providing an epigenetic mechanism by which maternal care affects the gene expression of the offspring (Weaver et al. 2004).

Overall, adverse life conditions during prenatal or early postnatal life may be highly detrimental to the function and responsiveness of the HPA axis to stress. This could ultimately lead to a more persistent exposure of the brain to elevated levels of glucocorticoids that may reduce cellular resiliency and lead to damage of function in certain brain regions (McEwen 2000).

Concluding Remarks

Prenatal stress has been linked to abnormal outcomes in rodents, non-human primates, and humans (Charil et al. 2010; Huizink et al. 2004; Weinstock 2001). The data described in this chapter provide evidence that maternal stress at critical periods of development may alter the programming of foetal brain areas controlling and processing important behaviours, thereby increasing susceptibility to psychopathology. The increase in stress hormones during critical windows of brain development is detrimental for normal neuronal differentiation and function. Depending on intensity and the time of gestation the stress takes place, the behavioural effects on the offspring range from learning and attention deficits, anxiety- and depressive-like behaviour to abnormal stress response, among others. It is clear from animal studies that prenatal stress affects the morphology of the offspring brains, but the mechanisms that lead to these effects remain poorly understood. Nevertheless, the maternal and foetal HPA axis and the placenta seem to represent the main candidates for these mechanisms.

As early as 1941, Sontag (1941) drew attention to the long-lasting consequences of gestational stress on infant development and behaviour. Extrapolation of experimental research results from prenatal stress studies in animals to gestational disturbances in humans has been attempted. Although we are aware that direct comparison between experimental animals and humans is a complex issue, understanding of brain mechanism underlying the link between prenatal stress and adult psychopathologies still relies on animal studies. In this context, it can be speculated that the impairments observed in limbic areas in animal studies after prenatal stress might induce behavioural effects in animals that correspond to similar effects in human subjects. Therefore, animal studies could provide a neurochemical and morphological basis to the observed human psychopathologies associated with gestational stress.

References

Adrover E, Berger MA, Perez AA, Tarazi FI, Antonelli MC (2007) Effects of prenatal stress on dopamine D2 receptor asymmetry in rat brain. Synapse 61(6):459–462

Alonso SJ, Arevalo R, Afonso D, Rodriguez M (1991) Effects of maternal stress during pregnancy on forced swimming test behavior of the offspring. Physiol Behav 50:511–517

Archer JE, Blackman DE (1971) Prenatal psychological stress and offspring behavior in rats and mice. Dev Psychobiol 4(3):193–248. doi:10.1002/dev.420040302

Baier CJ, Katunar MR, Adrover E, Pallares ME, Antonelli MC (2012) Gestational restraint stress and the developing dopaminergic system: an overview. Neurotox Res. doi:10.1007/s12640-011-9305-4

Barbazanges A, Piazza PV, Le Moal M, Maccari S (1996) Maternal glucocorticoid secretion mediates long-term effects of prenatal stress. J Neurosci 16(12):3943–3949

Barker DJ (1998) In utero programming of chronic disease. Clin Sci (Lond) 95(2):115–128

Barros VG, Duhalde-Vega M, Caltana L, Brusco A, Antonelli MC (2006a) Astrocyte-neuron vulnerability to prenatal stress in the adult rat brain. J Neurosci Res 83(5):787–800

Barros VG, Rodriguez P, Martijena ID, Perez A, Molina VA, Antonelli MC (2006b) Prenatal stress and early adoption effects on benzodiazepine receptors and anxiogenic behavior in the adult rat brain. Synapse 60(8):609–618

Bayer SA, Altman J, Russo RJ, Zhang X (1993) Timetables of neurogenesis in the human brain based on experimentally determined patterns in the rat. Neurotoxicology 14(1):83–144

Berger MA, Barros VG, Sarchi MI, Tarazi FI, Antonelli MC (2002) Long-term effects of prenatal stress on dopamine and glutamate receptors in adult rat brain. Neurochem Res 27(11):1525–1533

Buynitsky T, Mostofsky DI (2009) Restraint stress in biobehavioral research: recent developments. Neurosci Biobehav Rev 33(7):1089–1098. doi:10.1016/j.neubiorev.2009.05.004. S0149-7634(09)00074-8 [pii]

Charil A, Laplante DP, Vaillancourt C, King S (2010) Prenatal stress and brain development. Brain Res Rev 65(1):56–79. doi:10.1016/j.brainresrev.2010.06.002. S0165-0173(10)00072-X [pii]

Cintra A, Solfrini V, Bunnemann B, Okret S, Bortolotti F, Gustafsson JA, Fuxe K (1993) Prenatal development of glucocorticoid receptor gene expression and immunoreactivity in the rat brain and pituitary gland: a combined in situ hybridization and immunocytochemical analysis. Neuroendocrinology 57(6):1133–1147

Connors SL, Levitt P, Matthews SG, Slotkin TA, Johnston MV, Kinney HC, Johnson WG, Dailey RM, Zimmerman AW (2008) Fetal mechanisms in neurodevelopmental disorders. Pediatr Neurol 38(3):163–176. doi:10.1016/j.pediatrneurol.2007.10.009. S0887-8994(07)00542-5 [pii]

Cottrell EC, Seckl JR (2009) Prenatal stress, glucocorticoids and the programming of adult disease. Front Behav Neurosci 3:19. doi:10.3389/neuro.08.019.2009

Del Giudice M (2012) Fetal programming by maternal stress: insights from a conflict perspective. Psychoneuroendocrinology 37(10):1614–1629. S0306-4530(12)00191-6 [pii]. doi:10.1016/j.psyneuen.2012.05.014

Fride E, Weinstock M (1988) Prenatal stress increases anxiety related behavior and alters cerebral lateralization of dopamine activity. Life Sci 42(10):1059–1065

Fumagalli F, Molteni R, Racagni G, Riva MA (2007) Stress during development: impact on neuroplasticity and relevance to psychopathology. Prog Neurobiol 81(4):197–217. S0301-0082(07)00014-7 [pii]. doi:10.1016/j.pneurobio.2007.01.002

Harris A, Seckl J (2011) Glucocorticoids, prenatal stress and the programming of disease. Horm Behav 59(3):279–289. S0018-506X(10)00167-4 [pii]. doi:10.1016/j.yhbeh.2010.06.007

Hayashi A, Nagaoka M, Yamada K, Ichitani Y, Miake Y, Okado N (1998) Maternal stress induces synaptic loss and developmental disabilities of offspring. Int J Dev Neurosci 16(3–4):209–216

Henry C, Kabbaj M, Simon H, Le Moal M, Maccari S (1994) Prenatal stress increases the hypothalamo-pituitary-adrenal axis response in young and adult rats. J Neuroendocrinol 6(3):341–345

Huizink AC, de Medina PG, Mulder EJ, Visser GH, Buitelaar JK (2002) Psychological measures of prenatal stress as predictors of infant temperament. J Am Acad Child Adolesc Psychiatry 41(9):1078–1085

Huizink AC, Mulder EJ, Buitelaar JK (2004) Prenatal stress and risk for psychopathology: specific effects or induction of general susceptibility? Psychol Bull 130(1):115–142

Katunar M, Saez T, Brusco A, Antonelli M (2010) Ontogenetic expression of dopamine-related transcription factors and tyrosine hydroxylase in prenatally stressed rats. Neurotox Res 18(1):69–81

Kawamura T, Chen J, Takahashi T, Ichitani Y, Nakahara D (2006) Prenatal stress suppresses cell proliferation in the early developing brain. Neuroreport 17(14):1515–1518. doi:10.1097/01. wnr.0000236849.53682.6d. 00001756-200610020-00012[pii]

Koehl M, Barbazanges A, Le Moal M, Maccari S (1997) Prenatal stress induces a phase advance of circadian corticosterone rhythm in adult rats which is prevented by postnatal stress. Brain Res 759(2):317–320

Koenig JI, Elmer GI, Shepard PD, Lee PR, Mayo C, Joy B, Hercher E, Brady DL (2005) Prenatal exposure to a repeated variable stress paradigm elicits behavioral and neuroendocrinological changes in the adult offspring: potential relevance to schizophrenia. Behav Brain Res 156(2):251–261. S0166-4328(04)00216-5 [pii]. doi:10.1016/j.bbr.2004.05.030

Kraszpulski M, Dickerson PA, Salm AK (2006) Prenatal stress affects the developmental trajectory of the rat amygdala. Stress 9(2):85–95. J740525J80R55214 [pii]. doi:10.1080/10253890600798109

Lehmann J, Stohr T, Feldon J (2000) Long-term effects of prenatal stress experiences and postnatal maternal separation on emotionality and attentional processes. Behav Brain Res 107(1–2):133–144

Lemaire V, Koehl M, Le Moal M, Abrous DN (2000) Prenatal stress produces learning deficits associated with an inhibition of neurogenesis in the hippocampus. Proc Natl Acad Sci USA 97(20):11032–11037. 97/20/11032 [pii]

Lemaire V, Lamarque S, Le Moal M, Piazza PV, Abrous DN (2006) Postnatal stimulation of the pups counteracts prenatal stress-induced deficits in hippocampal neurogenesis. Biol Psychiatry 59(9):786–792. S0006-3223(05)01400-9 [pii]. doi:10.1016/j.Biopsych.2005.11.009

Levine S (1967) Maternal and environmental influences on the adrenocortical response to stress in weanling rats. Science 156(3772):258–260

Liu D, Diorio J, Day JC, Francis DD, Meaney MJ (2000) Maternal care, hippocampal synaptogenesis and cognitive development in rats. Nat Neurosci 3(8):799–806. doi:10.1038/77702

Maccari S, Piazza PV, Kabbaj M, Barbazanges A, Simon H, Le Moal M (1995) Adoption reverses the long-term impairment in glucocorticoid feedback induced by prenatal stress. J Neurosci 15(1 Pt 1):110–116

Maccari S, Darnaudery M, Morley-Fletcher S, Zuena AR, Cinque C, Van Reeth O (2003) Prenatal stress and long-term consequences: implications of glucocorticoid hormones. Neurosci Biobehav Rev 27(1–2):119–127. S0149763403000149 [pii]

Mastorci F, Vicentini M, Viltart O, Manghi M, Graiani G, Quaini F, Meerlo P, Nalivaiko E, Maccari S, Sgoifo A (2009) Long-term effects of prenatal stress: changes in adult cardiovascular regulation and sensitivity to stress. Neurosci Biobehav Rev 33(2):191–203. S0149-7634(08)00122-X [pii]. doi:10.1016/j.neubiorev.2008.08.001

Matthews SG (2001) Antenatal glucocorticoids and the developing brain: mechanisms of action. Semin Neonatol 6(4):309–317. doi:10.1053/siny.2001.0066. S1084-2756(01)90066-1[pii]

McEwen BS (2000) Effects of adverse experiences for brain structure and function. Biol Psychiatry 48(8):721–731. S0006-3223(00)00964-1 [pii]

Meaney MJ (2001) Maternal care, gene expression, and the transmission of individual differences in stress reactivity across generations. Annu Rev Neurosci 24:1161–1192. doi:10.1146/annurev.neuro.24.1.1161. 24/1/1161[pii]

Meaney MJ, Szyf M (2005) Maternal care as a model for experience-dependent chromatin plasticity? Trends Neurosci 28(9):456–463. S0166-2236(05)00189-X [pii]. doi:10.1016/j. tins.2005.07.006

Monteleone MC, Adrover E, Pallares ME, Antonelli MC, Frasch AC, Brocco MA (2014) Prenatal stress changes the glycoprotein GPM6A gene expression and induces epigenetic changes in rat offspring brain. Epigenetics 9(1):152–160. 25925 [pii]. doi:10.4161/epi.25925

Morley-Fletcher S, Darnaudery M, Koehl M, Casolini P, Van Reeth O, Maccari S (2003) Prenatal stress in rats predicts immobility behavior in the forced swim test. Effects of a chronic treatment with tianeptine. Brain Res 989(2):246–251. S0006899303032931 [pii]

Murmu MS, Salomon S, Biala Y, Weinstock M, Braun K, Bock J (2006) Changes of spine density and dendritic complexity in the prefrontal cortex in offspring of mothers exposed to stress during pregnancy. Eur J Neurosci 24(5):1477–1487. EJN5024 [pii]. doi:10.1111/j.1460-9568.2006.05024.x

Pallares ME, Scacchi Bernasconi PA, Feleder C, Cutrera RA (2007) Effects of prenatal stress on motor performance and anxiety behavior in Swiss mice. Physiol Behav 92(5):951–956. S0031-9384(07)00267-3 [pii]. doi:10.1016/j.physbeh.2007.06.021

Pallares ME, Adrover E, Baier CJ, Bourguignon NS, Monteleone MC, Brocco MA, Gonzalez-Calvar SI, Antonelli MC (2013a) Prenatal maternal restraint stress exposure alters the reproductive hormone profile and testis development of the rat male offspring. Stress 16(4):429–440. doi:10.3109/10253890.2012.761195

Pallares ME, Baier CJ, Adrover E, Monteleone MC, Brocco MA, Antonelli MC (2013b) Age-dependent effects of prenatal stress on the corticolimbic dopaminergic system development in the rat male offspring. Neurochem Res 38(11):2323–2335. doi:10.1007/s11064-013-1143-8

Rajakumar N, Leung LS, Ma J, Rajakumar B, Rushlow W (2004) Altered neurotrophin receptor function in the developing prefrontal cortex leads to adult-onset dopaminergic hyperresponsivity and impaired prepulse inhibition of acoustic startle. Biol Psychiatry 55(8):797–803. doi:10.1016/j.biopsych.2003.12.015. S0006322304000551[pii]

Rondo PH, Ferreira RF, Nogueira F, Ribeiro MC, Lobert H, Artes R (2003) Maternal psychological stress and distress as predictors of low birth weight, prematurity and intrauterine growth retardation. Eur J Clin Nutr 57(2):266–272. doi:10.1038/sj.ejcn.1601526. 1601526[pii]

Salm AK, Pavelko M, Krouse EM, Webster W, Kraszpulski M, Birkle DL (2004) Lateral amygdaloid nucleus expansion in adult rats is associated with exposure to prenatal stress. Brain Res Dev Brain Res 148(2):159–167. doi:10.1016/j.devbrainres.2003.11.005. S0165380603003468[pii]

Sandman CA, Davis EP, Buss C, Glynn LM (2011) Prenatal programming of human neurological function. Int J Pept 2011:837596. doi:10.1155/2011/837596

Silvagni A, Barros VG, Mura C, Antonelli MC, Carboni E (2008) Prenatal restraint stress differentially modifies basal and stimulated dopamine and noradrenaline release in the nucleus accumbens shell: an 'in vivo' microdialysis study in adolescent and young adult rats. Eur J Neurosci 28(4):744–758. EJN6364 [pii]. doi:10.1111/j.1460-9568.2008.06364.x

Son GH, Geum D, Chung S, Kim EJ, Jo JH, Kim CM, Lee KH, Kim H, Choi S, Kim HT, Lee CJ, Kim K (2006) Maternal stress produces learning deficits associated with impairment of NMDA receptor-mediated synaptic plasticity. J Neurosci 26(12):3309–3318. 26/12/3309 [pii]. doi:10.1523/JNEUROSCI.3850-05.2006

Sontag LD (1941) The significance of fetal environmental differences. Am J Obstet Gynecol 42(6):996–1003

Sun H, Guan L, Zhu Z, Li H (2013) Reduced levels of NR1 and NR2A with depression-like behavior in different brain regions in prenatally stressed juvenile offspring. PLoS One 8(11):e81775. doi:10.1371/journal.pone.0081775. PONE-D-13-37163[pii]

Thompson WR (1957) Influence of prenatal maternal anxiety on emotionality in young rats. Science 125(3250):698–699

Uno H, Eisele S, Sakai A, Shelton S, Baker E, DeJesus O, Holden J (1994) Neurotoxicity of glucocorticoids in the primate brain. Horm Behav 28(4):336–348. S0018-506X(84)71030-0 [pii]. doi:10.1006/hbeh.1994.1030

Valenzuela CF, Morton RA, Diaz MR, Topper L (2012) Does moderate drinking harm the fetal brain? Insights from animal models. Trends Neurosci 35(5):284–292. S0166-2236(12)00018-5 [pii]. doi:10.1016/j.tins.2012.01.006

Vallee M, Mayo W, Dellu F, Le Moal M, Simon H, Maccari S (1997) Prenatal stress induces high anxiety and postnatal handling induces low anxiety in adult offspring: correlation with stress-induced corticosterone secretion. J Neurosci 17(7):2626–2636

Van den Hove DL, Steinbusch HW, Scheepens A, Van de Berg WD, Kooiman LA, Boosten BJ, Prickaerts J, Blanco CE (2006) Prenatal stress and neonatal rat brain development. Neuroscience 137(1):145–155. S0306-4522(05)00985-1[pii]. doi:10.1016/j.neuroscience.2005.08.060

Ward IL, Weisz J (1984) Differential effects of maternal stress on circulating levels of corticosterone, progesterone, and testosterone in male and female rat fetuses and their mothers. Endocrinology 114(5):1635–1644. doi:10.1210/endo-114-5-1635

Weaver IC, Cervoni N, Champagne FA, D'Alessio AC, Sharma S, Seckl JR, Dymov S, Szyf M, Meaney MJ (2004) Epigenetic programming by maternal behavior. Nat Neurosci 7(8):847–854. doi:10.1038/nn1276. nn1276[pii]

Weinstock M (2001) Alterations induced by gestational stress in brain morphology and behaviour of the offspring. Prog Neurobiol 65(5):427–451. S0301-0082(01)00018-1 [pii]

Weinstock M (2005) The potential influence of maternal stress hormones on development and mental health of the offspring. Brain Behav Immun 19(4):296–308. S0889-1591(04)00132-1 [pii]. doi:10.1016/j.bbi.2004.09.006

Weinstock M (2008) The long-term behavioural consequences of prenatal stress. Neurosci Biobehav Rev 32(6):1073–1086

Welberg LA, Seckl JR, Holmes MC (2000) Inhibition of 11beta-hydroxysteroid dehydrogenase, the foeto-placental barrier to maternal glucocorticoids, permanently programs amygdala GR mRNA expression and anxiety-like behaviour in the offspring. Eur J Neurosci 12(3):1047–1054. ejn958 [pii]

Yang J, Han H, Cao J, Li L, Xu L (2006) Prenatal stress modifies hippocampal synaptic plasticity and spatial learning in young rat offspring. Hippocampus 16(5):431–436. doi:10.1002/hipo.20181

Zuena AR, Mairesse J, Casolini P, Cinque C, Alema GS, Morley-Fletcher S, Chiodi V, Spagnoli LG, Gradini R, Catalani A, Nicoletti F, Maccari S (2008) Prenatal restraint stress generates two distinct behavioral and neurochemical profiles in male and female rats. PLoS One 3(5):e2170. doi:10.1371/journal.pone.0002170

Chapter 8
Early Postnatal Development of Somastostatinergic Systems in Brainstem Respiratory Network

Isabel Llona, Paula Farías, and Jennifer L. Troc-Gajardo

Abstract Somatostatin is a peptide able to stop breathing, acting in the neural network that generates and control the respiratory rhythm. In this chapter, we present data on the early postnatal development of somatostatinergic systems in the mouse brainstem and summarize evidence for their influence on the generation and control of the respiratory rhythm.

Keywords Somatostatin • Respiration • Ontogeny • SST

Abbreviations

7	Facial nucleus
10	Dorsal motor nucleus of vagus
12 or XII	Hypoglossal nucleus
Amb	Ambiguus nucleus
AVC	Anteroventral cochlear nucleus
BötC	Bötzinger nucleus
CIC	Central nucleus inferior colliculus
CPG	Central pattern generator
Cu	Cuneate nucleus
cVRG	Caudal ventral respiratory group
DMTg	Dorsomedial tegmental area
DPGi	Dorsal paragigantocellular nucleus
DRC	Dorsal respiratory column
E	Embryonic
ECIC	External cortex of the inferior colliculus

I. Llona, PhD (✉) • P. Farías, PhD • J.L. Troc-Gajardo
Laboratorio de Sistemas Neurales, Departamento de Biología, Facultad de Química y Biología, Universidad de Santiago de Chile, Postal Code 9170022 Santiago, Chile
e-mail: isabel.llona@usach.cl

© Springer International Publishing AG 2017
R. von Bernhardi et al. (eds.), *The Plastic Brain*, Advances in Experimental Medicine and Biology 1015, DOI 10.1007/978-3-319-62817-2_8

ECu	External cuneate nucleus
Gi	Gigantocellular reticular nucleus
GiV	Gigantocellular reticular nucleus, ventral part
GrC	Granular layer of the cochlear nucleus
IO	Inferior olive
IRt	Intermediate reticular nucleus
KF	Kölliker-Fuse nucleus
LC	Locus coeruleus
LPBE	Lateral parabrachial nucleus, external part
LPBS	Lateral parabrachial nucleus, superior part
LPGi	Lateral paragigantocellular nucleus
LRt	Lateral reticular nucleus
LVPO	Lateral ventral periolivary nucleus
MdD	Medullary reticular nucleus, dorsal part
ml	Medial lemniscus
Mo5	Motor trigeminal nucleus
MPB	Medial parabrachial nucleus
MVe	Medial vestibular nucleus
MVeMC	Medial vestibular nucleus, mediocaudal part
MVePC	Medial vestibular nucleus, parvicellular part
MVPO	Medioventral periolivary nucleus
NK1R	Neurokinin 1 receptor
NTS	Nucleus of solitary tract (labeled Sol in figures)
P	Postnatal
P7	Perifacial zone
PCRtA	Parvicellular reticular formation
Pert	Parvicellular reticular nucleus
pFRG	Parafacial respiratory group
PGi	Paragigantocellular nucleus
ppy	Peripyramidal nucleus
preBötC	PreBötzinger complex
PRG	Pontine respiratory group
py	Pyramidal tract
RIA	Radioimmunoassay
RTN	Retrotrapezoid nucleus
RtTg	Reticulotegmental nucleus of the pons
RVL	Rostroventrolateral reticular nucleus
rVRG	Rostral ventral respiratory group
scp	Superior cerebelar peduncule
SIDS	Sudden infant death syndrome
Sol DL	Dorsolateral sub nucleus of NTS
Sol V	Ventral NTS
Sol VL	Ventrolateral sub nucleus of NTS
sp5	Spinal trigeminal tract

SpVe   Spinal vestibular nucleus
SST   Somatostatin
Su5   Supratrigeminal nucleus
VCP   Ventral cochlear nucleus, posterior part
VRC   Ventral respiratory column

Introduction

Since the neuropeptide somatostatin (SST) was discovered as an inhibitor of growth hormone secretion (Brazeau et al. 1973), many other biological functions have been described for this peptide. These functions include inhibition of secretion of insulin, glucagon, gastrin and other gastrointestinal hormones (Corleto 2010), a potent anti-proliferative action on various cancer cell lines (Keri et al.1996; Pyronnet et al. 2008), and anxiolytic and antidepressant-like effects (Engin et al. 2008; Butler et al. 2012). In addition, SST has been implicated in cognitive and neurological diseases such as Parkinson's and Huntington's diseases (Tuboly and Vecsei 2013).

The expression of SST has been described in various extra hypothalamic brain areas, including the anterior olfactory nucleus, cortex, hippocampus, striatum, brainstem (Viollet et al. 2008; Martel et al. 2012), and in such sensory systems as the visual (Cervia et al. 2008) and the olfactory systems (Lepousez et al. 2010). SST is expressed in many nuclei of the brainstem related to generation and control of respiration (Llona and Eugenín 2005), and it is currently used as an anatomical marker for the preBötzinger complex (preBötC) (Stornetta et al. 2003), which is considered fundamental for generating inspiration (Smith et al. 1991, 2013; Feldman et al. 2013; Feldman and Kam 2015).

However, little information has been published regarding the expression of SST in the early neonatal period in the mouse. As the mouse is suitable for in vitro preparations for the study of the generation and control of respiration, we decided to investigate the presence of SST in the mouse brainstem during the early postnatal weeks. SST binding site abnormalities in the brainstem have been detected in infants who died from Sudden Infant Death Syndrome (SIDS), a leading cause of mortality in human infants aged between 1 and 12 month in the USA (Kochanek et al. 2014). As further discussed in Chap. 11, a defect in the neuroregulation of the cardiorespiratory system is strongly suspected in this syndrome (Kinney et al. 1992; Kinney and Thach 2009). A high incidence of prolonged central apneas has been reported in infants who subsequently died of SIDS (Schechtman et al. 1991), and some studies have shown an increased density of SST binding sites in the brainstem of victims of SIDS (Carpentier et al. 1998).

The aim of this chapter is to describe recent data on the development of the somatostatinergic systems in the brainstem respiratory network and their function in respiratory rhythm generation and control.

The Neural Respiratory Network

Breathing in mammals is a vital function based on an autonomic and rhythmic motor behavior that allows the O_2/CO_2 exchange in alveoli. The respiratory rhythm is commanded by the respiratory central pattern generator (CPG), a neuronal network localized in the brainstem (Feldman et al. 2013; Feldman and Kam 2015; Smith et al. 2013). This network generates and controls the synchronous and rhythmic discharge of spinal and cranial motoneurons innervating respiratory muscles (Bianchi et al. 1995; Feldman et al. 2013). Sensory information such as chemoreception (Nattie and Li 2012), mechanoreception and thermoreception is integrated at the CPG to adjust the respiratory rhythm to different physiological situations (Molkov et al. 2013).

The anatomical organization of respiratory network in the brainstem is complex, with the respiratory neurons being located in the so-called dorsal (DRC) and ventral (VRC) respiratory columns (Bianchi et al. 1995). The DRC corresponds to the nucleus of the solitary tract (NTS), which receives, among others, afferents from the peripheral chemoreceptors (carotid and aortic bodies) and vagal pulmonary mechanoreceptors, and projects to premotor and phrenic motoneurons (Bianchi et al. 1995). At the level of the pons and as a continuation of the DRC lies the pontine respiratory group (PRG), which contains the Kölliker-Fuse (KF) and the parabrachial nuclei. It has been proposed that PRG participates in the respiratory response to smell, temperature, and defense reactions and transmits information from the hypothalamus to the respiratory centers (Bianchi et al. 1995). Destruction of the DRC, the PRG or both does not prevent the generation of the respiratory rhythm (Smith et al. 1991; Infante et al. 2003).

The VRC contains the Bötzinger nuclei (BötC), the parafacial respiratory group (pFRG), retrotrapezoid nucleus (RTN), the preBötC and the rostral ventral respiratory group (rVRG). These nuclei project at different spinal levels to premotor and motor neurons that innervate the muscles of the respiratory pump and control the upper airway resistance (Bianchi et al. 1995). Experiments using electrophysiological recordings, pharmacological manipulations, and lesions, both in vivo and in vitro, confirmed that the preBötC is critically involved in the generation of the respiratory rhythm (Smith et al. 1991; Ramirez et al. 1998; Gray et al. 2001; Tan et al. 2008). The preBötC, originally identified by Smith et al. (1991), is considered the minimum nucleus needed for the generation of the respiratory rhythm. It is located ventral to the semi compact division of the nucleus ambiguus (Amb), caudal to the compact division of the Amb and rostral to the lateral reticular nucleus (LRt). The preBötC contains a heterogeneous population of neurons with a subset of glutamatergic neurons functioning as pacemaker neurons fundamental for the generation of the respiratory rhythm (Feldman et al. 2013).

The respiratory rhythm is a set of finely tuned rhythmic motor patterns (Feldman and Kam 2015) (see also Chaps. 9, 10, 11, and 12). It is generated by the coupling of two oscillators (Feldman et al. 2003; Mellen et al. 2003) located in the VRC: the preBötC and pFRG (Smith et al. 2013; Onimaru and Homma 2003; Feldman et al. 2013), which interact through their projections (Tan et al. 2010; Feldman et al. 2013, Feldman and Kam 2015). The neurons that discharge during inspiratory phase and are inactive during expiratory phase (Smith et al. 1991) express neurokinin 1 receptor (NK1R) (Gray et al. 1999, 2001), SST (Stornetta et al. 2003; Gray et al. 2010) and the glycoprotein Reelin (Tan et al. 2012). Killing these neurons with NK1R toxins results in an ataxic respiratory rhythm (Gray et al. 2001), and silencing SST containing neurons induces a persistent apnea (Tan et al. 2008). However, SST null mutant mice are viable and breathe normally (Zeyda and Hochgeschwender 2008). On the other hand, neurons located in the pFRG are silent during inspiratory phase and discharge during active expiration (Onimaru and Homma 2003; Mellen et al. 2003).

The respiratory rhythm is modulated by many substances, including Substance P, SST, serotonin, and TRH (thyrotropin releasing hormone), synthesized by neurons located in various nuclei of the respiratory neural network (Bianchi et al. 1995; Alheid and McCrimmon 2008; Doi and Ramirez 2008; Feldman et al. 2003). Modulation by SST has been explored using various types of preparations, resulting always in a depression of the respiratory activity by SST. When SST was injected intracisternally in adult rats, the respiratory frequency and amplitude decreased until an irreversible apnea developed (Kalia et al. 1984a; Harfstrand et al. 1984). Soon after, it was demonstrated that SST acting at the VRC produced apnea and suppressed the ventilatory response to hypercapnia (Yamamoto et al. 1988; Chen et al. 1990). This evidence points to multiple sites of interaction between SST and the respiratory network. Burke et al. (2010) microinjected SST into the BötC, pre-BötC and caudal VRG (cVRG) in anesthetized rats. SST microinjections powerfully inhibited respiratory neurons throughout the VRC. In the BötC, SST eliminated post inspiratory activity in phrenic nerve recordings, an effect similar to inhibition of the KF nucleus in the pons. Injection in the preBötC produced depression of respiration or apnea. Injection into the cVRG reduced phrenic nerve amplitude (Burke et al. 2010). Pantaleo et al. (2011) investigated the respiratory responses to SST microinjected into the BötC and preBötC in anesthetized rabbits. SST in the BötC decreased respiratory frequency. However, when SST was microinjected in the preBötC, respiratory frequency increased. This unexpected result can be attributed to the high dose of SST injected in rabbits (150–250 pmol) or to species differences. Besides, the microinjection of the antagonist cyclosomatostatin decreased respiratory frequency, suggesting a tonic release of SST. Ramirez-Jarquín et al. (2012) showed in neonatal mice that blockade of type-2 SST receptors in vivo and in vitro increased respiratory, frequency pointing to a tonic release of the peptide in the preBötC modulating respiration. This hypothesis is supported by results obtained by Gray et al. (2010) in an *en bloc* preparation, where quantum slowing by SST was observed.

In sum, the respiratory network may be modulated by tonic release of SST. The relevance of the somatostatinergic tone to the control of respiratory rhythm during development remains to be determined.

Localization of SST-Positive Cells and Terminals in Respiratory Nuclei

Initial studies showed that expression of SST increased in the rat brainstem from embryonic day 15 (E15) until birth (Shiosaka et al. 1981). Pagliardini et al. (2003) dated the birth of SST positive neuron of preBötC between E12.5 and E13.5. During the perinatal period from E19 to postnatal day 5 (P5) expression of SST was higher, and only a few SST positive structures, such as the NTS and the parabrachial area, were labeled in adults (Inagaki et al. 1982). RIA assays confirmed that the SST levels in the brainstem were maximal at P7 (McGregor et al. 1982), with the highest level in the NTS (Douglas and Palkovits 1982). The SST mRNA showed a similar pattern of expression (Lowe et al. 1987; Shiraishi et al. 1993) to that of the peptide. Later, Tan et al. (2010) traced the SST pathway from the preBötC to other respiratory nuclei in the adult rat brainstem and showed a vast array of projections to respiratory nuclei. SST neurons project to contralateral preBötC, ipsi and contralateral BötC, cVRG, pFRG/RTN, NTS, the parabrachial and KF nucleus. Wei et al. (2012) studied the synaptic relationship between SST and NK1R expressing neurons in the pre-BötC. They found SST highly expressed in terminals, somata and primary dendrites. Large neurons expressing NK1R were not positive for SST, but received somatosta-tinergic inputs. Some SST-positive terminals were glutamatergic and the others GABAergic. Shimada and Ishikawa (1989) compared the distribution of SST-containing neurons of adult rat and mouse and found basically the same pattern of SST-immunoreactivity; however, these authors did not examine the lower brainstem. In mice, SST expression can be dated to early embryonic stages. Gray et al. (2010) showed that SST-positive neurons in the preBötC derived from the Dbx1 cell lineage and were born between E9.5 and E11.5. However, little has been reported on the early postnatal development of the SST neuronal systems of the mouse lower brainstem.

We mapped SST expression in the brainstem of neonatal mice by immunohisto-chemistry. From birth (P0) until P15 we found not only respiratory nuclei labeled, but also non-respiratory nuclei, as shown in Fig. 8.1. Labeling of somata and termi-nals did not change noticeably between P0 and P15, but the density of terminals showed a tendency to decrease (Table 8.1).

In the DRC we found only a few (less than 10 cells) SST immunolabelled cells in the subnucleus dorsolateral of NTS (SolDL) and an extensive network of fibers and terminals in various sub nuclei of the NTS (Fig. 8.1a–d). In the adult rat, neu-rons and nerve terminals containing SST were described in the ventrolateral (SolVL) and ventral (SolV) sub nuclei of the NTS (Johansson et al. 1984; Kalia et al. 1984b; Vincent et al. 1985). We did not find labeled cells in the ventrolateral or ventral sub nuclei of the NTS. This discrepancy in the distribution reported for the adult rat can be attributed to the use of colchicine pre-treatment in these previous studies, the age of the animal or species differences. At least three origins can be postulated for the SST terminals present in the NTS. In the rat, an SST pathway has been traced between the paragigantocellularis nucleus (PGi) and the NTS. These neurons often co-express methionine-enkephalin and also project to the spinal cord (Millhorn et al. 1987; Johnson et al. 2002). We also found SST-positive neurons in the DPGi.

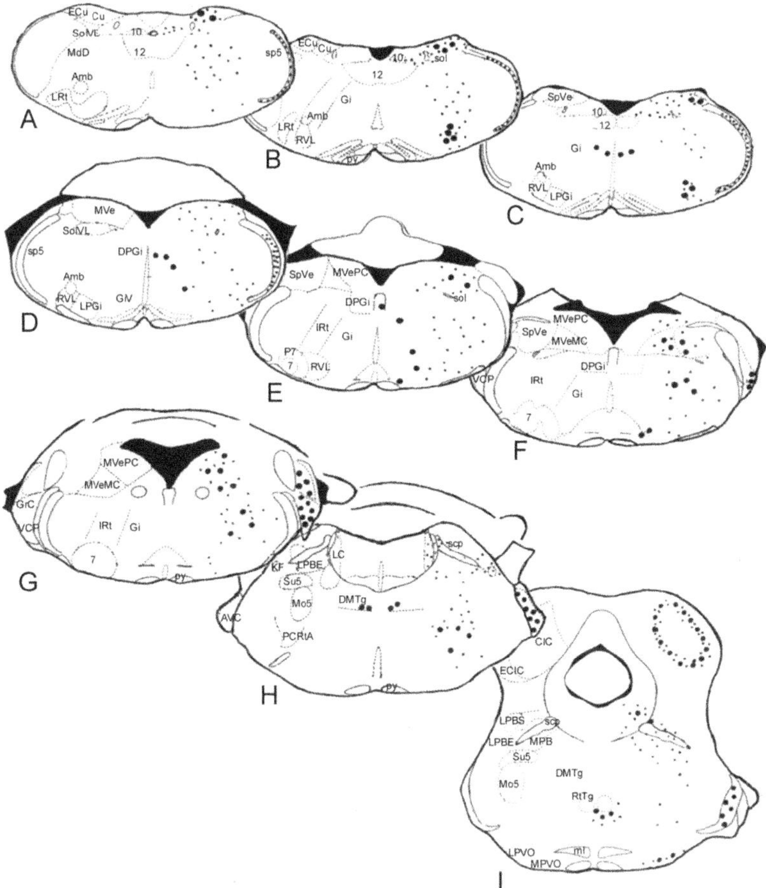

Fig. 8.1 Somatostatin expression in mouse neonatal brainstem. Serial frontal sections through the lower brainstem of neonatal mouse, showing the distribution of SST immunolabelled structures. *Large dots* (●) indicate somata and *small dots* (•) terminals and fibers. The density of dots was drawn according to the abundance of labelling at P0. The drawings are modifications of the atlas representations of the adult mouse brainstem corresponding to co-ordinates between −7.48 (A) and −4.98 (I) from bregma (Franklin and Paxinos 1997). *Abbreviations*: *7* facial nucleus, *10* dorsal motor nucleus of vagus, *12* or *XII* hypoglossal nucleus, *Amb* ambiguus nucleus, *AVC* anteroventral cochlear nucleus, *CIC* central nucleus inferior colliculus, *Cu* cuneate nucleus, *DMTg* dorsomedial tegmental area, *DPGi* dorsal paragigantocellular nucleus, *ECIC* external cortex of the inferior colliculus, *ECu* external cuneate nucleus, *Gi* gigantocellular reticular nucleus, *GiV* gigantocellular reticular nucleus, ventral part, *GrC* granular layer of the cochlear nucleus, *IO* inferior olive, *IRt* intermediate reticular nucleus, *KF* Kolliker Fuse nucleus, *LC* locus coeruleus, *LPBE* lateral parabrachial nucleus, external part, *LPBS* lateral parabrachial nucleus, superior part, *LPGi* lateral paragigantocellular nucleus, *LRt* lateral reticular nucleus, *LVPO* lateral ventral periolivary nucleus, *MdD* medullary reticular nucleus, dorsal part, *ml* medial lemniscus, *Mo5* motor trigeminal nucleus, *MPB* medial parabrachial nucleus, *MVe* medial vestibular nucleus, *MVeMC* medial vestibular nucleus, mediocaudal part, *MVePC* medial vestibular nucleus, parvicellular part, *MVPO* medioventral periolivary nucleus, *P7* perifacial zone, *PCRtA* parvicellular reticular formation, *Pert* parvicellular reticular nucleus, *ppy* peripyramidal nucleus, *py* pyramidal tract, *RtTg* reticulotegmental nucleus of the pons, *RVL* rostroventrolateral reticular nucleus, *scp* superior cerebelar peduncule, *Sol* nucleus of solitary tract, *Sol DL* dorsolateral sub nucleus of NTS, *Sol VL* ventrolateral sub nucleus of NTS, *sp5* spinal trigeminal tract, *SpVe* spinal vestibular nucleus, *Su5* supratrigeminal nucleus, *VCP* ventral cochlear nucleus, posterior part

Table 8.1 Somatostatin immunoreactive structures in the brainstem of the neonatal mouse

Nuclei		Postnatal age (days)							
		P0 (4)	P1 (3)	P2 (4)	P3 (3)	P4 (5)	P5 (2)	P7 (5)	P15 (2)
Cuneate (Cu)	S	##	##	##	##	##	##	#	#
	T	++	++	++	++	++	++	+	+
Dorsal medullary reticular (MdD)	S	–	–	–	–	–	–	–	–
	T	++	++	++	++	++	++	++	++
Intermediate reticular (IRt)	S					#			
	T	+++	+++	+++	+++	+++	+++	+++	+++
Spinal trigeminal tract (sp5)	S	–	–	–	–	–	–	–	–
	T	+++	+++	+++	+++	+++	+++	+++	+++
PreBötzinger Complex (preBötC)	S	#	#	#	#	#	#	#	#
	T	++	++	++	++	++	++	–	–
Solitary tract (Sol)	S	–	–	–	–	–	–	–	–
	T	+++	+++	+++	+++	+++	+++	+++	++
Gigantocellular reticular (Gi)	S	#	#	#	#	#	#	#	#
	T	++	++	++	++	++	+	+	+
Dorsal paragiganto cellular (DPGi)	S	##	##	##	##	##	##	##	##
	T	–	–	–	–	–	–	–	–
Spinal vestibular (SpVe)	S	##	##	##	##	##	##	##	#
	T	+++	+++	+++	+++	+++	+++	+++	++
Medullary vestibular (MVeMC, MVePC)	S	#	#	#	#	#	#	#	#
	T	+++	+++	+++	+++	+++	+++	+++	++
Cochlear (VCP)	S	###	###	###	###	###	###	##	##
	T	–	–	–	–	–	–	–	–
Peripyramidal (PPy)	S	#	#	#	#	#	#	#	#
	T	–	–	–	–	–	–	–	–
Dorsomedial tegmental (DMTg)	S	#	#	#	#	#	#	#	#
	T	–	–	–	–	–	–	–	–
Lateral parabraquial (LPBS, LPBE)	S	#	#			#	#	#	
	T	++	++			+++	+++	++	

Somatostatin was detected by immunohistochemistry in free-floating 20–30 µm sections using a rabbit anti SST antibody (1/4000, Santa Cruz, CA), a goat anti rabbit biotinylated secondary antibody (1/1000, Vector Labs, CA) and avidin-biotin-peroxidase complex (1/500, Vector Labs, CA) revealed with DAB-Ni. S = Somata (#); T = terminals or fibres (+). The numbers of animals used are indicated under the respective ages. The number of symbols indicates the number of labelled cells per section: #, few (less than 15); # #, moderate (15–30); # # #, abundant (> 30 cells). Terminals and fibres were estimated according to the intensity of labelling: weak (+), moderate (++) and intense (+++). (–) Indicates that no SST labelled structures were observed. Absence of symbols indicates an area not studied. For abbreviations see legend of Fig. 8.1

Thus, the origin of SST terminals in the NTS could be in part the PGi nucleus. Also, some of the SST terminals in the NTS in the neonatal mouse may originate in the periventricular SST immunoreactive neurons of the hypothalamus, as described for the adult rat (Krisch 1981). In adult rats, Tan et al. (2010) have shown that SST neurons in the preBötC project to the NTS.

The ventrolateral and ventral sub nuclei of the NTS receive afferents from lung stretch receptors and play a role in the inspiratory off-switch mechanism (Kalia 1981; Kalia et al. 1984b). In addition, the NTS responds to moderate hypoxia with an increase in the hypoxia-inducible factor-1α (HIF-1α) (Pascual et al. 2001). SST secreted from terminals in the NTS may act as an inhibitory transmitter of respiration. Jacquin et al. (1988) have provided direct evidence for an inhibitory effect of SST in the NTS. The activation of SST receptors in neurons of the NTS depresses excitability through hyperpolarization, result of augmentation of a voltage-dependent outward current blocked by muscarinic agonists (Jacquin et al. 1988). On the other hand, in the adult cat, microinjection of SST in the region of the lateral PGi consistently causes apnea (Yamamoto et al. 1988), as well as in the adult rat (Chen et al. 1990). The information converging in the NTS from peripheral reflexes, the CPG, and hypothalamus may use SST as signaling molecule.

In the PRG, scarce but intensely labeled SST positive-neurons and abundant terminals were found in the parabrachial nucleus (Fig. 8.1i). These terminals could arise from the preBötC, as shown by Tan et al. (2010) in adult rats. We have previously demonstrated, in the brainstem spinal cord preparation of the newborn mouse that a pontobulbar lesion increases the frequency of fictive respiration recorded from ventral roots (Infante et al. 2003). In the adult rat, both SST and SST receptors have been localized to A5 (pontine noradrenergic neurons) (Moyse et al. 1992; Wynne and Robertson 1997); it has been shown these A5 neurons are inhibitory (Errchidi et al. 1991; Hilaire and Duron 1999). Thus, SST may in part mediate the inhibitory control of the PRG on the respiratory rhythm generator.

The most significant nucleus with SST labeled neurons in the VRC is the rostro-ventrolateral medulla, ventral to the nucleus Amb (Figs. 8.1b, c and 8.2c–f) corresponding to preBötC. These neurons are fusiform, have a diameter of 15 μm and are found from P0 to P15 (Fig. 8.2d–f). Between P0 and P15, no differences have been observed in their distribution and number (10–15 cells per section; Table 8.1). Besides, a densely labeled network of SST positive fibers and terminals is observed in the preBötC (Fig. 8.2d). The density of terminals decreases with age, and at P7, labeled terminals are barely detected (Table 8.1). The exact origin of these terminals is unknown at the moment, but they may originate in the contralateral preBötC, the BötC or some other brainstem nuclei (Stornetta et al. 2003; Tan et al. 2010; Wei et al. 2012). Early postnatal immunolabeling for SST in the preBötC is similar to that in adult rats (Stornetta et al. 2003; Tan et al. 2010). In the preBötC, SST neurons are glutamatergic (Wang et al. 2001; Guyenet et al. 2002). The decrease in the SST terminals reported in this study may be related to changes in the organization of the rhythmogenic neurons during this period, as reported for glycinergic mechanisms (Paton et al. 1994). Our findings of a decrease in the SST terminals in the preBötC during the first week of life are interesting in the view of the suggested role for SST in the SIDS (Carpentier et al. 1998). High density of SST-containing perikarya and fibers in respiratory nuclei of the adult human and infant brainstem has also been shown (Bouras et al. 1987; Chigr et al. 1989). SIDS most frequently occurs between 1 and 4 months, a period of development during which major cardiorespiratory changes takeplace. In fact, increased density of SST binding sites in the medial and

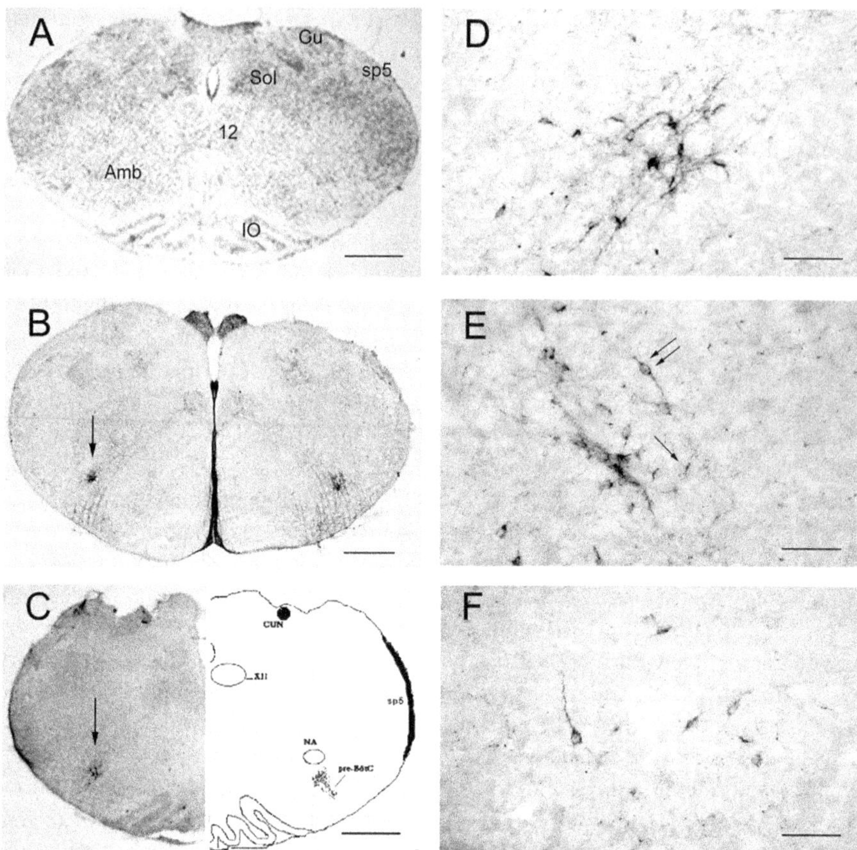

Fig. 8.2 Somatostatin in the preBötzinger complex. Neonatal brainstem coronal sections at the level of the preBötC: (**a**) Cresyl violet staining of the brainstem at P2 showing some anatomical landmarks such as the inferior olive (IO), the nucleus ambiguous (Amb), the hypoglossal nucleus (12), the solitary tract nucleus (Sol), cuneate nucleus (Cu), and the spinal trigeminal tract (sp5); (**b**) NK1-R staining at P1 with the *arrow* pointing to the Amb; (**c**) SST staining at P0, where the *arrow* points to the preBötC, ventral to the Amb. At the right is a drawing at the rostrocaudal level corresponding to −7.08 from bregma in adult mouse (Franklin and Paxinos 1997), and plate 24 in the neonatal mouse atlas (Jacobowitz and Abbott 1998) showing the localization of the preBötC; (**d–f**) SST-labelled cells and terminals in the preBötC at three different postnatal ages: P0 (**d**), P3 (**e**) and P7 (**f**). SST positive somata (*double arrows*) and terminals (*single arrow*) are labeled in panel **e**. Calibration bars: 400 μm (**a–c**) 50 μm (**d–f**)

lateral parabrachial nuclei of the brainstem in SIDS victims compared to matched controls has been reported by Carpentier et al. (1998).

The expression of SST in glutamatergic neurons is not restricted to the preBötC, but is also the case in the cuneate nucleus (Wang et al. 2000). It is tempting to hypothetize that SST may be part of a mechanism to avoid over excitation by released glutamate. However, it has not been determined if SST can modulate glutamate secretion from this population of neurons, as has been demonstrated in retina (Dal Monte et al. 2003) and hippocampus (Kozhemyakin et al. 2013). On the other hand, it is not known how glutamate can modulate SST release. Although SST receptor types 2 and 4 have been described in the preBötC, the pre- or postsynaptic localization of SST receptors has not been determined. It remains to be determined how the interaction between excitatory and inhibitory synapses in the preBötC is regulated by tonic release of SST.

In sum, SST is expressed from early development in the brainstem respiratory network, and increasing data on the anatomy and physiology of the peptide on the respiratory network have been obtained. However, a clear understanding of the mechanisms controlling SST release in the respiratory network has not been obtained.

Acknowledgements This article is dedicated to the memory of Dr. Silvia Araneda. This work was funded by grants from the National Fund for Scientific and Technological Development (FONDECYT # 1000025) and Research Division of the University of Santiago (DICYT # 029843ILL # 020643LLR). JL Troc-Gajardo is a PhD candidate at the PhD Program in Neuroscience, Universidad de Santiago de Chile, and she was financed by Programa de Formación de Capital Humano Avanzado # 21130284, Comisión Nacional de Investigación Científica y Tecnológica, Chile.

References

Alheid G, McCrimmon D (2008) The chemical neuroanatomy of breathing. Respir Physiol Neurobiol 164:3–11. doi:10.1016/j.resp.2008.07.014

Bianchi A, Denavit-Saubié M, Champagnat J (1995) Central control of breathing in mammals: neuronal circuitry, membrane properties, and neurotransmitters. Physiol Rev 75:1–45

Bouras C, Magistretti PJ, Morrison JH, Constantinidis J (1987) An immunohistochemical study of pro-somatostatin-derived peptides in the human brain. Neuroscience 22:781–800

Brazeau P, Vale W, Burgus R, Ling N, Butcher M, Rivier J, Guillemin R (1973) Hypothalamic polypeptide that inhibits the secretion of immunoreactive pituitary growth hormone. Science 179:77–79

Burke P, Abbott S, McMullan S, Goodchild A, Pilowsky P (2010) Somatostatin selectively ablates post-inspiratory activity after injection into the Bötzinger complex. Neuroscience 167(2):528–539. doi:10.1016/j.neuroscience.2010.01.065

Butler RK, White LC, Frederick-Duus D, Kaigler KF, Fadel JR, Wilson MA (2012) Comparison of the activation of somatostatin and neuropeptide Y containing neuronal populations of the rat amygdala following two different anxiogenic stressors. Exp Neurol 238:52–63. doi:10.1016/j.expneurol.2012.08.002

Carpentier V, Vaudry H, Mallet E, Laquerrière A, Leroux P (1998) Increased density of somatostatin binding sites in respiratory nuclei of the brainstem in sudden infant death syndrome. Neuroscience 86:159–166

Cervia D, Casini G, Bagnoli P (2008) Physiology and pathology of somatostatin in the mammalian retina: a current view. Mol Cel Endocrinol 286:112–122. doi:10.1016/j.mce.2007.12.009

Chen Z, Hedner T, Hedner J (1990) Local application of somatostatin in the rat ventrolateral brain medulla induces apnea. J Appl Physiol 69:2233–2238

Chigr F, Najimi M, Leduque P, Charnay Y, Jordan D, Chiyvialle J, Tohyama M, Kopp N (1989) Anatomical distribution of somatostatin immunoreactivity in the infant brainstem. Neuroscience 29:615–628

Corleto V (2010) Somatostatin and the gastrointestinal tract. Curr Opin Endocrinol Diabetes Obes 17(1):63–68. doi:10.1097/MED.0b013e32833463ed

Dal Monte M, Petrucci C, Cozzi A, Allen JP, Bagnoli P (2003) Somatostatin inhibits potassium-evoked glutamate release by activation of the sst(2) somatostatin receptor in the mouse retina. Naunyn Schmiedeberg's Arch Pharmacol 367(2):188–192

Doi A, Ramirez JM (2008) Neuromodulation and the orchestration of the respiratory rhythm. Respir Physiol Neurobiol 164(1–2):96–104. doi:10.1016/j.resp.2008.06.007

Douglas FL, Palkovits M (1982) Distribution and quantitative measurements of somatostatin-like immunoreactivity in the lowerbrainstem of the rat. Brain Res 242(2):369–373

Engin E, Stellbrink J, Treit D, Dickson C (2008) Anxiolytic and antidepressant effects of intra-cerebroventricularly administered somatostatin: behavioral and neurophysiological evidence. Neuroscience 157:666–676. doi:10.1016/j.neuroscience.2008.09.037

Errchidi S, Monteau R, Hilaire G (1991) Noradrenergic modulation of the medullary respiratory rhythm generator in the newborn rat: an in vitro study. J Physiol Lond 443:477–498

Feldman J, Kam K (2015) Facing the challenge of mammalian neural microcircuits: taking a few breaths may help. J Physiol 593(1):3–23. doi:10.1113/jphysiol.2014.277632

Feldman J, Mitchell GS, Nattie EE (2003) Breathing: rhythmicity, plasticity, chemosensitivity. Annu Rev Neurosci 26:239–266

Feldman J, Del Negro CA, Gray PA (2013) Understanding the rhythm of breathing: so near, yet so far. Annu Rev Physiol 75:423–452. doi:10.1146/annurev-physiol-040510-130049

Franklin K, Paxinos G (1997) The mouse brain in stereotaxic coordinates. Academic, San Diego

Gray P, Rekling J, Bocchiaro C, Feldman J (1999) Modulation of respiratory frequency by pepti-dergic input to rhythmogenic neurons in the preBötzinger complex. Science 286:1566–1568

Gray P, Janczewski W, Mellen N, McCrimmon D, Feldman J (2001) Normal breathing requires preBötzinger complex neurokinin-1 receptor-expressing neurons. Nat Neurosci 4:927–930

Gray PA, Hayes JA, Ling GY, Llona I, Tupal S, Picardo MC, Ross SE, Hirata T, Corbin JG, Eugenín J, Del Negro CA (2010) Developmental origin of preBötzinger complex respiratory neurons. J Neurosci 30:14883–14895. doi:10.1523/JNEUROSCI.4031-10.2010

Guyenet PG, Sevigny P, Weston MC, Stornetta RL (2002) Neurokinin-1 receptor-expressing cells of the ventral respiratory group are functionally heterogeneous and predominantly glutamater-gic. J Neurosci 22:3806–3816

Harfstrand A, Kalia M, Fuxe K, Kaijser L, Agnati LF (1984) Somatostatin-induced apnea: interaction with hypoxia and hypercapnea in the rat. Neurosci Lett 50:37–42

Hilaire G, Duron B (1999) Maturation of the mammalian respiratory system. Physiol Rev 79:325–360

Inagaki S, Shiosaka S, Takatsuki K, Sakanaka M, Takagi H, Senba E, Kawai Y, Minagawa H, Matsuzaki T, Tohyama M (1982) Regional distribution of somatostatin-containing neuron system in the lower brain stem of the neonatal rat. J Hirnforsch 23(1):77–85

Infante CD, von Bernhardi R, Rovegno M, Llona I, Eugenín JL (2003) Respiratory responses to pH in the absence of pontine and dorsal medullary areas in the newborn mouse in vitro. Brain Res 984(1–2):198–205

Jacobowitz DM, Abbott LC (1998) Chemoarchitectonic atlas of the developing mouse brain. CRC Press LLC, Boca Raton

Jacquin T, Champagnat J, Madamba S, Denavit-Saubié M, Siggins GR (1988) Somatostatin depresses excitability in neurons of the solitary tract complex through hyperpolarization and augmentation of I_M, a non-inactivating voltage-dependent outward current blocked by muscarinic agonist. Proc Natl Acad Sci U S A 85:948–952

Johansson O, Hökfelt T, Elde R (1984) Immunohistochemical distribution of somatostatin-like immunoreactivity in the central nervous system of the adult rat. Neuroscience 13:265–339

Johnson AD, Peoples J, Stornetta RL, Van Bockstaele EJ (2002) Opioid circuits originating from the nucleus paragigantocellularis and their potential role in opiate withdrawal. Brain Res 955:72–84

Kalia M (1981) Anatomical organization of central respiratory neurons. Annu Rev Physiol 43:105–120

Kalia M, Fuxe K, Agnati L, Hökfelt T, Härfstrand A (1984a) Somatostatin produces apnea and is localized in medullary respiratory nuclei: a possible role in apneic syndromes. Brain Res 296:339–344

Kalia M, Fuxe K, Hökfelt T, Johansson O, Lang R, Ganten D, Cuello C, Terenius L (1984b) Distribution of neuropeptide immunoreactive nerve terminals within the subnuclei of the nucleus tractus solitarius of the rat. J Comp Neurol 222:409–444

Keri G, Erchegyi J, Horvath A, Mezo I, Idei M, Vantus T, Balogh A, Vadasz Z, Bokonyi G, Seprodi J, Teplan I, Csuka O, Tejeda M, Gaal D, Szegedi Z, Szende B, Roze C, Kalthoff H, Ullrich A (1996) A tumor-selective somatostatin analog (TT-232) with strong in vitro and in vivo antitumor activity. Proc Natl Acad Sci U S A 93:12513–12518

Kinney HC, Thach BT (2009) The sudden infant death syndrome. N Engl J Med 361(8):795–805. doi:10.1056/NEJMra0803836

Kinney HC, Filiano JJ, Harper RM (1992) The neuropathology of the sudden infant death syndrome. A review. J Neuropathol Exp Neurol 51:115–126

Kochanek KD, Murphy SL, Xu J, Arias E (2014) Mortality in the United States, 2013. NCHS Data Brief 178:1–8

Kozhemyakin M, Rajasekaran K, Todorovic MS, Kowalski SL, Balint C, Kapur J (2013) Somatostatin type-2 receptor activation inhibits glutamate release and prevents status epilepticus. Neurobiol Dis 54:94–104. doi:10.1016/j.nbd.2013.02.015

Krisch B (1981) Somatostatin-immunoreactive fibber projections into the brain stem and the spinal cord of the rat. Cell Tissue Res 217(3):531–552

Lepousez G, Csaba Z, Bernard V, Loudes C, Videau C, Lacombe J, Epelbaum J, Viollet C (2010) Somatostatin interneurons delineate the inner part of the external plexiform layer in the mouse main olfactory bulb. J Comp Neurol 518:1976–1994. doi:10.1002/cne.22317

Llona I, Eugenin J (2005) Central actions of somatostatin in the generation and control of breathing. Biol Res 38:347–352

Lowe WL Jr, Schaffner AE, Roberts CT Jr, Le Roith D (1987) Developmental regulation of somatostatin gene expression in the brain is region specific. Mol Endocrinol 1(2):181–187

Martel G, Dutar P, Epelbaum J, Viollet C (2012) Somatostatinergic systems: an update on brain functions in normal and pathological aging. Front Endocrinol 3:154. doi:10.3389/fendo.2012.00154

McGregor GP, Woodhams PL, O'Shaughnessy DJ, Ghatei MA, Polak JM, Bloom SR (1982) Developmental changes in bombesin, substance P, somatostatin and vasoactive intestinal polypeptide in the rat brain. Neurosci Lett 28(1):21–27

Mellen N, Janczewski W, Bocchiaro CM, Feldman JL (2003) Opioid-induced quantal slowing reveals dual networks for respiratory rhythm generation. Neuron 37:821–826

Millhorn DE, Seroogy K, Hökfelt T, Schmued LC, Terenius L, Buchan A, Brown JC (1987) Neurons of the ventral medulla oblongata that contain both somatostatin and enkephalinimmunoreactivities project to nucleus tractussolitarii and spinal cord. Brain Res 424(1):99–108

Molkov YI, Bacak BJ, Dick TE, Rybak I (2013) A control of breathing by interacting pontine and pulmonary feedback loops. Front Neural Circuits 7:16. doi:10.3389/fncir.2013.00016

Moyse E, Beaudet A, Bertherat J, Epelbaum J (1992) Light microscopic radioautographic localization of somatostatin binding sites in the brainstem of the rat. J Chem Neuroanat 5(1):75–84

Nattie E, Li A (2012) Central chemoreceptors: locations and functions. Compr Physiol 2(1):221–254. doi:10.1002/cphy.c100083

Onimaru H, Homma I (2003) A novel functional neuron group for respiratory rhythm generation in the ventral medulla. J Neurosci 23(4):1478–1486

Pagliardini S, Ren J, Greer J (2003) Ontogeny of the pre-Bötzinger complex in perinatal rats. J Neurosci 23(29):9575–9584

Pantaleo T, Mutolo D, Cinelli E, Bongianni F (2011) Respiratory responses to somatostatin microinjections into the Bötzinger complex and the pre-Bötzinger complex of the rabbit. Neurosci Lett 498:26–30. doi:10.1016/j.neulet.2011.04.054

Pascual O, Denavit-Saubié M, Dumas S, Kietzmann T, Ghilini G, Mallet J, Pequignot JM (2001) Selective cardiorespiratory and catecholaminergic areas express the hypoxia-inducible factor-1α (HIF-1α) under in vivo hypoxia in rat brain. Eur J Neurosci 14:1981–1991

Paton JF, Ramirez JM, Richter DW (1994) Mechanisms of respiratory rhythm generation change profoundly during early life in mice and rats. Neurosci Lett 170:167–170

Pyronnet S, Bousquet C, Najib S, Azar R, Lakla H, Susini C (2008) Antitumor effects of somatostatin. Mol Cel Endocrinol 286:230–237. doi:10.1016/j.mce.2008.02.002

Ramirez JM, Schwarzacher SW, Pierrefiche O, Olivera BM, Richter DW (1998) Selective lesioning of the cat pre-Bötzinger complex in vivo eliminates breathing but not gasping. J Physiol 507(Pt 3):895–907

Ramírez-Jarquín J, Lara-Hernández S, López-Guerrero J, Aguileta M, Rivera-Angulo A, Sampieric A, Vacac L, Ordaza B, Peña-Ortega F (2012) Somatostatin modulates generation of inspiratory rhythms and determines asphyxia survival. Peptides 34:360–372. doi:10.1016/j.peptides.2012.02.011

Schechtman VL, Harper RM, Wilson AJ, Southall DP (1991) Sleep apnea in infants who succumb to the sudden infant death syndrome. Pediatrics 87:841–846

Shimada O, Ishikawa H (1989) Somatostatin-containing neurons in the mouse brain: an immunohistochemical study and comparison with the rat brain. Arch Histol Cytol 52:201–212

Shiosaka S, Takatsuki K, Sakanaka M, Inagaki S, Takagi H, Senba E, Kawai Y, Tohyama M (1981) Ontogeny of somatostatin-containing neuron system of the rat: immunohistochemical observations. J Comp Neurol 203(2):173–188

Shiraishi S, Kuriyama K, Saika T, Yoshida S, Lin I, Kitajiri M, Yamashita T, Kumazawa T, Shiosaka S (1993) Autoradiographic localization of somatostatin mRNA in the adult rat lower brainstem: observation by the double illumination technique. Neuropeptides 24:71–79

Smith J, Ellenberger H, Ballanyi K, Richter D, Feldman J (1991) Pre-Bötzinger complex: a brainstem region that may generate respiratory rhythm in mammals. Science 254:726–729

Smith J, Abdala A, Borgmann A, Rybak I, Paton J (2013) Brainstem respiratory networks: building blocks and microcircuits. Trends Neurosci 36(3):152–162. doi:10.1016/j.tins.2012.11.004

Stornetta R, Rosin D, Wang H, Sevigny C, Weston M, Guyenet M (2003) A group of glutamatergic interneurons expressing high levels of both neurokinin-1 receptors and somatostatin identifies the region of the pre-Botzinger complex. J Comp Neurol 455:499–512

Tan W, Janczewski W, Yang P, Shao X, Callaway E, Feldman J (2008) Silencing preBötzinger complex somatostatin-expressing neurons induces persistent apnea in awake rat. Nat Neurosci 11:538–540. doi:10.1038/nn.2104

Tan W, Pagliardini S, Yang P, Janczewski W, Feldman J (2010) Projections of preBötzinger complex neurons in adult rats. J Comp Neurol 518:1862–1878. doi:10.1002/cne.22308

Tan W, Sherman D, Turesson J, Shao XM, Janczewski WA, Feldman JL (2012) Reelin demarcates a subset of pre-Bötzinger complex neurons in adult rat. J Comp Neurol 520(3):606–619. doi:10.1002/cne.22753

Tuboly G, Vécsei L (2013) Somatostatin and cognitive function in neurodegenerative disorders. Mini Rev Med Chem 13:34–46

Vincent SR, McIntosh CHS, Buchan AMJ, Brown JC (1985) Central somatostatin systems revealed with monoclonal antibodies. J Comp Neurol 238:169–186

Viollet C, Lepousez G, Loudes C, Videau C, Simon A, Epelbaum J (2008) Somatostatinergic systems in brain: networks and functions. Mol Cell Endocrinol 286:75–87

Wang TJ, Lue JH, Shieh JY, Wen CY (2000) Somatostatin-IR neurons are a major subpopulation of the cuneothalamic neurons in the rat cuneate nucleus. Neurosci Res 38(2):199–207

Wang H, Stornetta R, Rosin D, Guyenet P (2001) Neurokinin-1 receptor-immunoreactive neurons of the central respiratory group in the rat. J Comp Neurol 434:128–146

Wei XY, Zhao Y, Wong-Riley MT, Ju G, Liu YY (2012) Synaptic relationship between somatostain- and neurokinin-1 receptor immunoreactive neurons in the preBötzinger complex of rats. J Neurochem 122:923–933. doi:10.1111/j.1471-4159.2012.07862.x

Wynne B, Robertson D (1997) Somatostatin and substance P-like immunoreactivity in the auditory brainstem of the adult rat. J Chem Neuroanat 12(4):259–266

Yamamoto Y, Runold M, Prabhakar N, Pantaleo T (1988) Somatostatin in the control of respiration. Acta Physiol Scand 134:529–533

Zeyda T, Hochgeschwender U (2008) Null mutant mouse models of somatostatin and cortistatin, and their receptors. Mol Cel Endocrinol 286:18–25. doi:10.1016/j.mce.2007.11.029

Part III
Neural Plasticity in the Respiratory Rhythm

Chapter 9
Respiratory Rhythm Generation: The Whole Is Greater Than the Sum of the Parts

Consuelo Morgado-Valle and Luis Beltran-Parrazal

Abstract Breathing is a continuous behavior essential for life in mammals and one of the few behaviors that can be studied in vivo in intact animals awake, anesthetized or decerebrated and in highly reduced in vitro and in situ preparations. The preBötzinger complex (preBötC) is a small nucleus in the brainstem that plays an essential role in normal breathing and is widely accepted as the site necessary and sufficient for generation of the inspiratory phase of the respiratory rhythm. Substantial advances in understanding the anatomical and cellular basis of respiratory rhythmogenesis have arisen from in vitro and in vivo studies in the past 25 years; however, the underlying cellular mechanisms remain unknown.

Keywords Rhythm generation • preBötzinger Complex • Pacemaker • Inspiratory rhythm • Central pattern generator

Abbreviations

ACSF	Artificial cerebrospinal fluid
AMPA	α- amino-3-hydroxy-5-methyl-4-isoxazolepropionic acid
AP	Action potential
APV	2-amino-5-phosphonovaleric acid
Ca^{2+}	Calcium
Cd^{2+}	Cadmium
Cl^-	Chloride
CNQX	6-cyano-7-nitroquinoxaline-2,3-dione
I_{CAN}	Ca^{2+} activated non-selective cation current
I_{NaP}	Persistent-Na^+ current

C. Morgado-Valle, PhD (✉) • L. Beltran-Parrazal, PhD
Centro de Investigaciones Cerebrales, Universidad Veracruzana,
Berlin 7, Fracc., Montemagno Animas, C.P. 91190 Xalapa, Veracruz, Mexico
e-mail: comorgado@uv.mx

© Springer International Publishing AG 2017
R. von Bernhardi et al. (eds.), *The Plastic Brain*, Advances in Experimental
Medicine and Biology 1015, DOI 10.1007/978-3-319-62817-2_9

K^+	Potassium
Mg^{2+}	Magnesium
mRNA	Messenger ribonucleic acid
Na^+	Sodium
NK1R	Neurokinin 1 receptor
NMDA	N-methyl-D-aspartate
NMDAR	NMDA receptor
preBötC	PreBötzinger complex
Sr^{2+}	Strontium
TRPC	Transient receptor potential cation channels
TRPM	Transient receptor potential melastatin
VGCC	Voltage-gated Ca^{2+} channels
V_m	Membrane potential
XIIn	Hypoglossal nerve

Introduction

We breathe from cradle to grave. Respiration is a complex process of gas exchange that sustains life. When we think about our breathing, we think of how necessary the lungs, nose, trachea, mouth and maybe even the stomach are. We rarely think of the role the brain plays in the process of breathing.

Breathing is a behavior generated in the brainstem. Like all behaviors, breathing can be measured and can be modulated by external stimuli. In fact, a group of neuroscientists has assumed the task of understanding (mainly in mammals) how the brain generates and modulates the respiratory rhythm. As mentioned in Chap. 8, there have been great advances in the field of neuroscience studying the neural control of breathing in the last two decades. These advances range from describing the anatomical location in the brainstem of the respiratory rhythm generation core -the preBötC (Smith et al. 1991)- to the development of new methodologies for studying breathing in the laboratory in preparations known as "in vitro", all in order to understand and possibly cure some diseases of respiration in which the brain plays a key role.

The study of respiratory rhythm generation is a fascinating area of neurobiology, not only for being a critical-for-life behavior, but also for its plasticity. In mammals, the neural mechanisms of respiratory rhythmogenesis must be functional even before birth to promote lung development. In humans, fetal respiratory movements can be detected in the uterus around the 30th week of pregnancy (Florido et al. 2005), suggesting that the neural network involved in the generation of breathing is functional at that age, unlike other neural networks such as those involved in locomotion. Throughout life, plasticity of this essential neuronal network is a must. Adaptations must be made "on the fly", i.e., while still functioning, and there is little margin for error. The respiratory rhythm must be modulated in the short-term during such processes as locomotion, exercise, and sleep, and in the long-term by illness, changes in altitude, and aging. Also, respiration must be integrated with speech, swallowing, chewing and defecation.

Breathing is very similar in non-primate mammals and humans, therefore the mechanisms underlying respiratory rhythm generation are studied mainly in rodents while assuming that the findings can be extrapolated to higher mammals including humans. Here we review some of the most controversial hypotheses that have emerged in the last two decades of studying the mechanisms underlying the respiratory rhythmogenesis in vitro, including the current data supporting or rejecting them.

Studying the Respiratory Rhythm Generation *In Vitro*: Going Back to the Origins

The preBötC is a very limited portion of the ventral respiratory column that in the adult rat extends 300–400 µm rostro-caudally (Fig. 9.1). The critical region for respiratory rhythm generation extends from the caudal end of retrofacial nucleus ~200 µm towards the obex (Smith et al. 1991). The preBötC is slightly caudal to the Bötzinger complex, the most rostral portion of the ventral respiratory column. The preBötC generates the inspiratory phase of the respiratory rhythm. This information is simple and obvious. However, as there is not just one single method for obtaining transverse slices, research groups around the world use a wide variety of in vitro slice preparations containing the preBötC. Such preparation variability is a source of confusion in the field of respiratory rhythm generation. Adding to this confusion are the differences attributable to the use of different species of rodents, i.e., rats and mice.

Brainstem slices of a wide range of thicknesses display rhythmic activity. The only requirement is that the preBötC is contained within a chunk of tissue. In transverse slices, using a suction electrode, rhythmic activity can be recorded from either hypoglossal nerve or from the surface of the ventral respiratory column. As a general rule, the thicker the chunk of tissue, the lower the concentration of K^+ needed in the artificial cerebrospinal fluid (ACSF) to obtain robust rhythmic activity. The presence of more structures, mainly rostral to the preBötC, changes the frequency and pattern of the inspiratory rhythm. This must be interpreted as the presence of modulatory inputs to the preBötC and not as a reconfiguration of the rhythmogenic network or as a change of the neuronal core properties.

The rhythmically active brainstem slice preparation described by Smith et al. (1991) contains the preBötC, the hypoglossal nucleus and the hypoglossal nerve (XIIn) from which a motor output in phase with preBötC neurons can be recorded (Fig. 9.1b). Advantages and limitations of this preparation, as well as a methodology to obtain reproducible rhythmic slices have been extensively reviewed (Funk and Greer 2013). Efforts have been made to illustrate the variability of rhythmic activity in slices that contain the preBötC either in the rostral or caudal surface or enclosed in

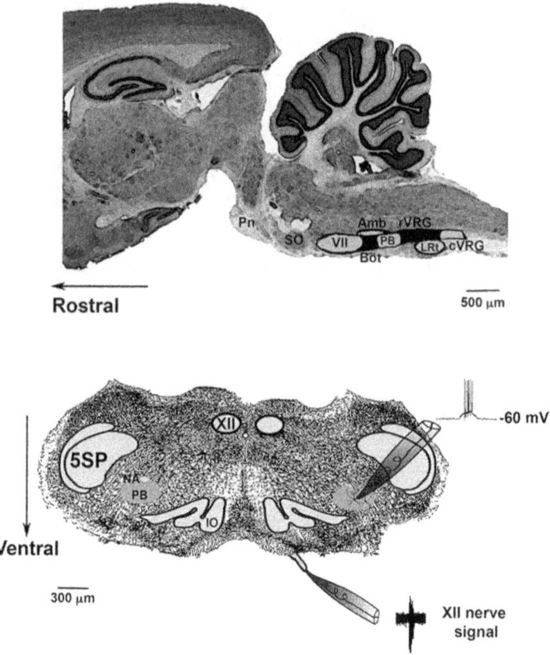

Fig. 9.1 Respiratory network compartments in rat brainstem. *Top* Parasagittal brainstem section showing nuclei that comprise the ventral respiratory column. Rostral to caudal: Bötzinger complex (*BötC*), preBötzinger complex (*PB*), rostral and caudal ventral respiratory group (*rVRG* and *cVRG*). Other anatomical landmarks: Pons (*Pn*), superior olivary complex (*SO*), (*VII*) facial nucleus, nucleus ambiguus (*Amb*), and lateral reticular nucleus (*LRt*). *Bottom* Coronal brainstem section showing the anatomical position of PB and hypoglossal (*XII*) nerve. We show a representation of recording electrodes and electrophysiological traces. Single inspiratory *PB* neurons can be recorded in current-clamp mode using patch-clamp electrodes, whereas hypoglossal nerve population activity in phase with *PB* can be obtained using suction electrodes. Anatomical landmarks: spinal trigeminal nucleus (*5SP*), hypoglossal motor nucleus (*XII*). inferior olive (*IO*) and nucleus ambiguus (*NA*)

the middle of a chunk of tissue 200 up to 700 µm thick (Ruangkittisakul et al. 2006, 2008). According to these authors, in neonatal rats the preBötC extends rostro-caudally $\leq$200 µm and, despite their thickness, slices containing the preBötC are rhythmic in ACSF containing physiological K$^+$ concentrations, i.e. 3 mM, and 1 mM Ca^{2+}. This result contrasts with the original preBötC-containing slice description (Smith et al. 1991) that states that slices thinner than 500 µm do not generate motor output at 3 mM K$^+$ but generate rhythmic output at K$^+$ concentrations of 9–11 mM.

It is important to emphasize that the preBötC network generates the inspiratory phase of the respiratory rhythm (Fig. 9.2); there are not expiratory neurons in the preBötC. Only two electrophysiological signatures can be found when recording neurons from the preBötC: inspiratory and, silent or with few action potentials (APs). Slices containing more rostral structures associated with expiration, such as the Bötzinger complex, may have expiratory neurons in the rostral surface.

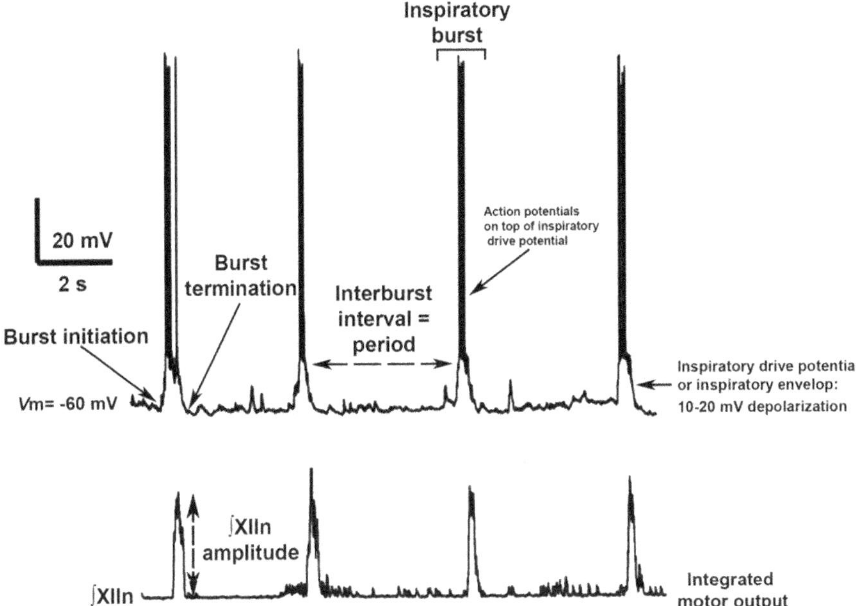

Fig. 9.2 Nomenclature used in the field of respiratory rhythm generation. *Top trace:* Whole-cell patch-clamp recording in current-clamp mode from an inspiratory preBötC neuron maintained at a $V_m \approx -60$ mV. When the network is active, as result of synaptic interactions preBötC neurons fire synchronous inspiratory bursts that consist of APs on top of a 10–20 mV depolarization lasting 0.3–0.8 s, called the inspiratory drive potential or inspiratory envelope. Some important questions in the field are those aimed at understanding the mechanisms for initiation and termination of the inspiratory burst. Pharmacological or ionic manipulations that modify the duration of the inspiratory envelope or the interburst interval are considered relevant for establishing rhythmogenic mechanisms. *Bottom trace:* Integrated hypoglossal nerve (∫XIIn) motor activity in phase with inspiratory neurons. Pharmacological or ionic manipulations that modify the amplitude of the ∫XIIn motor output suggest modulatory mechanisms

A different set of preparations has been developed to studying respiratory rhythm generation and modulation: the tilted sagittal slice (Paton et al. 1994), the tilted sagittal slab (Barnes et al. 2007), the working heart-brainstem preparation (Paton 1996a, b), and the lung-attached brainstem-spinal cord preparation (Mellen and Feldman 2000). Common characteristics of these preparations are: (1) the preservation of more intact circuits e.g. the whole preBötC, the whole ventral respiratory column, chemosensitive areas, efferent and/or afferent nuclei, and (2) their ability to generate rhythmic activity in ACSF containing physiological concentrations of K$^+$, i.e., 3 mM.

The convenience in using these preparations depends on the question to be answered: whether addressing mechanisms for rhythm generation or for modulation. As these preparations contain more structures, preBötC neurons are not always readily available on the surface for visualized patch-clamp recording and, often, blind-patch recordings must be performed, which adds a degree of difficulty to the interpretation of data.

Many Questions, But Few Conclusive Answers

After more than two decades of studying the properties of preBötC neurons, many facts have emerged, but the sum of the available information is not enough to explain how the respiratory rhythm is generated. Ionic currents such as the persistent-Na$^+$ current (I_{NaP}) and the Ca^{2+} activated non-selective cation current (I_{CAN}) co-exist in preBötC neurons. The channels responsible for those currents co-exist with fast voltage-gated Na$^+$ channels, voltage-gated Ca^{2+} channels (VGCC), Ca^{2+} activated K$^+$ channels, inward-rectifying K$^+$ channels (Papadopoulos et al. 2008) and pumps. Furthermore, it is clear now that the preBötC is a heterogeneous network in which some neurons express neurokinin 1 receptor (NK1R) while others produce somatostatin (see Chap. 8); some neurons produce glutamate while others produce glycine; some neurons express metabotropic glutamate, AMPA- and NMDA receptors (although in basal conditions the NMDA receptors (NMDAR) seem to be silent); some neurons respond to agonists or antagonists of acetylcholine-, norepinephrine-, serotonin-, or purine receptors, among others. And last but not least, the anatomical region named preBötC also contains neurons that remain silent at all times, or should we ignore those neurons?

Unfortunately, pharmacology has not been very useful and can sometimes lead to confounding results. For instance, in addition to blocking the I_{NaP}, riluzole also blocks other Na$^+$ channels, activates K$^+$ channels, and activates small conductance Ca^{2+}-activated K$^+$ channels (Beltran-Parrazal and Charles 2003; Dimitriadi et al. 2013). In addition to blocking the I_{CAN}, flufenamic acid also inhibits voltage-gated Na$^+$ channels (Yau et al. 2010), modulates transient receptor potential cation (TRPC) channels (Kolaj et al. 2014), modulates the transient outward K$^+$ current (Zhao et al. 2007), closes gap junctions, and causes a sustained increase in intracellular [Ca^{2+}] and transient increases in I_{CAN} and a Ca^{2+}-activated slow outward Cl$^-$ current (Lee et al. 1996). Then, the questions emerge: should we use such nonspecific drugs to discard the pacemaker hypothesis, to propose a new mechanism of pacemaking, or to attribute to a type of channel(s) a necessary role for rhythmogenesis?

Pacemaker Neurons or Not?

A pacemaker neuron is one that when isolated synaptically has the ability to fire a burst of APs. This means that such a neuron expresses a set of currents that allows the membrane potential to spontaneously "jump" from a given potential to a more depolarized potential and fire a burst of APs and then return to the initial potential. A pacemaker can be voltage dependent or voltage-independent. A voltage-dependent pacemaker needs to be at a specific potential to fire a burst of APs, whereas in a voltage-independent pacemaker the burst of APs occurs independently of the membrane potential. It is very important to remember that to clearly identify a neuron as a pacemaker, the bursts of APs must occur when the neuron is synaptically isolated.

A very common mistake is the use as synonyms of the terms 'bursting neuron' and 'pacemaker neuron'. All preBötC neurons are bursting neurons, so that when the network is active and, as a result of synaptic interactions, preBötC neurons fire a synchronous burst of APs on top of a 10–20 mV, 0.3–0.8 s depolarization known as inspiratory drive potential or inspiratory envelope (Fig. 9.2). All preBötC neurons express pacemaking-promoting currents such as I_{NaP} and I_{CAN} (Del Negro et al. 2002; Peña et al. 2004). However, fewer than 10% are pacemaker neurons (Del Negro et al. 2005) and the vast majority are non-pacemaker neurons requiring excitatory synaptic input to burst rhythmically. Therefore, pacemaker and bursting cannot be synonymous.

Fact: both pacemaker and non-pacemaker neurons co-exist in the preBötC. Fact: based on their neurotransmitter phenotype, within the pacemaker neuron group there are at least two types: excitatory putatively glutamatergic and inhibitory glycinergic (Morgado-Valle et al. 2010). A type of pacemaker neuron named a "conditional" pacemaker has been identified (Viemari and Ramirez 2006). Conditional pacemaker neurons do not burst in synaptic isolation but burst in the presence of a neuromodulator such as norepinephrine.

Based on their pacemaking mechanisms, at least two types of pacemaker neurons have been described: I_{NaP}-mediated, which are voltage-dependent, and I_{CAN}-mediated, which are putatively voltage-independent (Peña et al. 2004; but see Del Negro et al. 2005). To identify the I_{NaP}-mediated voltage-dependent pacemaker neurons, one must block fast synaptic transmission, either by adding antagonists of neurotransmitter receptors such as 6-cyano-7-nitroquinoxaline-2,3-dione (CNQX), 2-amino-5-phosphonovaleric acid (APV), bicuculline and strychnine to the ACSF or by lowering neurotransmission using a low Ca^{2+}, high Mg^{2+} ACSF (Fig. 9.3). Once synaptic transmission is blocked, one must, with applied current set the membrane potential to ~ -50 mV, a potential at which the I_{NaP} activates. If the neuron is a pacemaker, membrane potential will be unstable and eventually will depolarize further, fire a burst of APs, spontaneously return to ~ -50 mV, and repeat the cycle. If one hyperpolarizes the membrane potential below ~ -50 mV, the neuron will remain silent and will not fire bursts of APs; and if one depolarizes from -50 mV, tonic firing of APs will be established (Fig. 9.3a). This was the methodology used to identify pacemaker neurons in a then controversial experiment that found that in the absence of I_{NaP}-mediated voltage-dependent pacemakers the respiratory rhythm persists (Fig. 9.3b) (Del Negro et al. 2002).

I_{CAN}-mediated voltage-independent pacemaker neurons were identified using a different approach. The use of low Ca^{2+} high Mg^{2+} ACSF as a method to identify pacemakers was challenged. The notion that lowering the extracellular Ca^{2+} would prevent the identification of pacemaker neurons with an ionic mechanism dependent on Ca^{2+} was pushed forward (Peña et al. 2004). In order to test such a possibility, several ionic manipulations were done. For instance, a neuron identified as non-pacemaker in a cocktail of fast-transmission receptor blockers, expressed pacemaker properties when Ca^{2+} was completely eliminated from the ACSF, in a solution the authors referred as low Ca^{2+} high Mg^{2+}. However, according to their methodology, such a solution contained no Ca^{2+} and 2.5 mM Mg^{2+}. This sole manipulation,

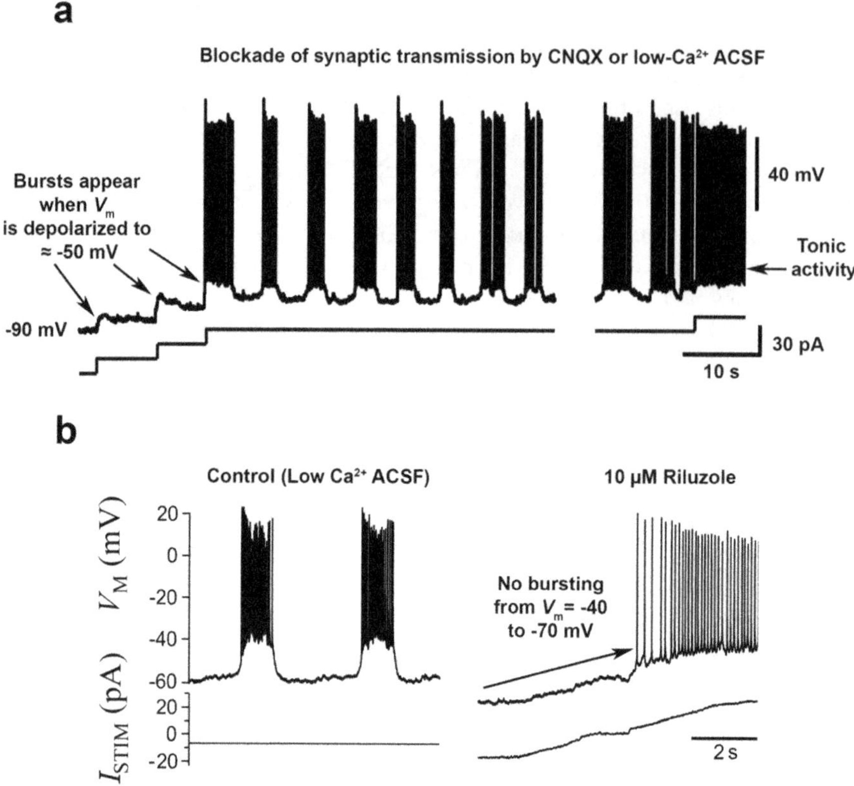

Fig. 9.3 Method to identify preBötC I_{NaP}-mediated pacemaker neurons. (**a**) Pacemaking in I_{NaP}-mediated pacemaker neurons is voltage-dependent. After blocking fast synaptic transmission either by adding antagonists of neurotransmitter receptors to the artificial cerebrospinal fluid (ACSF) or, by lowering neurotransmission using a low Ca²⁺, high Mg²⁺ ACSF, membrane potential is set to ~ − 50 mV, the potential in which the I_{NaP} activates. If the neuron is a pacemaker, bursts of APs will appear and the membrane potential will spontaneously return to ~ − 50 mV and repeat the cycle. If the membrane potential is hyperpolarized, the neuron will remain silent, i.e., will not fire burst of APs; if the membrane potential is depolarized beyond −50 mV, tonic firing of APs will be established. (**b**) Pacemaking activity of I_{NaP}-mediated voltage-dependent pacemakers identified in low Ca²⁺, high Mg²⁺ ACSF is abolished in the presence of riluzole

viz. removing extracellular Ca²⁺, can induce periodic spontaneous synchronized activity in the absence of synaptic transmission (Shuai et al. 2003).

In a different set of experiments, the nonspecific Ca²⁺-channel blocker Cd²⁺ was added to putatively voltage-dependent pacemaker neurons. Some of those pacemaker neurons ceased their activity, leading to the conclusion that in those Cd²⁺-sensitive pacemakers, pacemaking emerged by an inward current activated by Ca²⁺ influx during APs (Peña et al. 2004). I_{CAN} was then proposed as such a pacemaking-promoting current. However, a question is pertinent here: why does a pacemaking-promoting current require APs to become activated? One must assume that the Ca²⁺

influx necessary to activate I_{CAN} occurs during the APs in the immediate preceding burst. Therefore, the question remains: is activation of I_{CAN} a cause or consequence of bursting?

A Role for I_{CAN}?

The notion that a Ca^{2+}-activated inward current was relevant for rhythm generation came from indirect evidence: Cd^{2+} abolishes bursting in a class of riluzole-insensitive voltage-dependent pacemakers (Peña et al. 2004). This finding occurred at an historical moment for the field: the hypothesis that the respiratory rhythm is generated by a pacemaker-driven mechanism had been challenged (Del Negro et al. 2002). Therefore, the presence in the preBötC of two types of pacemaker neurons suggested that the pacemaker hypothesis should not be discarded. By use of flufenamic acid, a compound that was thought at the time to be a specific I_{CAN} blocker, the putative role of I_{CAN} as a pacemaking-promoting current in preBötC neurons was established (Peña et al. 2004).

There was a substantial leap from the idea that I_{CAN} is a pacemaking-promoting current to the notion that inspiratory bursts in the preBötC neurons depend on I_{CAN} (Pace et al. 2007). Further studies established the presence in the preBötC neurons of mRNA for two members of the transient receptor potential melastatin (TRPM) family of ion channels, TRPM4 and TRPM5, which are the known to give rise to I_{CAN} (Crowder et al. 2007). Though interesting, the presence of TRPM channels in preBötC neurons did not demonstrate their necessity for rhythm generation.

Mediated by TRPM4 and TRPM5 channels, I_{CAN} was suggested as necessary for respiratory rhythmogenesis (Pace et al. 2007; Mironov 2008). Other than the presence of TRPM4/5 channels, the alleged necessity of I_{CAN} for inspiratory drive was based on indirect evidence using nonspecific pharmacology targeting many players in signal transduction cascades. The functional properties and regulation of TRPM4 channels are well characterized in cell lines (Launay et al. 2002; Nilius et al. 2004a, b; Voets and Nilius, 2007). However, the role and physiology of endogenous TRPM4 channels in preBötC neurons and, thus, in respiratory rhythm generation was not clear. In heterologous expression systems TRPM4 is activated by internal Ca^{2+} with a K_d ranging from 0.4 to 9.8 μM and inhibited by intracellular free adenosine triphosphate (ATP) in the 10 μM range (Launay et al. 2002; Nilius et al. 2004a, b). However, in slice preparations physiological intracellular Ca^{2+} and free ATP concentrations are difficult to determine. Intracellular standard solutions used by most groups to study endogenous inspiratory activity in preBötC neurons in control conditions involve perfusion of the neuron with ≥300 μM free ATP and ~5 nM free Ca^{2+} (http://maxchelator.stanford.edu/CaMgATPEGTA-NIST.htm; Temperature = 27 °C; pH = 7.4; Ionic = 0.08). This suggests that under control conditions at resting potential, TRPM4 channels are not active and therefore I_{CAN} is not participating in the initiation of the inspiratory drive potential. Furthermore, at resting potential cytoplasmic Ca^{2+} concentration is presumably in equilibrium with the intracellular solution and is in the low nM range, far from the concentration needed in heterologous systems to activate TRPM4 channels. In heterologous systems, TRPM5 is insensitive to ATP concentrations up to 1 mM (Ullrich et al. 2005). PreBötC neurons express both TRPM4 and TRPM5 channels (Crowder et al. 2007), suggesting that in experimental conditions intracellular free ATP is not inhibiting TRPM5-mediated I_{CAN}.

Despite of the evidence supporting the "I_{CAN} necessity" hypothesis, a definitive test of the non-essential role of I_{CAN} for breathing would be the existence of TRPM4, TRPM5 or TRPM4/TRPM5 knockout mice that are viable with no signs of respiratory problems (Damak et al. 2006; Barbet et al. 2008; Ohkuri et al. 2009; Lei et al. 2014).

Evidence for the Role of Ca^{2+} in Rhythm Generation

We suggest that, other than for synaptic transmission, Ca^{2+} influx does not participate in the initiation or termination of inspiratory drive. This is based on four experimental observations:

1. NMDARs, putatively an important source of Ca^{2+} influx in preBötC neurons, do not appear to be important. The respiratory rhythm is virtually identical in slices from knockout mice lacking the NR1 subunit gene necessary for assembly of functional NMDAR and, in slices from control mice (Funk et al. 1997). Also, blockade of NMDAR with the pore blocker MK-801 does not abolish rhythm generation in the slice preparation (Morgado-Valle and Feldman 2007).
2. Although VGCC are major contributors to Ca^{2+} influx in neurons, the depolarized potentials required for activation of VGCC are only reached during APs, which in preBötC neurons occur only on top of the inspiratory envelope. The inspiratory envelope is a 10–20 mV synchronous depolarization from a $V_m \approx -60$ mV. Therefore, the inspiratory envelope per se does not reach the potential for activation of VGCC. Somatic Ca^{2+} influx is a consequence of APs during the inspiratory phase and does not contribute substantively to the inspiratory envelope (Fig. 9.4) (Morgado-Valle et al. 2008). Furthermore, individual blockade of L- N- or P/Q type VGCC does not significantly affect rhythm generation (Morgado-Valle et al. 2008). However, since excitatory synaptic transmission depends on N-type and P/Q-type VGCC, they are required for stable breathing (Koch et al. 2013).
3. Blocking release of Ca^{2+} from intracellular stores, which is proposed as a source of Ca^{2+} to activate I_{CAN} in preBötC neurons (Pace et al. 2007), does not affect the rhythm. Inhibition of the sarco/endoplasmic reticulum Ca^{2+} ATPases and depletion of the intracellular Ca^{2+} stores does not affect respiratory rhythm generation (Fig. 9.5), suggesting that intracellular Ca^{2+} stores, without regard to their somatic or dendritic location, do not significantly contribute to rhythm generation in the preBötC in vitro (Beltran-Parrazal et al. 2012).

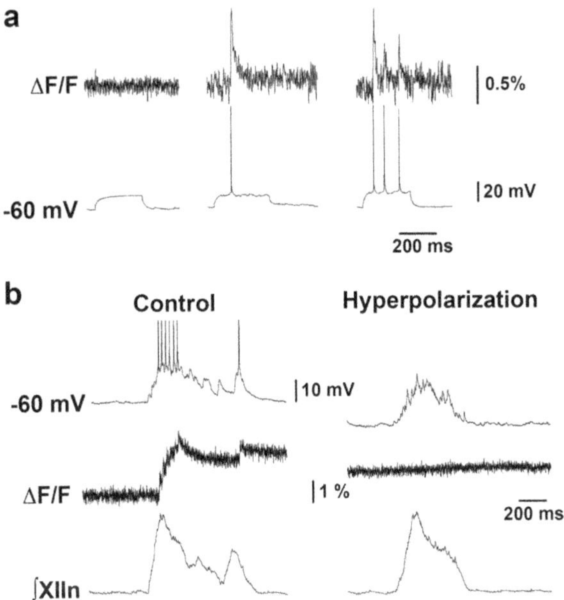

Fig. 9.4 Ca^{2+} transients in preBötC inspiratory neurons loaded with a Ca^{2+}-sensitive dye. (**a**) Somatic Ca^{2+} transients (*upper traces*) in response to evoked APs (*lower traces*) in the interburst interval of an inspiratory neuron ($V_\mathrm{m} \approx -60$ mV). (**b**) Somatic Ca^{2+} transients in active inspiratory preBötC neurons are seen only in response to APs. Traces are: *upper*, current-clamped inspiratory neuron; *middle*, Fluo-4 Ca^{2+} signal expressed as ΔF/F; *lower*, respiratory-related motor output recorded from XIIn. ΔF/F did not change prior to the first AP, suggesting that there are not Ca^{2+} transients associated with the inspiratory envelope. Note that hyperpolarization to eliminate APs produces a flat ΔF/F trace with no somatic Ca^{2+} transients, despite the presence of inspiratory envelope

4. Although several groups have proposed that the inspiratory burst depends on I_CAN, which is mediated by TRPM4 and TRPM5 channels (Pace et al. 2007; Mironov 2008), there is evidence against this. PreBötC neurons express both TRPM4 and TRPM5 channels (Crowder et al. 2007). However, TRPM4 and TRPM5 knockout and TRPM4/TRPM5 double knockout mice are viable with no apparent signs of respiratory problems (Damak et al. 2006; Barbet et al. 2008; Ohkuri et al. 2009; Lei et al. 2014).

In neonatal rat slices containing the preBötC, substitution of extracellular Ca^{2+} by Sr^{2+} significantly increases inspiratory drive duration without affecting AP properties (Fig. 9.6) (Morgado-Valle et al. 2014). Sr^{2+} supports synaptic transmission and desynchronizes neurotransmitter release. Our findings suggest that in the presence of Sr^{2+} in the preBötC network, the fast release of glutamate is less efficient whereas the slow release is prolonged and, therefore, the inspiratory burst lasts longer. We propose that the duration of inspiratory burst is determined by the kinetics of the presynaptic molecular machinery for neurotransmitter release, which is ultimately determined by the kinetics of Ca^{2+} influx and endogenous buffering.

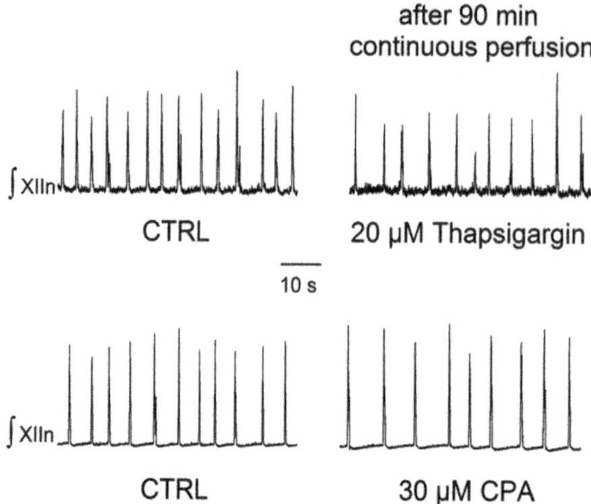

Fig. 9.5 Inhibition of the sarco/endoplasmic reticulum Ca^{2+} ATPases (SERCA) does not affect respiratory rhythm generation. Continuous 90 min bath perfusion of the SERCA blockers thapsigargin or cyclopiazonic acid (CPA) does not interfere with rhythm generation mechanisms, suggesting that the intracellular Ca^{2+} stores, without regard to their somatic or dendritic location, do not significantly contribute to rhythm generation in the preBötC in vitro. Compare ∫XIIn traces in control conditions (CTRL) (*left*) and after blockade of SERCA (*right*)

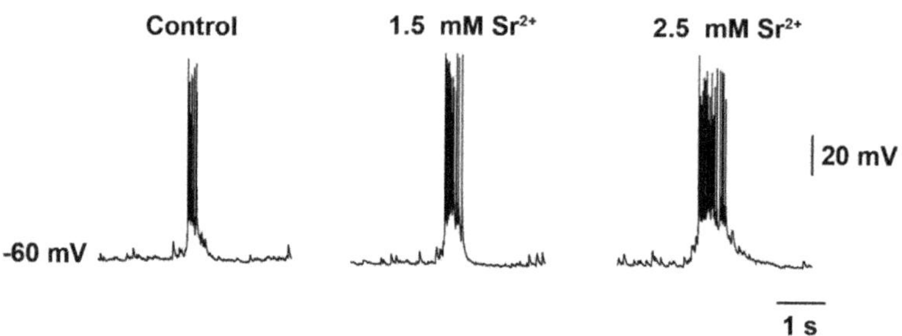

Fig. 9.6 Substitution of extracellular Ca^{2+} by Sr^{2+} significantly prolongs the inspiratory burst. Examples of inspiratory bursts from a current-clamped preBötC neuron under control conditions (*left*), in 1.5 mM Sr^{2+}-ACSF (*middle*) and in 2.5 mM Sr^{2+}-ACSF (*right*). Calibration bars apply to all traces

Concluding Remarks

The rhythmic activity of the neuronal network that we call the preBötC depends on glutamatergic transmission, which in turn is modulated by a wide variety of neurotransmitters and neuropeptides from nuclei that send afferents to both left and right preBötCs, which are synchronous. Such modulatory connectivity gives the preBötC the necessary plasticity to generate distinctive, state-dependent rhythmic patterns that translate into motor output (Fig. 9.7). The possibility exists that those

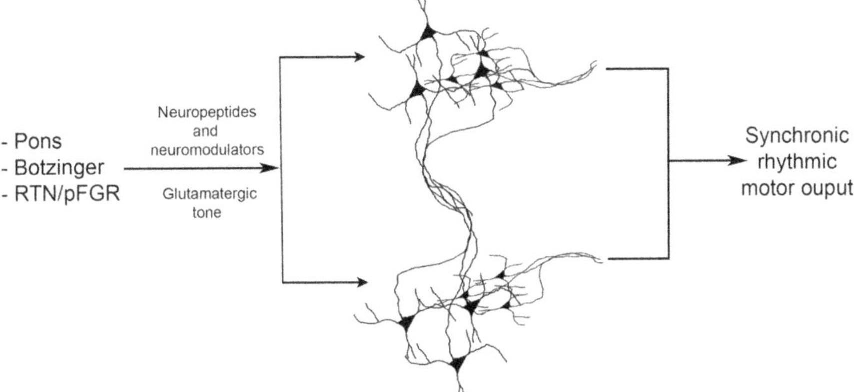

Fig. 9.7 The rhythmic activity of the preBötC is modulated by neurotransmitters and neuropeptides from nuclei that send afferents to both preBötCs. Such modulatory connectivity gives to the preBötC the necessary plasticity to generating distinctive, state-dependent rhythmic patterns that are translated into motor output. Are those modulatory afferents involved in respiratory rhythm generation by maintaining a glutamatergic tone that allows preBötC neurons to be bistable? Perhaps, given the fact that thin slices cannot generate rhythmic activity in a physiological concentration of K^+

modulatory afferents are involved in respiratory rhythm generation by maintaining a glutamatergic tone that allows the membrane of preBötC neurons to be bistable. This is likely, since thin slices cannot generate rhythmic activity in a physiological concentration of K^+. However, when structures anterior to the preBötC are included, in vitro preparations in 3 mM K^+ become spontaneously rhythmical. This is not indicative of a "reconfiguration" of the network, but reflects that afferents from anterior nuclei in the medulla modulate the preBötC.

The mechanism for respiratory rhythm generation remains unknown, despite much information collected both at cellular and network levels. Nonetheless, several conclusions can be drawn. First, while more intact preparations might not have added to understanding the mechanisms for rhythm generation, they have given us a glimpse of how intricate the modulation of the preBötC network can be. Second, that a non-physiological concentration of K^+ is needed to obtain rhythmicity in a very reduced preparation does not mean that the network has been reconfigured. The most parsimonious explanation is that by removing rostral structures, a tonic excitatory input is lost, or that one or several modulatory inputs, either excitatory or inhibitory were removed. Third, pacemaking is not synonymous with bursting. In the absence of synaptic transmission, pacemaking is determined by intrinsic cellular properties. Fourth, it is probable that Ca^{2+} is not playing a role in rhythm generation other than enabling synaptic transmission.

Acknowledgements This research was supported by grants CONACYT-128392, CONACYT-15 3627 and the Beltran-Morgado Foundation for the Advancement and Communication of Neuroscience in Veracruz.

References

Barbet G, Demion M, Moura IC, Serafini N, Leger T, Vrtovsnik F, Monteiro RC, Guinamard R, Kinet JP, Launay P (2008) The calcium-activated nonselective cation channel TRPM4 is essential for the migration but not the maturation of dendritic cells. Nat Immunol 9:1148–1156

Barnes BJ, Tuong CM, Mellen NM (2007) Functional imaging reveals respiratory network activity during hypoxic and opioid challenge in the neonate rat tilted sagittal slab preparation. J Neurophysiol 97:2283–2292

Beltran-Parrazal L, Charles A (2003) Riluzole inhibits spontaneous Ca2+ signaling in neuroendocrine cells by activation of K+ channels and inhibition of Na+ channels. Br J Pharmacol 140:881–888

Beltran-Parrazal L, Fernandez-Ruiz J, Toledo R, Manzo J, Morgado-Valle C (2012) Inhibition of endoplasmic reticulum Ca^{2+} ATPase in preBötzinger complex of neonatal rat does not affect respiratory rhythm generation. Neuroscience 224:116–124

Crowder EA, Saha MS, Pace RW, Zhang H, Prestwich GD, Del Negro CA (2007) Phosphatidylinositol 4,5-bisphosphate regulates inspiratory burst activity in the neonatal mouse preBotzinger complex. J Physiol Lond 582:1047–1058

Damak S, Rong M, Yasumatsu K, Kokrashvili Z, Perez CA, Shigemura N, Yoshida R, Mosinger B Jr, Glendinning JI, Ninomiya Y, Margolskee RF (2006) Trpm5 null mice respond to bitter, sweet, and umami compounds. Chem Senses 31:253–264

Del Negro CA, Morgado-Valle C, Feldman JL (2002) Respiratory rhythm: an emergent network property? Neuron 34:821–830

Del Negro CA, Morgado-Valle C, Hayes JA, Mackay DD, Pace RW, Crowder EA, Feldman JL (2005) Sodium and calcium current-mediated pacemaker neurons and respiratory rhythm generation. J Neurosci 25:446–453

Dimitriadi M, Kye MJ, Kalloo G, Yersak JM, Sahin M, Hart AC (2013) The neuroprotective drug riluzole acts via small conductance Ca2+-activated K+ channels to ameliorate defects in spinal muscular atrophy models. J Neurosci 33:6557–6562

Florido J, Cortés E, Gutiérrez M, Soto VM, Miranda MT, Navarrete L (2005) Analysis of fetal breathing movements at 30-38 weeks of gestation. J Perinat Med 33:38–41

Funk GD, Greer JJ (2013) The rhythmic, transverse medullary slice preparation in respiratory neurobiology: contributions and caveats. Respir Physiol Neurobiol 186:236–253

Funk GD, Johnson SM, Smith JC, Dong XW, Lai J, Feldman JL (1997) Functional respiratory rhythm generating networks in neonatal mice lacking NMDAR1 gene. J Neurophysiol 78:1414–1420

Koch H, Zanella S, Elsen GE, Smith L, Doi A, Garcia AJ 3rd, Wei AD, Xun R, Kirsch S, Gomez CM, Hevner RF, Ramirez JM (2013) Stable respiratory activity requires both P/Q-type and N-type voltage-gated calcium channels. J Neurosci 33:3633–3645

Kolaj M, Zhang L, Renaud LP (2014) Novel coupling between TRPC-like and KNa channels modulates low threshold spike-induced afterpotentials in rat thalamic midline neurons. Neuropharmacology 86:88–96

Launay P, Fleig A, Perraud AL, Scharenberg AM, Penner R, Kinet JP (2002) TRPM4 is a Ca2+-activated nonselective cation channel mediating cell membrane depolarization. Cell 109:397–407

Lee RJ, Shaw T, Sandquist M, Partridge LD (1996) Mechanism of action of the non-steroidal anti-inflammatory drug flufenamate on [Ca2+]i and Ca(2+)-activated currents in neurons. Cell Calcium 19:431–438

Lei YT, Thuault SJ, Launay P, Margolskee RF, Kandel ER, Siegelbaum SA (2014) Differential contribution of TRPM4 and TRPM5 nonselective cation channels to the slow afterdepolarization in mouse prefrontal cortex neurons. Front Cell Neurosci 8:267. doi:10.3389/fncel.2014.00267

Mellen NM, Feldman JL (2000) Phasic lung inflation shortens inspiration and respiratory period in the lung-attached neonate rat brain stem spinal cord. J Neurophysiol 83:3165–3168

Mironov SL (2008) Metabotropic glutamate receptors activate dendritic calcium waves and TRPM channels which drive rhythmic respiratory patterns in mice. J Physiol Lond 586:2277–2291

Morgado-Valle C, Feldman JL (2007) NMDA receptors in preBotzinger complex neurons can drive respiratory rhythm independent of AMPA receptors. J Physiol 582:359–368

Morgado-Valle C, Beltran-Parrazal L, DiFranco M, Vergara JL, Feldman JL (2008) Somatic Ca2+ transients do not contribute to inspiratory drive in preBotzinger Complex neurons. J Physiol 586:4531–4540

Morgado-Valle C, Baca SM, Feldman JL (2010) Glycinergic pacemaker neurons in preBötzinger complex of neonatal mouse. J Neurosci 30:3634–3639

Morgado-Valle C, Fernandez-Ruiz J, Lopez-Meraz L, Beltran-Parrazal L (2014) Substitution of extracellular calcium (Ca2+) by strontium (Sr2+) prolongs inspiratory burst in preBötzinger Complex (preBötC) inspiratory neurons. J Neurophysiol 113(4):1175–1183. doi:10.1152/jn.00705.2014

Nilius B, Prenen J, Janssens A, Voets T, Droogmans G (2004a) Decavanadate modulates gating of TRPM4 cation channels. J Physiol 560:753–765

Nilius B, Prenen J, Voets T, Droogmans G (2004b) Intracellular nucleotides and polyamines inhibit the Ca2+-activated cation channel TRPM4b. Pflugers Arch 448:70–75

Ohkuri T, Yasumatsu K, Horio N, Jyotaki M, Margolskee RF, Ninomiya Y (2009) Multiple sweet receptors and transduction pathways revealed in knockout mice by temperature dependence and gurmarin sensitivity. Am J Physiol Regul Integr Comp Physiol 296:R960–R971

Pace RW, Mackay DD, Feldman JL, Del Negro CA (2007) Inspiratory bursts in the preBotzinger complex depend on a calcium-activated non-specific cation current linked to glutamate receptors in neonatal mice. J Physiol Lond 582:113–125

Papadopoulos N, Winter SM, Härtel K, Kaiser M, Neusch C, Hülsmann S (2008) Possible roles of the weakly inward rectifying k+ channel Kir4.1 (KCNJ10) in the pre-Bötzinger complex. Adv Exp Med Biol 605:109–113

Paton JF (1996a) A working heart-brainstem preparation of the mouse. J Neurosci Methods 65:63–68

Paton JF (1996b) The ventral medullary respiratory network of the mature mouse studied in a working heart-brainstem preparation. J Physiol 493:819–831

Paton JF, Ramirez JM, Richter DW (1994) Functionally intact in vitro preparation generating respiratory activity in neonatal and mature mammals. Pflugers Arch 428:250–260

Peña F, Parkis MA, Tryba AK, Ramirez JM (2004) Differential contribution of pacemaker properties to the generation of respiratory rhythms during normoxia and hypoxia. Neuron 43:105–117

Ruangkittisakul A, Schwarzacher SW, Secchia L, Poon BY, Ma Y, Funk GD, Ballanyi K (2006) High sensitivity to neuromodulator-activated signaling pathways at physiological [K+] of confocally imaged respiratory center neurons in on-line-calibrated newborn rat brainstem slices. J Neurosci 26:11870–11880

Ruangkittisakul A, Schwarzacher SW, Secchia L, Ma Y, Bobocea N, Poon BY, Funk GD, Ballanyi K (2008) Generation of eupnea and sighs by a spatiochemically organized inspiratory network. J Neurosci 28:2447–2458

Shuai J, Bikson M, Hahn PJ, Lian J, Durand DM (2003) Ionic mechanisms underlying spontaneous CA1 neuronal firing in Ca2+-free solution. Biophys J 84:2099–2111

Smith JC, Ellenberger HH, Ballanyi K, Richter DW, Feldman JL (1991) Pre-Botzinger complex: a brainstem region that may generate respiratory rhythm in mammals. Science 254:726–729

Ullrich ND, Voets T, Prenen J, Vennekens R, Talavera K, Droogmans G, Nilius B (2005) Comparison of functional properties of the Ca2+-activated cation channels TRPM4 and TRPM5 from mice. Cell Calcium 37:267–278

Viemari JC, Ramirez JM (2006) Norepinephrine differentially modulates different types of respiratory pacemaker and nonpacemaker neurons. J Neurophysiol 95:2070–2082

Voets T, Nilius B (2007) Modulation of TRPs by PIPs. J Physiol 582:939–944

Yau HJ, Baranauskas G, Martina M (2010) Flufenamic acid decreases neuronal excitability through modulation of voltage-gated sodium channel gating. J Physiol 588:3869–3882

Zhao ZG, Zhang M, Zeng XM, Fei XW, Liu LY, Zhang ZH, Mei YA (2007) Flufenamic acid bi-directionally modulates the transient outward K(+) current in rat cerebellar granule cells. J Pharmacol Exp Ther 322:195–204

Chapter 10
The Onset of the Fetal Respiratory Rhythm: An Emergent Property Triggered by Chemosensory Drive?

Sebastián Beltrán-Castillo, Consuelo Morgado-Valle, and Jaime Eugenín

Abstract The mechanisms responsible for the onset of respiratory activity during fetal life are unknown. The onset of respiratory rhythm may be a consequence of the genetic program of each of the constituents of the respiratory network, so they start to interact and generate respiratory cycles when reaching a certain degree of maturation. Alternatively, generation of cycles might require the contribution of recently formed sensory inputs that will trigger oscillatory activity in the nascent respiratory neural network. If this hypothesis is true, then sensory input to the respiratory generator must be already formed and become functional before the onset of fetal respiration. In this review, we evaluate the timing of the onset of the respiratory rhythm in comparison to the appearance of receptors, neurotransmitter machinery, and afferent projections provided by two central chemoreceptive nuclei, the raphe and locus coeruleus nuclei.

Keywords Respiratory rhythm • Fetal breathing • Central chemoreception • Raphe nuclei • Locus coeruleus nuclei • Brainstem • Ontogeny

Abbreviations

5-HT	5-hydroxytryptamine (serotonin)
5-HT1AR	Serotonin receptor 1A
5-HT2AR	Serotonin receptor 2A

S. Beltrán-Castillo, PhD • J. Eugenín, MD, PhD (✉)
Laboratorio de Sistemas Neurales, Facultad de Química y Biología, Departamento de Biología, Universidad de Santiago de Chile, USACH, PO 9170022 Santiago, Chile
e-mail: jaime.eugenin@usach.cl; jeugenin@gmail.com

C. Morgado-Valle, PhD (✉)
Centro de Investigaciones Cerebrales, Universidad Veracruzana,
Campus Xalapa, Berlin 7, Fracc., Monte Magno Animas, C.P. 91190
Xalapa, Veracruz, Mexico
e-mail: comorgado@uv.mx

© Springer International Publishing AG 2017
R. von Bernhardi et al. (eds.), *The Plastic Brain*, Advances in Experimental Medicine and Biology 1015, DOI 10.1007/978-3-319-62817-2_10

5-HT2BR	Serotonin receptor 2B
5-HT4R	Serotonin receptor 4
5-HT7R	Serotonin receptor 7
5-HTergic	Serotonergic
6-OH DA	6-hydroxy dopamine
8-OH DPAT	8-hydroxy-diproplaminotetralin
A5	Noradrenergic neurons in pons region near A6
A6	Noradrenergic neurons in locus coeruleus
ACh	Acetylcholine
aCSF	Artificial cerebrospinal fluid
AMPA	α-amino-3-hydroxy-5-methyl-4-isoxazolepropionic acid
Ascl1	Achaete-scute complex-like 1
Atoh1	Protein atonal homolog 1
BMP	Bone morphogenetic protein
CNO	Clozapine-N-oxide
CNS	Central nervous system
CO_2	Carbon dioxide
DBH-SAP	β-hydroxilase-saporin
Dbx1	Developing brain homeobox 1
DRC	Dorsal ventral respiratory column
DRG	Dorsal root ganglion
E	Embryonic day
Egr2	Early growth response 2
ePF	Embryonic parafacial oscillator
GABA	γ-aminobutyric acid
Gata2	GATA binding protein 2
Gata3	GATA binding protein 3
GFP	Green fluorescent protein
H^+	Protons
Hoxa2	Homeobox protein Hox-2A
I_{CAN}	Calcium-activated non-specific cationic current
I_h	Hyperpolarization activated current
I_{NaP}	Persistent Na^+ current
Lbx1	Ladybird homeobox 1
LC	Locus Coeruleus
LF	Low frequency
Lmx1b	LIM homeobox transcription factor 1β
Lmx1b^{f/f/p}	Lmx1b conditional knockout
MAO-A	Monoamine oxidase A
Mash1	Mammalian achaete scute homolog-1
NA	Noradrenaline
NAergic	Noradrenergic
NALCN	Non-selective cationic cannel
NK1R	Neurokinin 1 receptor
NTS	Nucleus tractus solitaries

Pet-1	Pheochromocytoma 12 ETS factor-1
Pet-1$^{-/-}$	Pheochromocytoma 12 ETS factor-1 knockout mice
Phox2a	Paired-like homeobox 2a
Phox2b	Paired-like homeobox 2b
preBötC	Prebötzinger complex
RN	Raphe nucleus
RPG	Respiratory pattern generator
RTN/pFRG	Retrotrapezoide nucleus/parafacial respiratory group
SERT	Serotonin transporter
SNpc	Substantia nigra pars compacta
SP	Substance P
SSTR	Somatostatin receptor
Tg8	Monoamine oxidase A deficient mice
TH	Tyrosine hydroxylase
TpOH	Tryptophan hydroxylase
TRH	Thyrotropin-releasing hormone
VRC	Ventral respiratory column

Introduction

Breathing is a vital homeostatic motor behavior whose main function is to achieve proper ventilation of lung alveoli for an optimal alveolar arterial exchange of carbon dioxide and oxygen. Breathing is possible thanks to the rhythmic and coordinated contraction of respiratory muscles generating changes in airflow resistance and a gradient of pressure along the airway pathway (Feldman et al. 2013). As already discussed in Chaps. 8 and 9, respiratory motoneurons receive a timed motor command generated by the respiratory pattern generator (RPG) in the brainstem. The RPG includes respiratory neurons distributed in several nuclei along the dorsal and ventral respiratory columns (DRC and VRC, respectively), with the VRC being necessary for generating the rhythm. The preBötzinger complex (preBötC), which is the generator of inspiration (Smith et al. 1991; Feldman et al. 2013), the postinspiratory complex (PiCo) (Anderson et al. 2016) and the retrotrapezoid nucleus/parafacial respiratory group (RTN/pFRG), which generate the post-inspiratory activity and the pre-inspiratory/expiratory activities, respectively (Onimaru and Homma 2003; Janczewski and Feldman 2006), are the principal nuclei in the VRC with oscillatory properties. Several brainstem regions show intrinsic chemosensitivity influencing the respiratory rhythm. These regions include the locus coeruleus (A6), the nucleus tractus solitarius (NTS) in the dorsal respiratory column, the preBötC and the RTN/pFRG in the ventral respiratory column, the ventrolateral medullary region, and the medullary raphe nucleus (RN) (Ballantyne and Scheid 2001; Erlichman et al. 1998; Gourine et al. 2010; Wenker et al. 2010; Coates et al. 1993; Feldman et al. 2003; Nattie 1999;

Oyamada et al. 1998; Solomon 2003; Wang et al. 1998). Several neuronal phenotypes are observed in these chemosensitive regions including, among others, populations of glutamatergic, serotonergic, noradrenergic, and cholinergic neurons (Li et al. 1999; Bernard et al. 1996; Biancardi et al. 2008; Bianchi et al. 1995; Soulier et al. 1997).

Developmentally, the differentiation of the neural progenitors that originate the main neuromodulatory sensory inputs to the RPG begins before the onset of the respiratory rhythm. However, it is not clear whether the sensory inputs are functional then. These inputs might not only modify the activity of the RPG from the very onset of the rhythm, but they also can influence RPG development and maturation (Abreu-Villaca et al. 2011; Cermak et al. 1998; Erickson et al. 2007; Fujii and Arata 2010; Hodges and Richerson 2008; Shideler and Yan 2010; Yew and Chan 1999).

Absence of respiratory movements before the onset of the respiratory rhythmic activity in the brainstem makes improbable the influence of lung and rib cage mechanoreceptors to trigger the rhythm. By contrast, central chemoreception seems to be fundamental to trigger the respiratory rhythm during ontogeny. Although in vitro preparations can sustain a respiratory-like rhythm in "deafferented" conditions (for example isolated preBötC "islands"), the autonomy of the rhythm is not complete and its functional expression depends on chemosensory signals. Chemosensitivity has a central role in the maintenance of the respiratory rhythm in postnatal life. A small increase in carbon dioxide/proton (CO_2/H^+) levels produces a large increase in respiratory frequency and tidal volume (hence, in ventilation), and a small decrease from normal CO_2/H^+ levels produces the opposite effects, including apnea during sleep or anesthesia (Nattie 1999). Furthermore, alkalosis (pH 7.6 or higher) can stop the respiratory rhythm (Eugenin et al. 2006; Eugenin and Nicholls 1997; Infante et al. 2003). During fetal life, the fetus is maintained in a more acidic environment than that after the birth (Kubli et al. 1969). This slightly more acidic environment may be crucial to induce the appearance of the respiratory rhythm (Fig. 10.1). In fact, a respiratory-like rhythm can be recorded in mice from phrenic roots at embryonic day 13 (E13) only if the brainstem-spinal cord preparation is superfused in acidified medium (Eugenin et al. 2006). This suggests that as soon as E13, a bulbospinal respiratory command can affect phrenic motoneurons (Eugenin et al. 2006) and that the rhythm onset may occur well before E15, as determined by other authors (Thoby-Brisson et al. 2005). Therefore, it is reasonable to propose that the existence of a chemosensory drive could be relevant for the ontogenic onset of the respiratory rhythm.

Here we review available data, mostly obtained in rodents, to ascertain the time course of developmental events associated with the emergence of a functional respiratory pattern generator and its modulation by central chemoreception input. We focus our analysis on the locus coeruleus and the RN, the main central chemoreceptor nuclei sending distinct projections to the respiratory network. We further analyze the time correspondence between the onset of the respiratory rhythm and the appearance of receptors, neurotransmitter machinery, and afferent projections at the level of the respiratory network.

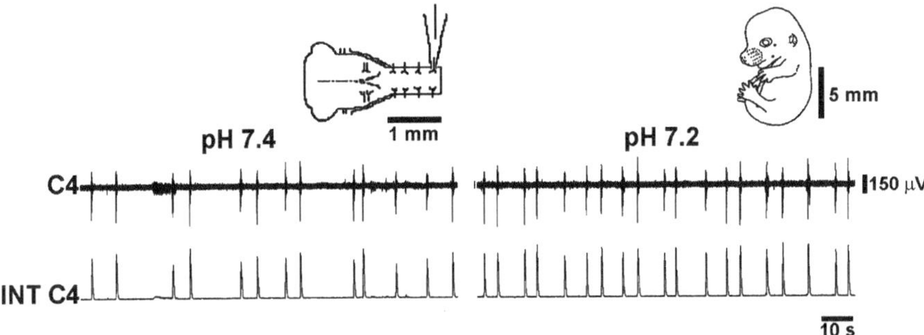

Fig. 10.1 Acidification of the superfusion medium increases the frequency of fictive respiration in *en bloc* preparation obtained from E15 fetal mouse. *Upper drawings*: *en bloc* preparation with a suction electrode placed onto the C4 ventral root (*left*) and an E15 mouse fetus from which it was taken (*right*). At E15 the fetus is ~1 cm and the explant ~2 mm long. Note that spontaneous activity recorded from a C4 ventral root is irregular at pH 7.4 and is faster and more regular at pH 7.2. Acidification was done by switching the mixture of gas from 5% CO_2/95% O_2 to 10% CO_2/90% O_2 at 22 °C. *Upper traces*, raw inspiratory bursts; *lower traces*, integrated inspiratory bursts processing raw recordings with a full wave rectifier (time constant is 100 ms)

Onset of Brainstem Rhythms

Mouse hindbrain is segmented in rhombomeres during E8 to E12. Rhombomeres correspond to transient cell-specific segments of the hindbrain defined transversely by arranged bands of gene expression (Lumsden 1999; Chambers et al. 2009). Development of rhombomeres in this brief embryonic period is a key process for the formation of central pattern generators within the brainstem (Champagnat and Fortin 1997; Champagnat et al. 2009; Fortin and Thoby-Brisson 2009).

Spontaneous activity appears in different brain territories during fetal periods in which synaptogenesis is occurring (Fig. 10.2). Such activity seems to be required for correct wiring of circuits and networks (Bosma 2010), and some of this corresponds to rhombomere progression and milestones. Spontaneous activity in the form of single asynchronous neuronal events is seen in the E9.5 mouse embryo; the duration of the events shortens by E10.5. In contrast, at the end of hindbrain segmentation, by E11.5, neuronal activity becomes highly synchronized, driven by serotonergic neurons located in the rostral raphe at the former rhombomeres r2 and r3 (Bosma 2010; Hunt et al. 2005) (see Chap. 11). Calcium imaging shows that the activity of this serotonergic "pacemaker" propagates into rostral and caudal hindbrain along the midline via serotonin receptor 2A (5-HT2AR) activation, and into lateral hindbrain. Blockade of serotonin receptor 2A (5-HT2AR) slows or abolishes the synchronized activity. In contrast, antagonists of several neurotransmitters including glutamate, glycine, γ-aminobutyric acid (GABA), acetylcholine (ACh), and noradrenaline (NA) do not block the propagation of this activity (Hunt et al. 2005). Synchronized activity persists to E13.5 (Bosma 2010; Hunt et al. 2005).

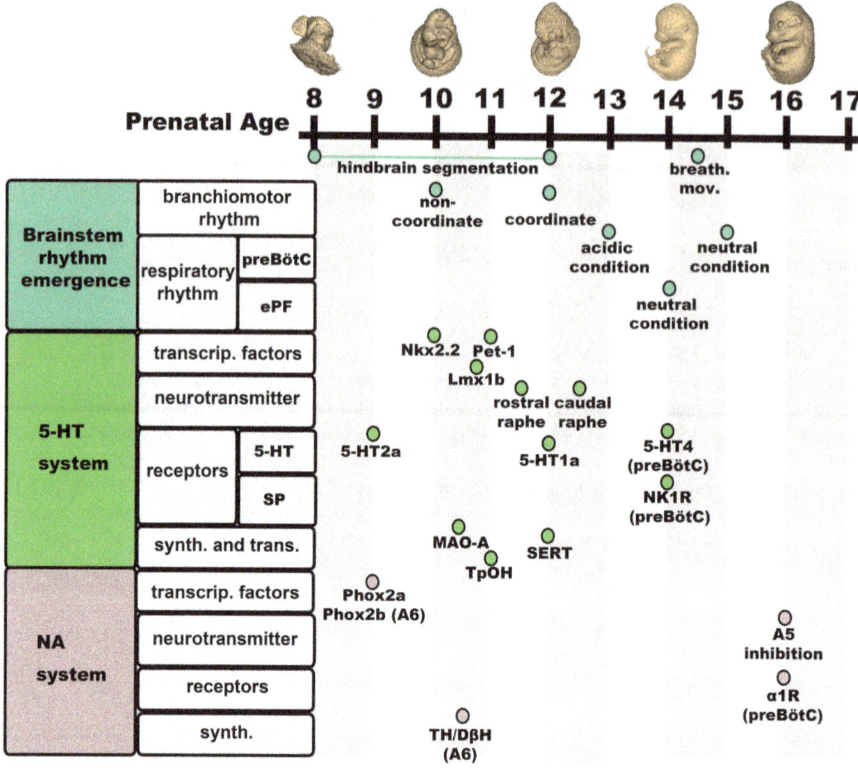

Fig. 10.2 Ontogeny of respiratory-like rhythm, serotonergic and inputs during mouse fetal development. Gestational days are indicated in the *upper scale*. Over this, drawings of E8, E10, E12, E14, and E16 mouse embryos obtained from EMAP eMouse Atlas Project (http://www.emouseatlas.org) (Richardson et al. 2014). Initiation of the specific expression of activity or molecule is indicated by a *dot*

The serotonergic neurons derived from rhombomeres r2 and r3 later develop into the RN of the reticular formation. Whether they contribute some drive to the nascent respiratory network is unknown.

Other primordial spontaneous low frequency rhythms (LF) can be detected with calcium imaging. They initiate at multiple sites: the rostrolateral medulla, the medulla at the level of the hypoglossal nucleus or in spinal cord segments. For a brief period up to E16, the medullary LF co-exists with the respiratory-like rhythm produced by the preBötC (Thoby-Brisson et al. 2005). The LF burst can interfere with the respiratory-like rhythm in vitro. Thus, when the LF burst coincides with the respiratory-like rhythm, the rhythm frequency increases. And the end of each burst of LF activity is followed by a delay in the onset of the next respiratory-like burst (Meillerais et al. 2010). These effects suggest that LF activity propagates into and influences the nascent respiratory network. However, the LF is generated in a different circuit with a different site of origin and different pharmacological properties.

Furthermore, the LF can be selectively abolished without compromising the generation of the respiratory rhythm. Therefore, the LF is not a precursor of the respiratory rhythm circuit (Meillerais et al. 2010; Champagnat et al. 2011).

The development of the RPG circuit begins early, and signs of its activity can be detected many days before birth (Figs. 10.1 and 10.2). Oscillatory LF branchiomotor activity can be recorded in vitro from trigeminal, facial, glossopharyngeal, and vagal nerve roots of mice at E10.5, whereas LF synchronized activity between adjacent branchiomotor oscillators can be recorded at E12.5 (Abadie et al. 2000). It is not clear whether early embryonic branchiomotor rhythms are direct precursors of the more mature high-frequency rhythms, like the respiratory rhythm (Thoby-Brisson et al. 2005; Champagnat et al. 2009). In fact, they differ in their susceptibility to blockers of gap junctions and glutamatergic transmission, suggesting the rhythms are based on different mechanisms (Thoby-Brisson et al. 2005). Bulbospinal projecting fibers can reach the thoracic level of the spinal cord of mouse at E14.5 (Ballion et al. 2002). Breathing movements are detected in murine fetuses from E14.5 (Abadie et al. 2000) and optimal synaptic connectivity between phrenic nerve and diaphragm muscle is already achieved around E17 in rats (Greer et al. 1999). Murine fetuses born by cesarean section before E18 cannot breathe spontaneously when exposed to air (Viemari et al. 2003). The impairment of breathing by prematurely born E15-E17 mouse fetuses is not due to a deficit in rhythm generation, but depends mainly on mechanical restrictions imposed by lung immaturity, such as those derived from insufficient production of lung surfactant (Viemari et al. 2003).

The Onset of the Fetal Respiratory Rhythm Evaluated In Vitro

The absence of breathing movements at a certain fetal age is not due to an absence of an active RPG but probably instead a lack of functional synapses on respiratory motoneurons and muscles. Researchers have used activity in the isolated brainstem-spinal cord (*"en bloc"*) preparation and the brainstem slice preparation to address when a respiratory generator emerges during development.

Thoby-Brisson et al. (2005) recorded a high frequency ("respiratory-like") rhythm either alone or co-existing with a low-frequency (LF, see above) rhythm in a small percentage of *en bloc* and slice preparations obtained from E14 fetal mice (Thoby-Brisson et al. 2005). The high frequency ("respiratory-like") rhythm became more sustained and appeared in a majority of the preparations at E15 and in all the preparations by E16 (Thoby-Brisson et al. 2005). Combining calcium imaging with electrophysiological recordings, these authors found that the generator of this respiratory-like rhythm was located in the ventral medulla in a position similar to the neonatal preBötC. Even more, as expected of preBötC neurons, the neurons forming the generator were connected bilaterally, expressed the neurokinin 1 receptor (NK1R), and their activities were modulated by substance P (SP) and opioids, while antagonists of α-amino-3-hydroxy-5-methyl-4-isoxazolepropionic acid (AMPA)/kainate glutamate receptors abolished their rhythm.

The preBötC is not the first respiratory nucleus showing respiratory-like activity (Fig. 10.2). Using calcium imaging, Thoby-Brisson et al. (2009) recorded rhythmical bursting activity from the embryonic parafacial oscillator (ePF), which is equivalent to the adult/neonatal RTN/pFRG, a day earlier than in the preBötC in brainstem slices under the same bath temperature and bath ionic composition (Thoby-Brisson et al. 2009). The ePF cells express the paired-like homeobox 2b (Phox2b), a homeodomain transcription factor required for maintaining the neuronal types that sustain reflex control of visceral autonomic functions and are derived from the early growth response 2 (Egr2, also known as Krox20) expressing precursors (Thoby-Brisson et al. 2009; Gray et al. 2010). *Egr2* null mice mutant (Krox20 mutant) not only shows ten times more apneas than wild type mice (Jacquin et al. 1996), but also lack ePF rhythm activity (Thoby-Brisson et al. 2009). The ePF oscillator shows a faster rhythm when it is free running, isolated from preBötC influence. In addition, it entrains the preBötC oscillator. Thus, silencing the Egr2 gene, which eliminates the ePF cells but not the preBötC neurons, or application of riluzole, which impairs ePF activity, slows the preBötC rhythm, as does tranverse section between both oscillators to separate them (Thoby-Brisson et al. 2009). In contrast to the preBötC rhythm, the ePF rhythm persists without glutamatergic transmission, but is impaired after gap junction blockade with carbenoxolone (Thoby-Brisson et al. 2009). In embryos, ePF pacemaking mechanisms seems to rely primarily on persistent Na^+ current (I_{NaP}), while the ePF rhythm is modulated by hyperpolarization activated (I_h) current (Thoby-Brisson et al. 2009).

The body of evidence indicates that ePF and preBötC oscillators emerge independently from different neural precursors and they differ in their underlying rhythm mechanisms. As mentioned before, the ePF activity precedes by 1 day the appearance of the preBötC activity, and the ePF oscillator couples and entrains the slower preBötC oscillator (Thoby-Brisson et al. 2009).

Development of RTN/pFRG as a Chemosensitive Nucleus

The RTN/pFRG is a group of neurons located rostral to preBötC and ventral to the facial nucleus and likely derived from ePF cells. The pFRG is postulated to be a second oscillator with pre-inspiratory activity (Onimaru and Homma 2003) and a conditional oscillator for active expiration (Fortuna et al. 2009; Feldman et al. 2013). The RTN is postulated to be an important site for central chemosensitivity (Stornetta et al. 2006; Guyenet et al. 2009; Takakura et al. 2006; Mulkey et al. 2007). Because both nuclei overlap, the researchers refer to this area as the RTN/pFRG. Genetic studies show that the RTN/pFRG contain neurons derived from *Phox2b*, protein atonal homolog 1 (*Atoh1*) and developing brain homeobox 1 (*Dbx1*) expressing progenitors (Gray 2008). Non-rhythmically-active glutamatergic neurons located in the medial RTN/pFRG (Feldman et al. 2003; Abbott et al. 2009) that express *Phox2b* and are sensitive to pH are important for the chemosensitive role of the RTN/pFRG (Guyenet et al. 2009; Stornetta et al. 2006; Dubreuil et al. 2008). Mice with mutations in *Phox2b* show a blunted response to CO_2 and a high death rate due to central apnea soon after birth. Interestingly, the *Phox2b[27Ala/+]* mice show normal anatomical structure in catecholaminergic nuclei (A6, A5) and NTS (Dubreuil et al. 2008). Conditional mutant mice lacking the RTN, as consequence of the expression of *Phox2b[27Ala]* restricted to rhombomeres r3 and r5, breathe normally at rest but show decreased response to CO_2 during their early postnatal days. In adulthood, these mutant mice recover 60% of the wild type CO_2 sensitivity (Ramanantsoa et al. 2011), so there is some plasticity of their central chemoreception.

When the RTN/pFRG input to preBötC appears is not clear, but it is probably after E15.5 because a transverse section caudal to the facial nucleus after this age can alter the frequency of the breathing rhythm in vitro (Fortin and Thoby-Brisson 2009). Therefore, RTN/pFRG input is important for chemosensory reflex responses, but probably it does not have an essential role in the emergence of the preBötC inspiratory rhythm.

The exact embryonic day in which the respiratory network initiates oscillatory activity is unsettled, leaving uncertainty in the timing of related events. For each mammalian species, the day in which the respiratory rhythm emerges is different. Furthermore, the exact embryonic day in which a respiratory-like rhythm emerges among species from same genus, e.g. genus Mus, is still controversial and elusive. As mentioned before, a respiratory-like rhythm can be recorded from phrenic roots at E13 if the brainstem-spinal cord preparation is superfused in acidified medium, which is representative of the acidic environment found *in uterus* (Eugenin et al. 2006). In addition, independent of the embryonic age evaluated, acidification of the superfusion medium always triggers a respiratory rhythm response even in the youngest preparations (Eugenin et al. 2006). Indeed, at the time the embryonic parafacial oscillator (ePF) activity emerges (E14.5), its frequency increases with acidification, suggesting that ePF cells have an early role in central chemoreception.

How Is the Inspiratory Rhythm Generated in the preBötC?

The preBötC, considered essential for the inspiratory activity (Feldman et al. 2013; Smith et al. 1991) (see Chaps. 8, 9, 11, and 12), is located ventral to the nucleus ambiguus in the ventral medulla. A population of glutamatergic neurons that express NK1 receptor (NK1R[+]) and somatostatin receptor (SSTR[+]) forms the core of the preBötC (Gray et al. 2010). This glutamatergic NK1R[+]/SSTR[+] neuronal population appears around E14.5-E15.5 (Gray et al. 2010; Pagliardini et al. 2003; Thoby-Brisson et al. 2005) and is derived from progenitor cells expressing the homeobox gene *Dbx1*, a gene that controls the fate of glutamatergic interneurons. *Dbx1* is required for preBötC rhythm generation in the mouse embryo (Gray et al. 2010; Bouvier et al. 2010). Indeed, the *Dbx1* null mutant mice do not show any inspiratory or expiratory movement in vivo at birth, and in in vitro brainstem preparations do not spontaneously generate inspiratory-like motor output either in pH 7.4 or acidified aCSF or in the presence of serotonin (Gray et al. 2010).

Although is widely accepted that the inspiratory rhythm is generated by glutamatergic propiobulbar neurons in the preBötC (Funk et al. 1993; Wallen-Mackenzie et al. 2006), how this works is a matter of debate. It is not clear whether the inspiratory burst is driven by a small population of specialized pacemaker neurons (Ben-Mabrouk and Tryba 2010; Pena et al. 2004; Ramirez et al. 2011; Tryba et al. 2008; Viemari and Ramirez 2006; Zavala-Tecuapetla et al. 2008) or is the result of excitatory interactions among neurons that express a synaptically triggered burst-generating conductance (Feldman 2007; Rekling and Feldman 1998; Feldman et al. 2013) forming an oscillator network.

Models based on pacemaker neurons can generate a rhythm with a period and burst duration that match the respiratory rhythm in normal or modified conditions (Rekling and Feldman 1998; Smith et al. 2000; Feldman and Cleland 1982; Feldman et al. 2013). However, these models assume that all pacemaker neurons are glutamatergic/excitatory neurons. Evidence shows that around 23% of preBötC pacemaker neurons are glycinergic (Morgado-Valle et al. 2010). Furthermore, the pharmacological blockade of I_{NaP} or calcium-activated non-specific cationic current (I_{CAN}), the two main pacemaking currents (see below), do not always perturb or stop the rhythm (Del Negro et al. 2002a, b, 2005; St-John 2008).

Models based on a group-pacemaker or an oscillating network use a group of interconnected glutamatergic preBötC neurons firing tonically at low rate between inspiratory bursts that increase their activity via positive feedback. This positive feedback activity is the source of a pre-inspiratory phase, which starts around 300–400 ms before the inspiratory burst and leads to a network-wide, synchronous inspiratory burst phase (Feldman et al. 2013; Rekling et al. 2000). Such models are incomplete and are at present working hypotheses, for the mechanism for inspiratory burst termination is unknown. While there is evidence supporting both pacemaker and group-pacemaker models, it is probable that rhythm generation is not based on a single mechanism. The fact that different underlying mechanisms may provide identical network activity (Prinz et al. 2004; Ramirez et al. 2012) suggests there is a need for more investigation into the network properties and the underlying mechanisms of rhythm generation.

Nevertheless, most research groups agree on the importance of some components of inspiratory rhythm generation: the voltage sensitivity and subthreshold activation of I_{NaP} (Koizumi and Smith 2008; Butera et al. 1999; Del Negro et al. 2010, 2002b), the I_{CAN} (Del Negro et al. 2005; Pace et al. 2007; Thoby-Brisson and Ramirez 2001) (see Chap. 9), and a population of glutamatergic NK1R[+]/SST[+] neurons in the preBötC (Gray et al. 2010; Bouvier et al. 2010). Other cell types modify the relative contribution of each of these elements through neuromodulators.

Neuromodulators of the Respiratory Rhythm

In neonates and adults, neuromodulators such as neuropeptides, biogenic amines, amino acids, and nucleotides released by neurons or astrocytes in various nuclei act on rhythmogenic and other components of the RPG. These neuromodulators shape the respiratory pattern accordingly to physiological demands, and their effects are dependent on the functional state of the respiratory network (Doi and Ramirez 2008; Ramirez et al. 2012).

The fact that neuromodulators affect the inspiratory rhythm during prenatal ontogeny suggests that neuromodulators and the machinery for their synthesis and degradation are available during embryonic life. Moreover, it implies that there is expression of functional receptors capable of activating intracellular cascades to evoke differential effects on the respiratory rhythm. Presumably various neuromodulators converge on the respiratory network and can influence the rhythm. In addition, the RPG is a dynamic network and shows state-dependency, so eliminating the actions of a neuromodulator during development might be compensated by the actions of another to reach a new balance.

The presence of a phrenic respiratory-like rhythm in acidic conditions at early embryonic stages reveals that neuronal progenitors of the NK1R$^+$ population another neuronal population can generate a respiratory-like rhythm. This also suggests that a chemosensory stimulus induces a change on intrinsic properties of rhythmogenic neurons allowing the rhythm to emerge. This is a reasonable hypothesis considering that, likely, the synaptic plasticity of the RPG and its oscillatory activity depend on the interplay of intrinsic properties and synaptic input. We will discuss this hypothesis in next sections. Likewise, we will try to answer the following questions: When do serotonergic and noradrenergic sensory input to the RPG emerge? When they are able of influencing the respiratory rhythm? And, what happens to the respiratory rhythm when the normal development of these sensory inputs is disturbed or completely suppressed? We chose the serotonergic and noradrenergic systems for our analysis because they constitute central chemosensory inputs that appears early during development, and their receptors and nerve fibers are easily distinguishable from the RPG neural network.

Serotonergic System

Serotonergic raphe neurons play two main general roles for RPG. One is a trophic-like role, influencing the development of the neural network circuits (Gaspar et al. 2003). The other is a chemosensory role, contributing to the CO_2-sensing drive of the respiratory rhythm as well as to the central chemoreflexes induced by hypercapnia (Richerson 2004).

Ontogeny of 5-HT Containing Neurons and Its Role in Neural Network Development

As further discussed in Chap. 11, serotonergic raphe neurons (Fig. 10.2) originate from a rostral cluster of progenitors derived from rhombomeres r1-r3 and from a caudal cluster derived from rhombomeres r5- r7 (Gaspar and Lillesaar 2012). Once they migrate from dorsal to ventral positions, precursor cells switch from visceral motor to serotonergic neurons (Pattyn et al. 2003), induced by the activation of the homebox gene *Nkx2.2* at E10.5. This gene expresses a transcription factor that suppresses *Phox2b* expression, an important paired-class homeobox gene required for generation of the visceral motor nucleus (Pattyn et al. 2000; Pattyn et al. 2003). *Nkx2.2* also induces the expression of the transcription factors coded in the genes pheochromocytoma 12 ETS factor-1 (*Pet-1*) and LIM homeobox transcription factor 1β (*Lmx1b*) (Pattyn et al. 2003; Cheng et al. 2003; Hendricks et al. 1999), pivotal factors for the post-mitotic development of serotonergic neurons (Cheng et al. 2003; Ding et al. 2003; Hendricks et al. 1999). The presence of Pet-1 and Lmx1b transcription factors allows the detection of serotonergic neurons before they express 5-HT (Goridis and Rohrer 2002). Lmx1b expression is around E10.75 (Ding et al. 2003) while Pet1 expression, restricted to the RN, occurs around E11 (Alenina et al. 2006), 0.5–1 day before the serotonergic neurons can be identified (Goridis and Rohrer 2002; Gaspar and Lillesaar 2012; Alenina et al. 2006). The serotonergic neuronal differentiation process also involves the genes mammalian achaete-scute homolog 1/achaete-scute complex-like 1 (*Mash1/Ascl1*), Gata binding protein 2 (*Gata2*), and Gata binding protein 3 (*Gata3*) (Craven et al. 2004; van Doorninck et al. 1999; Pattyn et al. 2003).

The availability of *Pet-1* knockout (*Pet-1$^{-/-}$*) (Hendricks et al. 2003) and *Lmx1b* conditional knockout (*Lmx1b$^{f/f/p}$*) mice (Ding et al. 2003) has helped to elucidate the role played by serotoninergic innervation in development and modulation of the RPG. *Pet-1$^{-/-}$* and *Lmx1b$^{f/f/p}$* mice can be considered models for evaluating the effects of partial and complete serotonergic depletion, because *Pet-1$^{-/-}$* mice retain 20–30% of their serotonergic neurons, while in *Lmx1b$^{f/f/p}$* mice there are none (Hendricks et al. 2003; Ding et al. 2003) (Gaspar and Lillesaar 2012). In general terms, the postnatal survival of both types of knockout mice shows that the serotonergic system is not essential for building the respiratory rhythm oscillator. However, it also suggests that this system is important in the modulation of the rhythm. In vivo, *Pet-1$^{-/-}$* mice show a transient respiratory depression, with rhythm instability that includes long apneas (Erickson et al. 2007) and exacerbated bradycardia during hyperthermia (Cummings et al. 2010). Nonetheless, the breathing pattern is improved by 9.5 postnatal days (P9.5) when becomes closer to that seen in *wild type*.

The *Lmx1b$^{f/f/p}$* is a mouse in which both alleles encoding for *Lmx1b* transcription factor were excised in *Pet-1* expressing neurons. Since *Lmx1b* is expressed prior to *Pet-1* in the serotonergic neurons, and the *Lmx1b* gene is essential for maintaining the serotonergic phenotype, this mouse is useful to evaluate the effect on respiratory network development of a transient expression of serotonergic precursor neurons. In fact, these mice display normal development of central serotonergic neurons during early prenatal development, followed by down regulation of *Lmx1b* expression at E12.5 that leads to the almost complete loss of serotonergic neurons (Zhao et al. 2006). *Lmx1b$^{f/f/p}$* mice display frequent apneas and decreased ventilation at early postnatal ages. The improvement in the breathing pattern of 2–4 week old mice suggests a compensation for the serotonergic deficit by unknown mechanisms (Hodges et al. 2009). Interestingly, exogenous application of agonists acting on 5-HT2AR and NK1R, reverses the ventilatory deficit in *Lmx1b^{f/f/p}* in vivo and in vitro (Hodges et al. 2009), suggesting that the development of postsynaptic mechanisms within serotonergic circuits occurs earlier than E12.5 and that the presence of serotonergic neurons after E12.5 is not required for rhythm generation. These results also suggest that the 5-HT/SP input is important for shaping the usual postnatal breathing pattern. Furthermore, *Lmx1b^{f/f/p}* adult mice show a 50% reduction in the ventilatory response to CO_2 in vivo (Hodges et al. 2008), suggesting that chemosensitive responses strongly depend on the presence of serotonergic neurons; other mechanisms, if present, are unable to compensate.

Monoamine oxidase A (MAO-A), an enzyme that metabolizes biogenic amines, has been implicated in embryonic brain development controlling apoptosis and neural differentiation of murine embryonic stem cells (Wang et al. 2011a, b). The influence of MAO-A in the emergence of ventilatory activity was evaluated in MAO-A deficient mice (*Tg8*), which exhibit excess biogenic amines in perinatal brain. At birth, these mice have an irregular respiratory pattern. Functional changes are accompanied by the increase in the number of spines and varicosities in phrenic motoneurons (Bou-Flores et al. 2000), revealing that levels of biogenic amines, perhaps serotonin, are determining the structure of the respiratory network.

Serotonergic Drive of the Respiratory Rhythm and Contribution to Central Chemoreception

The raphe neurons containing serotonin project to several nuclei along the substantia nigra pars compacta (SNpc), acting not only at the levels of the "input" (NTS) and the "output" (hypoglossal and phrenic motor nuclei) of the RPG, but also on nuclei contributing directly to the respiratory rhythm generation, such as the preBötC and the RTN/pFRG (Fuxe 1965; Holtman 1988; Pilowsky et al. 1990; Hodges and Richerson 2008). The effect of serotonin on breathing pattern is complex and involves several pre- and post-synaptic receptor subtypes with differential inhibitory and excitatory activities along the RPG (Hodges and Richerson 2008). In addition, serotonergic terminals can co-release the excitatory neuropeptides SP and thyrotropin-releasing hormone (TRH) (Holtman 1988; Hodges and Richerson 2008).

Serotonin modulates the respiratory rhythm via serotonin receptors 1A, 2A, 2B (5-HT2BR), 4 (5-HT4R), and 7 (5-HT7R) (Ramirez et al. 2012; Richter et al. 2003). These receptors are coupled to G-protein (GPCRs), which mediates the changes in the open probability of various ion channels. When administered in vitro, serotonin has a net stimulatory effect on the respiratory activity in neonatal mouse and rat preparations (Di Pasquale et al. 1992; Richerson 2004). Activation of 5-HT1AR overcomes apneustic episodes induced by hypoxia (Richter et al. 2003). In pentobarbitone-anaesthetized cats, activation of 5-HT1AR/5-HT7R with 8-hydroxy-diproplaminotetralin (8-OH-DPAT) has depressant effects on breathing, increasing the membrane hyperpolarization of expiratory neurons during inspiratory and post-inspiratory phases (Lalley et al. 1994). The blockade of 5-HT2AR with altanserin hydrochloride or the activation of 5-HT2BR with the agonist BW-723C86 increases the respiratory frequency in rat brainstem preparations (Niebert et al. 2011). The activation of 5-HT4R using BIMU8 overcomes fentanyl-induced respiratory depression and reestablishes a stable respiratory rhythm without the loss of fentanyl's analgesic effect in rats (Manzke et al. 2003). Interestingly, the combined activation of 5-HT4R using zacopride and 5-HT1AR/5-HT7R with 8-OH-DPAT reverses opioid-induced respiratory depression and hypoxia in goats (Sahibzada et al. 2000; Meyer et al. 2006).

The serotonergic raphe neurons have an important role in central chemoreception (Richerson 1995; Wang et al. 2001; Kanamaru and Homma 2007; Bernard et al. 1996), in particular those located in caudal raphe (Nattie and Li 2001; Bernard et al. 1996; Richerson 1995). CO_2/H^+-sensitive neurons are located in midline medullary raphe (Richerson 1995, 2004; Wang et al. 1998). Focal acidification of the midline raphe by microinjection of acetazolamide in anesthetized rats or by reverse microdialysis of acidified CSF in conscious rats or goats increases ventilation (Nattie and Li 2001; Hodges et al. 2004a, b). Inhibition or lesion of raphe neurons reduces the response to hypercapnia (Dreshaj et al. 1998; Messier et al. 2002, 2004; Taylor et al. 2006; Corcoran et al. 2013). Elimination of serotonergic neurons, by application of a monoclonal antibody against the serotonin transporter (SERT) conjugated to saporin into RN, blunts the CO_2 response (Nattie et al. 2004). Likewise, adult knock-out mice ($Lmx1b^{f/f/p}$ and $Pet-1^{(-/-)}$) with nearly complete absence of central 5-HT neurons have a 50% lower ventilatory response to hypercapnia (Hodges et al. 2008, 2011). The synthetic $G_{i/o}$ protein–coupled receptor Di can be expressed selectively in 5-HT neurons, which hyperpolarize when clozapine-N-oxide (CNO) is administered. Such selective inhibition reduces the ventilatory response to hypercapnia (Ray et al. 2011, 2013).

Ontogeny of Fibers Projecting into the RPG Expressing Serotonergic Phenotype

Serotonin appears in the hindbrain with a rostrocaudal gradient such that first it is detected in the rostral RN at E11.5 and then in the caudal RN at E12.5 (Ding et al. 2003). At E12.5, serotonergic fibers can be detected at cervical segments of the spinal cord in fetal mouse (Ballion et al. 2002), where later, at this and more caudal levels, they will induce distinct types of spinal cord rhythmic motor activities (Branchereau et al. 2000).

Appearance of Machinery for Synthesis and Degradation of Serotonin

The existence of a binding site for Pet-1 transcription factor on the promoter region of genes encoding for 5-HT1aR, 5-HT transporter (SERT), tryptophan hydroxylase (TpOH) and aromatic L-amino acid decarboxylase (Hendricks et al. 1999) suggests that the emergence of these elements follows a similar localization and temporal pattern of emergence to that of the determinants for the serotonergic neuron phenotype. TpOH (the enzyme that catalyzes the initial and rate-limiting step in the synthesis of 5-HT), SERT (important for the uptake of 5-HT from the synaptic space), and MAO-A (the enzyme that degrades 5-HT) can be detected in the same regions that the 5-HT transcription factors Pet-1 or Lmx1b are detected earlier. TpOH and SERT are detected at E11 and E12, respectively, in the mouse hindbrain (Zhao et al. 2006; Bruning et al. 1997), while MAO-A is detected as early as E10.5 in the mouse brain (Wang et al. 2011a).

Serotonergic Receptors

The 5-HT receptors 1A, 2A, 4 and 7 are considered involved in modulation of breathing (Richter et al. 2003). 5-HT1AR mRNA is detected in rat brainstem by E12, reaching a peak of transcription at E15 (Hillion et al. 1993). 5-HT2A receptor mRNA is detected by E9 and E11 in rat embryos, with increasing levels thereafter (Wu et al. 1999). In preBötC, strong expression of 5-HT4R, co-expressed with mu-opioid receptors, is reported at E14 (Manzke et al. 2008). The 5-HT7R shows low expression levels in preBötC (Richter et al. 2003) and the course of its expression during ontogeny has not been evaluated. Functional NK1R, the receptor for SP, a co-released neurotransmitter from raphe neurons, is detected in preBötC at E14.5 (Thoby-Brisson et al. 2005).

Noradrenergic System

The noradrenergic system plays a role in the development of the respiratory neural network (Hilaire 2006) and provides excitatory and inhibitory input into the RPG (Hilaire et al. 2004). The main noradrenergic nucleus, A6, also known as locus coeruleus (LC), located in the dorsal pons at the floor of the fourth ventricle, provides, among other functions, chemosensory drive to the respiratory rhythm (Nattie and Li 2012). A5 is another noradrenergic group of cells in the vicinity of the superior olivary complex in the pontine tegmentum.

Noradrenergic Input into the RPG

Evidence of noradrenergic modulation of respiratory rhythm has been obtained in vitro using rat *en bloc* preparations. In preparations preserving the pons, bath application of noradrenaline either increases or decreases the frequency of the respiratory rhythm. When pons is eliminated, noradrenaline only decreases the rhythm frequency (Errchidi et al. 1990; Hilaire et al. 1989). This dual effect on respiratory rhythm frequency is mediated by two different noradrenergic receptor subtypes expressed in the ventrolateral medulla: the α1 adrenoreceptor, increasing (Errchidi et al. 1991; Hilaire et al. 1989; Al-Zubaidy et al. 1996) and the α2 adrenoreceptor, decreasing the respiratory rhythm frequency (Arata et al. 1998). Inhibitory noradrenergic input is provided by the A5 area (Hilaire et al. 1989; Viemari et al. 2004b), whereas the excitatory input is provided by the LC (Viemari et al. 2004a), an area that is also involved in the hypercarbic ventilatory response (Biancardi et al. 2008; Filosa et al. 2002). In mouse *en bloc* preparations, noradrenaline superfusion decreases the frequency of the respiratory rhythm regardless of whether the pons is included or not (Jean-Charles and Gerard 2002). In mouse medullary slices, A1/C1 and A2/C2 areas are identified as the origin for a tonic excitatory preBötC input that acts as a stabilizer of breathing through the α2 adrenoreceptor (Zanella et al. 2006; Corcoran and Milsom 2009; Hilaire 2006). The α1 and α2 adrenoreceptors are coupled to Gq/11 and Gi/o, respectively, affecting different ion channels. In other systems, activation of α1 adrenoreceptors may suppress cesium-sensitive potassium channels in human osteoblasts, barium and tetraethylammonium-sensitive heat-activated potassium channels in rat dorsal root ganglion neurons, and cardiac delayed rectifier potassium currents (Kodama and Togari 2010; Wang et al. 2009; Yamamoto et al. 2009). Activation of α2 adrenoreceptors modulates the N-type voltage-gated calcium channels in cultured cerebellar neurons and the delayed-rectifier potassium current and sodium current in rat sympathetic neurons (Chen et al. 2009; Li and Horn 2008). In preBötC, activation of α2 adrenoreceptors is observed in pacemakers with burst properties mediated by I_{NaP} current (Viemari et al. 2011).

Locus Coeruleus as a Central Chemosensitive Nucleus

Detection of c-Fos protein as marker of neuronal activity revealed that the number of c-Fos positive neurons at the LC increased after exposure to hypercapnia (Teppema et al. 1997), while LC neurons in vivo and in brainstem slices increased their firing rate in response to hypercapnia (Elam et al. 1981; Stunden et al. 2001). Intrinsic CO_2/H^+ responsiveness of LC neurons was demonstrated by perforated patch-clamp recording from dissociated LC neurons phenotypically identified in primary cultures by endogenous expression of green fluorescent protein (GFP) obtained from Prp57 transgenic mice (Johnson et al. 2008). Focal acidification of the LC by both local application of acetazolamide, an inhibitor of the carbonic anhydrase enzyme, increased ventilation (Coates et al. 1993). Reduction of central chemoreflexes induced by hypercapnia and reduction of wakefulness at expense of NREM sleep were caused by destruction of brainstem catecholamine-containing neurons by injecting antidopamine beta-hydroxylase-saporin (DBH-SAP) conjugate into the fourth ventricle (Li and Nattie 2006). On the other hand, selective destruction of noradrenergic or NK1R-expressing LC neurons using injection of 6-hydroxydopamine (6-OH DA) or SP-saporin conjugate, respectively, into the LC also reduced the ventilatory response to hypercapnia (de Carvalho et al. 2010; Biancardi et al. 2008).

Ontogeny of Respiratory Related Noradrenergic System (Fig. 10.2)

The LC is the main noradrenergic brainstem nucleus contributing to the chemosensory drive of the respiratory rhythm. It consists of a dense group of neurons located in the dorsolateral tegmentum of the pons and it is derived from progenitors in dorsal rhombomere 1 (Aroca et al. 2006). Noradrenergic neurons of the LC are generated and start to grow fibers from E14 in rats (Olson and Seiger 1972) and E9 to E10.5 in mice (Steindler and Trosko 1989; Goridis and Rohrer 2002).

LC early development and specification requires several factors, both extrinsic and intrinsic (Goridis and Rohrer 2002). For example, the dorsal fate of LC neurons depends on bone morphogenetic protein (BMP) secreted initially by dorsal ectoderm and later by the roof plate and dorsal-most neural tube (Goridis and Rohrer 2002). In addition, there have been four transcription factors identified in mouse apparently acting in a linear cascade during development: Mash1 → paired-like homeobox 2a (Phox2a) → Phox2b → Rnx (Goridis and Rohrer 2002; Hirsch et al. 1998; Pattyn et al. 2000). The main gene for generation of noradrenergic LC neurons is *Phox2a*, a gene coding for a homeodomain protein expressed in peripheral autonomic ganglia and in some hindbrain nuclei (Morin et al. 1997); it lies downstream *Mash1* and upstream to *Phox2b* (Hirsch et al. 1998). Knockout mice for *Phox2a* show impaired respiratory rhythm in vivo and in vitro, impaired respiratory response to hypoxia in vitro (Viemari et al. 2004a), and low probability of survival for more than a day (Morin et al. 1997). Phox2b and Phox2a can be detected in LC during E9.5, but Phox2b is no longer expressed in LC neurons by E12.5. *Rnx* is required for formation of A5 neurons (Shirasawa et al. 2000) and for reaching the regular number of LC catecholaminergic neurons (Qian et al. 2001). $Rnx^{(-/-)}$ mice have a higher respiratory rate, more apneas, and die earlier than wild type mice (Shirasawa et al. 2000). The adequate development of brainstem noradrenergic neurons termed A1/C1, A5, A2/C2 and C2/C3 requires the expression of ladybird homeobox 1 (*Lbx1*) (Pagliardini et al. 2008), because the loss of Lbx1 expression results in ectopic noradrenergic neuron expression. Another gene required for normal development of LC is homeobox protein Hox-A2 (*Hoxa2*); its loss causes expansion of the number of respiratory-related neurons in A6 (Chatonnet et al. 2007).

Prenatal Development of Noradrenergic Fibers Projecting into the RPG and Adrenergic Receptor Expression

Noradrenergic neurons in the LC apparently start to grow axons by E14 in the rat (Olson and Seiger 1972) and E9 in mice (Steindler and Trosko 1989). Indirect evidence points to an early functional role of these noradrenergic projections, at least those provided by pontine inhibitory A5 input to preBötC. We have reported an increase in respiratory rate after ponto-medullary transection in *en bloc* preparations by E16 in mice (Eugenin et al. 2006). However, no inhibition of respiratory frequency by A5 was observed at E16 in mouse preparations (brainstem containing intact pons and medulla) (Viemari et al. 2003).

Considering the responsiveness to noradrenaline of brainstem preparations at E16, we can speculate about the emergence of functional α1 excitatory adrenoreceptor in preBötC (Viemari et al. 2003). However, the contribution of different adrenergic receptors during early prenatal life, for example through selective pharmacological agonists, has not been evaluated.

Enzymes for Synthesis and Degradation of Noradrenaline

Expression of tyrosine hydroxylase (TH), an enzyme essential for synthesis of noradrenaline, is detected by E10.5 in the lateral floor of the fourth ventricle, the area that gives rise to the LC (A6) (Morin et al. 1997; Lauder and Bloom 1975). However, in rats the number of TH positive neurons is low at E16, and this only increase until E19 in areas like C1 (Foster et al. 1985). Dopamine beta hydroxylase, the enzyme that catalyzes the conversion of dopamine to noradrenaline, also is already detected at E10.5 in A6. As mentioned before, the enzyme for noradrenaline degradation, MAO-A, is detected as early as E10.5 in embryonic mouse brain (Wang et al. 2011a, b).

Concluding Remarks

Our search was centered on a probable role achieved by noradrenergic and serotonergic central chemoreception as triggers of the onset of the respiratory rhythm during fetal development. The probability that neurons generate action potentials depends on their excitability and the balance of excitatory/inhibitory synaptic currents. The preBötC may integrate incoming sensory information and changes its properties according to its biochemical environmental (Ramirez et al. 2012). For example, reconfiguration of the properties could be by an increase in I_{NaP} density that may induce the transition from a silent to an actively, autonomously spiking neuron (Butera et al. 1999; Del Negro et al. 2002a; Thoby-Brisson and Ramirez 2001) or by alteration of the contribution of voltage-independent, non-selective cationic channel NALCN, thus activating several receptors coupled to G-proteins, such as those associated with NK1R (Lu et al. 2009, 2010). Central chemosensory input, like that of other afferents to the respiratory network, can cause an increase in the ratio of excitatory/inhibitory synaptic currents in preBötC neurons, either by activing receptors acting on channels or by changing expression genes coding for channels responsible for rhythmogenesis properties. But first some questions should be addressed: Is chemosensory input already available at the time of the emergence of the rhythm in the preBötC (Fig. 10.3)? Is such a contribution strictly necessary to trigger the rhythm?

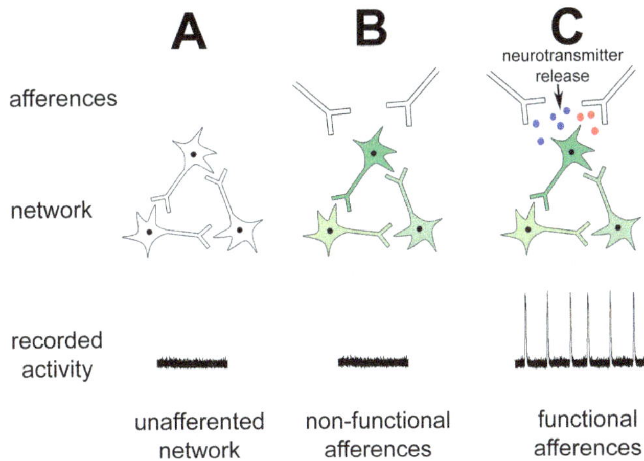

Fig. 10.3 Rhythmic activity may begin as result of concomitant changes in the intrinsic properties of the neural network and the arriving functional inputs. (**a**) immature network without input influences; (**b**) neurons forming the network change their properties, but they are still unable to generate a rhythm; (**c**) Functional afferents and neural network maturation are required for generating the rhythm

Our analysis, restricted only to the probable contribution of projections from RN and LC, does not allow us to have a clear answer for both questions. Serotonergic and noradrenergic fibers are able to extend to the cervical segments of the spinal cord at the fetal stage in which the respiratory rhythm starts. Therefore, although there is no direct evidence, developing fibers from RN and LC have enough time to reach neurons of the respiratory network in the brainstem and establishing synaptic connections with them, just as noradrenergic fibers do with cells in the embryonic hindbrain that generate spontaneous activity from E11.5 to E13.5 (Bosma 2010). In addition, most of the machinery for synthesis, uptake, and degradation of serotonin or noradrenaline are available at that time (Fig. 10.2). We know from the effects of experimentally altering the levels of 5-HT in the embryonic brain and from the early expression of 5-HT receptors and serotoninergic phenotype in brainstem that 5-HT exerts an early and strong influence on the emergence and maturation of circuitry, but not on the onset and maintenance of a functional respiratory rhythm.

At early postnatal life, almost the half of the LC neurons and less than a twentieth of RN neurons are already chemosensitive. In fact, ~3% of serotonergic raphe neurons are sensitive to CO_2 before P12 (Wang and Richerson 1999) and ~45% of LC neurons are (Stunden et al. 2001). Such results suggest that LC neurons are more likely contributing to chemosensory input to the respiratory network than raphe neurons during fetal life. However, future research should evaluate either single neuron chemosensitivity or the contribution of any brainstem nuclei to the overall chemosensory response during fetal life. Likewise, knowledge of the molecular mechanisms for chemosensitivity at each chemosensory site is vital to ascertain when various chemosensory structures are ready to affect the function of the respiratory network.

After genetic manipulation eliminating specific serotonergic or noradrenergic neuronal populations, the respiratory rhythm was still present in postnatal life. This reveals that the onset of the rhythm was achieved independently of the presence or absence of serotonergic or noradrenergic input. Whether the onset of the respiratory rhythm was delayed by the absence of input is an open question. On the other hand, elimination of one specific input involved in chemosensitivity induces early in postnatal life an inadequate breathing phenotype, which improves during adulthood (Hodges et al. 2009). The plasticity of the network and the redundancy of the effects of different neuromodulators on breathing may explain the improvement during adulthood. Different neuromodulators may activate different types of receptors, which in turn may activate the same or different intracellular cascades leading to similar effects on neuron properties, and hence on rhythmic activity. It is then reasonable to think that the lack of one input may be compensated by the presence of another. It is worth noting that the respiratory effects of altering serotonergic input may be caused by failures in this chemosensitive input (Corcoran et al. 2013) or deficit in the trophic actions that 5-HT exerts on neural circuits (Gaspar et al. 2003).

Contribution to the emergence of the respiratory rhythm provided by other chemosensitive nuclei should be considered. For example, RTN may have a more important role in chemosensitivity than RN during fetal life, because is known that mutant mice lacking RTN neurons can breathe normally in eucapnic normoxia, but show little response to CO_2 during the first postnatal days (Ramanantsoa et al. 2011). However, estimation of the relative contribution of chemosensory nuclei that themselves belong to the respiratory network, including the preBötC, RTN/pFRG and NTS, is a very difficult task. First, the same neurons that comprise the basic circuit for generating the rhythm might themselves be chemosensitive. Second, the molecular mechanisms for chemosensitivity are still an open question, and therefore, studies based on the early expression of particular molecules as "chemosensitivity markers" are not possible. And third, there is a total lack of studies evaluating the contribution of each chemosensitive site during the fetal period.

In summary, experimental data suggest that serotonergic and noradrenergic afferent input into the respiratory network is functional at the same time the respiratory rhythm emerges during development. However, data do not allow us to confirm or rule out the possibility that a chemosensory stimulus is required for triggering the onset of the respiratory rhythm. This is a reasonable hypothesis that should be considered for future research.

Acknowledgment Grants Fondo Nacional de Desarrollo Científico y Tecnológico (FONDECYT) # 1171434 (JE), Beltran-Morgado Foundation for the Advancement and Communication of Neuroscience in Veracruz, Comisión Nacional de Ciencia y Tecnología (CONICYT) #21120594 (S Beltrán-Castillo). DICYT-USACH (JE).

References

Abadie V, Champagnat J, Fortin G (2000) Branchiomotor activities in mouse embryo. Neuroreport 11(1):141–145

Abbott SB, Stornetta RL, Fortuna MG, Depuy SD, West GH, Harris TE, Guyenet PG (2009) Photostimulation of retrotrapezoid nucleus phox2b-expressing neurons in vivo produces long-lasting activation of breathing in rats. J Neurosci 29(18):5806–5819. doi:10.1523/JNEUROSCI.1106-09.2009. 29/18/5806 [pii]

Abreu-Villaca Y, Filgueiras CC, Manhaes AC (2011) Developmental aspects of the cholinergic system. Behav Brain Res 221(2):367–378. doi:10.1016/j.bbr.2009.12.049. S0166-4328(10)00017-3 [pii]

Alenina N, Bashammakh S, Bader M (2006) Specification and differentiation of serotonergic neurons. Stem Cell Rev 2(1):5–10. doi:10.1007/s12015-006-0002-2

Al-Zubaidy ZA, Erickson RL, Greer JJ (1996) Serotonergic and noradrenergic effects on respiratory neural discharge in the medullary slice preparation of neonatal rats. Pflugers Arch 431(6):942–949

Anderson TM, Garcia AJ, Baertsch NB, Pollak J, Bloom JC, Wei AD, Rai KG, Ramirez J-M (2016) A novel excitatory network for the control of breathing. Nature 536(7614):76–80

Arata A, Onimaru H, Homma I (1998) The adrenergic modulation of firings of respiratory rhythm-generating neurons in medulla-spinal cord preparation from newborn rat. Exp Brain Res 119(4):399–408

Aroca P, Lorente-Canovas B, Mateos FR, Puelles L (2006) Locus coeruleus neurons originate in alar rhombomere 1 and migrate into the basal plate: studies in chick and mouse embryos. J Comp Neurol 496(6):802–818. doi:10.1002/cne.20957

Ballantyne D, Scheid P (2001) Central chemosensitivity of respiration: a brief overview. Respir Physiol 129(1–2):5–12. S0034568701002973 [pii]

Ballion B, Branchereau P, Chapron J, Viala D (2002) Ontogeny of descending serotonergic innervation and evidence for intraspinal 5-HT neurons in the mouse spinal cord. Brain Res Dev Brain Res 137(1):81–88. S0165380602004145 [pii]

Ben-Mabrouk F, Tryba AK (2010) Substance P modulation of TRPC3/7 channels improves respiratory rhythm regularity and ICAN-dependent pacemaker activity. Eur J Neurosci 31(7):1219–1232. doi:10.1111/j.1460-9568.2010.07156.x. EJN7156 [pii]

Bernard DG, Li A, Nattie EE (1996) Evidence for central chemoreception in the midline raphe. J Appl Physiol 80(1):108–115

Biancardi V, Bicego KC, Almeida MC, Gargaglioni LH (2008) Locus coeruleus noradrenergic neurons and CO2 drive to breathing. Pflugers Arch 455(6):1119–1128. doi:10.1007/s00424-007-0338-8

Bianchi AL, Denavit-Saubie M, Champagnat J (1995) Central control of breathing in mammals: neuronal circuitry, membrane properties, and neurotransmitters. Physiol Rev 75(1):1–45

Bosma MM (2010) Timing and mechanism of a window of spontaneous activity in embryonic mouse hindbrain development. Ann N Y Acad Sci 1198:182–191. doi:10.1111/j.1749-6632.2009.05423.x

Bou-Flores C, Lajard AM, Monteau R, De Maeyer E, Seif I, Lanoir J, Hilaire G (2000) Abnormal phrenic motoneuron activity and morphology in neonatal monoamine oxidase A-deficient transgenic mice: possible role of a serotonin excess. J Neurosci 20(12):4646–4656. 20/12/4646 [pii]

Bouvier J, Thoby-Brisson M, Renier N, Dubreuil V, Ericson J, Champagnat J, Pierani A, Chedotal A, Fortin G (2010) Hindbrain interneurons and axon guidance signaling critical for breathing. Nat Neurosci 13(9):1066–1074. doi:10.1038/nn.2622. nn.2622 [pii]

Branchereau P, Morin D, Bonnot A, Ballion B, Chapron J, Viala D (2000) Development of lumbar rhythmic networks: from embryonic to neonate locomotor-like patterns in the mouse. Brain Res Bull 53(5):711–718. S0361-9230(00)00403-2 [pii]

Bruning G, Liangos O, Baumgarten HG (1997) Prenatal development of the serotonin transporter in mouse brain. Cell Tissue Res 289(2):211–221

Butera RJ Jr, Rinzel J, Smith JC (1999) Models of respiratory rhythm generation in the pre-Botzinger complex. I Bursting pacemaker neurons. J Neurophysiol 82(1):382–397

de Carvalho D, Bicego KC, de Castro OW, da Silva GS, Garcia-Cairasco N, Gargaglioni LH (2010) Role of neurokinin-1 expressing neurons in the locus coeruleus on ventilatory and cardiovascular responses to hypercapnia. Respir Physiol Neurobiol 172(1–2):24–31. doi:10.1016/j.resp.2010.04.016

Cermak JM, Holler T, Jackson DA, Blusztajn JK (1998) Prenatal availability of choline modifies development of the hippocampal cholinergic system. FASEB J 12(3):349–357

Chambers D, Wilson LJ, Alfonsi F, Hunter E, Saxena U, Blanc E, Lumsden A (2009) Rhombomere-specific analysis reveals the repertoire of genetic cues expressed across the developing hindbrain. Neural Dev 4:6. doi:10.1186/1749-8104-4-6

Champagnat J, Fortin G (1997) Primordial respiratory-like rhythm generation in the vertebrate embryo. Trends Neurosci 20(3):119–124

Champagnat J, Morin-Surun MP, Fortin G, Thoby-Brisson M (2009) Developmental basis of the rostro-caudal organization of the brainstem respiratory rhythm generator. Philos Trans R Soc Lond Ser B Biol Sci 364(1529):2469–2476. doi:10.1098/rstb.2009.0090

Champagnat J, Morin-Surun MP, Bouvier J, Thoby-Brisson M, Fortin G (2011) Prenatal development of central rhythm generation. Respir Physiol Neurobiol 178(1):146–155. doi:10.1016/j.resp.2011.04.013

Chatonnet F, Wrobel LJ, Mezieres V, Pasqualetti M, Ducret S, Taillebourg E, Charnay P, Rijli FM, Champagnat J (2007) Distinct roles of Hoxa2 and Krox20 in the development of rhythmic neural networks controlling inspiratory depth, respiratory frequency, and jaw opening. Neural Dev 2:19. doi:10.1186/1749-8104-2-19. 1749-8104-2-19 [pii]

Chen BS, Peng H, Wu SN (2009) Dexmedetomidine, an alpha2-adrenergic agonist, inhibits neuronal delayed-rectifier potassium current and sodium current. Br J Anaesth 103(2):244–254. doi:10.1093/bja/aep107. aep107 [pii]

Cheng L, Chen CL, Luo P, Tan M, Qiu M, Johnson R, Ma Q (2003) Lmx1b, Pet-1, and Nkx2.2 coordinately specify serotonergic neurotransmitter phenotype. J Neurosci 23(31):9961–9967. 23/31/9961 [pii]

Coates EL, Li A, Nattie EE (1993) Widespread sites of brain stem ventilatory chemoreceptors. J Appl Physiol 75(1):5–14

Corcoran AE, Milsom WK (2009) Maturational changes in pontine and medullary alpha-adrenoceptor influences on respiratory rhythm generation in neonatal rats. Respir Physiol Neurobiol 165(2–3):195–201. doi:10.1016/j.resp.2008.11.009. S1569-9048(08)00324-8 [pii]

Corcoran AE, Richerson GB, Harris MB (2013) Serotonergic mechanisms are necessary for central respiratory chemoresponsiveness in situ. Respir Physiol Neurobiol 186(2):214–220. doi:10.1016/j.resp.2013.02.015

Craven SE, Lim KC, Ye W, Engel JD, de Sauvage F, Rosenthal A (2004) Gata2 specifies serotonergic neurons downstream of sonic hedgehog. Development 131(5):1165–1173. doi:10.1242/dev.01024. 131/5/1165[pii]

Cummings KJ, Li A, Deneris ES, Nattie EE (2010) Bradycardia in serotonin-deficient pet-1-/- mice: influence of respiratory dysfunction and hyperthermia over the first 2 postnatal weeks. Am J Physiol Regul Integr Comp Physiol 298(5):R1333–R1342. doi:10.1152/ajpregu.00110.2010. 298/5/R1333 [pii]

Del Negro CA, Koshiya N, Butera RJ Jr, Smith JC (2002a) Persistent sodium current, membrane properties and bursting behavior of pre-botzinger complex inspiratory neurons in vitro. J Neurophysiol 88(5):2242–2250. doi:10.1152/jn.00081.2002

Del Negro CA, Morgado-Valle C, Feldman JL (2002b) Respiratory rhythm: an emergent network property? Neuron 34(5):821–830. S0896627302007122 [pii]

Del Negro CA, Morgado-Valle C, Hayes JA, Mackay DD, Pace RW, Crowder EA, Feldman JL (2005) Sodium and calcium current-mediated pacemaker neurons and respiratory rhythm generation. J Neurosci 25(2):446–453. doi:10.1523/JNEUROSCI.2237-04.2005. 25/2/446 [pii]

Del Negro CA, Hayes JA, Pace RW, Brush BR, Teruyama R, Feldman JL (2010) Synaptically activated burst-generating conductances may underlie a group-pacemaker mechanism for respiratory rhythm generation in mammals. Prog Brain Res 187:111–136. doi:10.1016/B978-0-444-53613-6.00008-3. B978-0-444-53613-6.00008-3 [pii]

Di Pasquale E, Morin D, Monteau R, Hilaire G (1992) Serotonergic modulation of the respiratory rhythm generator at birth: an in vitro study in the rat. Neurosci Lett 143(1–2):91–95

Ding YQ, Marklund U, Yuan W, Yin J, Wegman L, Ericson J, Deneris E, Johnson RL, Chen ZF (2003) Lmx1b is essential for the development of serotonergic neurons. Nat Neurosci 6(9):933–938. doi:10.1038/nn1104. nn1104[pii]

Doi A, Ramirez JM (2008) Neuromodulation and the orchestration of the respiratory rhythm. Respir Physiol Neurobiol 164(1-2):96–104. doi:10.1016/j.resp.2008.06.007. S1569-9048(08)00166-3[pii]

van Doorninck JH, van Der Wees J, Karis A, Goedknegt E, Engel JD, Coesmans M, Rutteman M, Grosveld F, De Zeeuw CI (1999) GATA-3 is involved in the development of serotonergic neurons in the caudal raphe nuclei. J Neurosci 19(12):RC12

Dreshaj IA, Haxhiu MA, Martin RJ (1998) Role of the medullary raphe nuclei in the respiratory response to CO2. Respir Physiol 111(1):15–23

Dubreuil V, Ramanantsoa N, Trochet D, Vaubourg V, Amiel J, Gallego J, Brunet JF, Goridis C (2008) A human mutation in Phox2b causes lack of CO2 chemosensitivity, fatal central apnea, and specific loss of parafacial neurons. Proc Natl Acad Sci U S A 105(3):1067–1072. doi:10.1073/pnas.0709115105. 0709115105 [pii]

Elam M, Yao T, Thoren P, Svensson TH (1981) Hypercapnia and hypoxia: chemoreceptor-mediated control of locus coeruleus neurons and splanchnic, sympathetic nerves. Brain Res 222(2):373–381

Erickson JT, Shafer G, Rossetti MD, Wilson CG, Deneris ES (2007) Arrest of 5HT neuron differentiation delays respiratory maturation and impairs neonatal homeostatic responses to environmental challenges. Respir Physiol Neurobiol 159(1):85–101. doi:10.1016/j.resp.2007.06.002. S1569-9048(07)00173-5 [pii]

Erlichman JS, Li A, Nattie EE (1998) Ventilatory effects of glial dysfunction in a rat brain stem chemoreceptor region. J Appl Physiol (1985) 85(5):1599–1604

Errchidi S, Hilaire G, Monteau R (1990) Permanent release of noradrenaline modulates respiratory frequency in the newborn rat: an in vitro study. J Physiol 429:497–510

Errchidi S, Monteau R, Hilaire G (1991) Noradrenergic modulation of the medullary respiratory rhythm generator in the newborn rat: an in vitro study. J Physiol 443:477–498

Eugenin J, Nicholls JG (1997) Chemosensory and cholinergic stimulation of fictive respiration in isolated CNS of neonatal opossum. J Physiol Lond 501(Pt 2):425–437

Eugenin J, von Bernhardi R, Muller KJ, Llona I (2006) Development and pH sensitivity of the respiratory rhythm of fetal mice in vitro. Neuroscience 141(1):223–231. doi:10.1016/j.neuroscience.2006.03.046. S0306-4522(06)00420-9 [pii]

Feldman SR (2007) Looking beyond the borders of our specialty: the 2006 Clarence S. Livingood MD lecture. Dermatol Online J 13(4):20

Feldman JL, Cleland C (1982) Possible roles of pacemaker neurons in mammalian respiratory rhythmogenesis. In: Carpenter D (ed) Cellular pacemarkers. Wiley, New York, pp 104–128

Feldman JL, Mitchell GS, Nattie EE (2003) Breathing: rhythmicity, plasticity, chemosensitivity. Annu Rev Neurosci 26:239–266. doi:10.1146/annurev.neuro.26.041002.131103. 041002.131103[pii]

Feldman JL, Del Negro CA, Gray PA (2013) Understanding the rhythm of breathing: so near, yet so far. Annu Rev Physiol 75:423–452. doi:10.1146/annurev-physiol-040510-130049

Filosa JA, Dean JB, Putnam RW (2002) Role of intracellular and extracellular pH in the chemosensitive response of rat locus coeruleus neurones. J Physiol 541(Pt 2):493–509. PHY_14142 [pii]

Fortin G, Thoby-Brisson M (2009) Embryonic emergence of the respiratory rhythm generator. Respir Physiol Neurobiol 168(1–2):86–91. doi:10.1016/j.resp.2009.06.013. S1569-9048(09)00179-7 [pii]

Fortuna MG, Stornetta RL, West GH, Guyenet PG (2009) Activation of the retrotrapezoid nucleus by posterior hypothalamic stimulation. J Physiol 587(Pt 21):5121–5138. doi:10.1113/jphysiol.2009.176875. jphysiol.2009.176875 [pii]

Foster GA, Schultzberg M, Goldstein M, Hokfelt T (1985) Ontogeny of phenylethanolamine N-methyltransferase- and tyrosine hydroxylase-like immunoreactivity in presumptive adrenaline neurones of the foetal rat central nervous system. J Comp Neurol 236(3):348–381. doi:10.1002/cne.902360306

Fujii M, Arata A (2010) Adrenaline modulates on the respiratory network development. Adv Exp Med Biol 669:25–28. doi:10.1007/978-1-4419-5692-7_5

Funk GD, Smith JC, Feldman JL (1993) Generation and transmission of respiratory oscillations in medullary slices: role of excitatory amino acids. J Neurophysiol 70(4):1497–1515

Fuxe K (1965) Evidence for the existence of monoamine neurons in the central nervous system. 3. The monoamine nerve terminal. Z Zellforsch Mikrosk Anat 65:573–596

Gaspar P, Lillesaar C (2012) Probing the diversity of serotonin neurons. Philos Trans R Soc Lond Ser B Biol Sci 367(1601):2382–2394. doi:10.1098/rstb.2011.0378

Gaspar P, Cases O, Maroteaux L (2003) The developmental role of serotonin: news from mouse molecular genetics. Nat Rev Neurosci 4(12):1002–1012. doi:10.1038/nrn1256

Goridis C, Rohrer H (2002) Specification of catecholaminergic and serotonergic neurons. Nat Rev Neurosci 3(7):531–541. doi:10.1038/nrn871

Gourine AV, Kasymov V, Marina N, Tang F, Figueiredo MF, Lane S, Teschemacher AG, Spyer KM, Deisseroth K, Kasparov S (2010) Astrocytes control breathing through pH-dependent release of ATP. Science 329(5991):571–575. doi:10.1126/science.1190721. science.1190721 [pii]

Gray PA (2008) Transcription factors and the genetic organization of brain stem respiratory neurons. J Appl Physiol (1985) 104(5):1513–1521. doi:10.1152/japplphysiol.01383.2007. 01383.2007 [pii]

Gray PA, Hayes JA, Ling GY, Llona I, Tupal S, Picardo MC, Ross SE, Hirata T, Corbin JG, Eugenin J, Del Negro CA (2010) Developmental origin of preBotzinger complex respiratory neurons. J Neurosci 30(44):14883–14895. doi:10.1523/JNEUROSCI.4031-10.2010. 30/44/14883 [pii]

Greer JJ, Allan DW, Martin-Caraballo M, Lemke RP (1999) An overview of phrenic nerve and diaphragm muscle development in the perinatal rat. J Appl Physiol 86(3):779–786

Guyenet PG, Bayliss DA, Stornetta RL, Fortuna MG, Abbott SB, DePuy SD (2009) Retrotrapezoid nucleus, respiratory chemosensitivity and breathing automaticity. Respir Physiol Neurobiol 168(1–2):59–68. doi:10.1016/j.resp.2009.02.001. S1569-9048(09)00029-9 [pii]

Hendricks T, Francis N, Fyodorov D, Deneris ES (1999) The ETS domain factor pet-1 is an early and precise marker of central serotonin neurons and interacts with a conserved element in serotonergic genes. J Neurosci 19(23):10348–10356

Hendricks TJ, Fyodorov DV, Wegman LJ, Lelutiu NB, Pehek EA, Yamamoto B, Silver J, Weeber EJ, Sweatt JD, Deneris ES (2003) Pet-1 ETS gene plays a critical role in 5-HT neuron development and is required for normal anxiety-like and aggressive behavior. Neuron 37(2):233–247

Hilaire G (2006) Endogenous noradrenaline affects the maturation and function of the respiratory network: possible implication for SIDS. Auton Neurosci 126-127:320–331. doi:10.1016/j.autneu.2006.01.021. S1566-0702(06)00048-8 [pii]

Hilaire G, Monteau R, Errchidi S (1989) Possible modulation of the medullary respiratory rhythm generator by the noradrenergic A5 area: an in vitro study in the newborn rat. Brain Res 485(2):325–332. 0006-8993(89)90577-5 [pii]

Hilaire G, Viemari JC, Coulon P, Simonneau M, Bevengut M (2004) Modulation of the respiratory rhythm generator by the pontine noradrenergic A5 and A6 groups in rodents. Respir Physiol Neurobiol 143(2–3):187–197

Hillion J, Milne-Edwards JB, Catelon J, de Vitry F, Gros F, Hamon M (1993) Prenatal developmental expression of rat brain 5-HT1A receptor gene followed by PCR. Biochem Biophys Res Commun 191(3):991–997

Hirsch MR, Tiveron MC, Guillemot F, Brunet JF, Goridis C (1998) Control of noradrenergic differentiation and Phox2a expression by MASH1 in the central and peripheral nervous system. Development 125(4):599–608

Hodges MR, Richerson GB (2008) Contributions of 5-HT neurons to respiratory control: neuromodulatory and trophic effects. Respir Physiol Neurobiol 164(1–2):222–232. doi:10.1016/j. resp.2008.05.014. S1569-9048(08)00139-0 [pii]

Hodges MR, Klum L, Leekley T, Brozoski DT, Bastasic J, Davis S, Wenninger JM, Feroah TR, Pan LG, Forster HV (2004a) Effects on breathing in awake and sleeping goats of focal acidosis in the medullary raphe. J Appl Physiol (1985) 96(5):1815–1824. doi:10.1152/japplphysiol.00992.2003

Hodges MR, Opansky C, Qian B, Davis S, Bonis J, Bastasic J, Leekley T, Pan LG, Forster HV (2004b) Transient attenuation of CO2 sensitivity after neurotoxic lesions in the medullary raphe area of awake goats. J Appl Physiol 97(6):2236–2247

Hodges MR, Tattersall GJ, Harris MB, McEvoy SD, Richerson DN, Deneris ES, Johnson RL, Chen ZF, Richerson GB (2008) Defects in breathing and thermoregulation in mice with near-complete absence of central serotonin neurons. J Neurosci 28(10):2495–2505. doi:10.1523/JNEUROSCI.4729-07.2008. 28/10/2495 [pii]

Hodges MR, Wehner M, Aungst J, Smith JC, Richerson GB (2009) Transgenic mice lacking serotonin neurons have severe apnea and high mortality during development. J Neurosci 29(33):10341–10349. doi:10.1523/JNEUROSCI.1963-09.2009. 29/33/10341 [pii]

Hodges MR, Best S, Richerson GB (2011) Altered ventilatory and thermoregulatory control in male and female adult Pet-1 null mice. Respir Physiol Neurobiol 177(2):133–140. doi:10.1016/j.resp.2011.03.020

Holtman JR Jr (1988) Immunohistochemical localization of serotonin- and substance P-containing fibers around respiratory muscle motoneurons in the nucleus ambiguus of the cat. Neuroscience 26(1):169–178

Hunt PN, McCabe AK, Bosma MM (2005) Midline serotonergic neurones contribute to widespread synchronized activity in embryonic mouse hindbrain. J Physiol 566(Pt 3):807–819. doi:10.1113/jphysiol.2005.089581

Infante CD, von Bernhardi R, Rovegno M, Llona I, Eugenin JL (2003) Respiratory responses to pH in the absence of pontine and dorsal medullary areas in the newborn mouse in vitro. Brain Res 984(1–2):198–205

Jacquin TD, Borday V, Schneider-Maunoury S, Topilko P, Ghilini G, Kato F, Charnay P, Champagnat J (1996) Reorganization of pontine rhythmogenic neuronal networks in Krox-20 knockout mice. Neuron 17(4):747–758

Janczewski WA, Feldman JL (2006) Distinct rhythm generators for inspiration and expiration in the juvenile rat. J Physiol 570(Pt 2):407–420. doi:10.1113/jphysiol.2005.098848. jphysiol.2005.098848 [pii]

Jean-Charles V, Gerard H (2002) Noradrenergic receptors and in vitro respiratory rhythm: possible interspecies differences between mouse and rat neonates. Neurosci Lett 324(2):149–153

Johnson SM, Haxhiu MA, Richerson GB (2008) GFP-expressing locus ceruleus neurons from Prp57 transgenic mice exhibit CO2/H+ responses in primary cell culture. J Appl Physiol (1985) 105(4):1301–1311. doi:10.1152/japplphysiol.90414.2008

Kanamaru M, Homma I (2007) Compensatory airway dilation and additive ventilatory augmentation mediated by dorsomedial medullary 5-hydroxytryptamine 2 receptor activity and hypercapnia. Am J Physiol Regul Integr Comp Physiol 293(2):R854–R860. doi:10.1152/ajpregu.00829.2006. 00829.2006 [pii]

Kodama D, Togari A (2010) Modulation of potassium channels via the alpha1B-adrenergic receptor in human osteoblasts. Neurosci Lett 485(2):102–106. doi:10.1016/j.neulet.2010.08.073. S0304-3940(10)01153-5 [pii]

Koizumi H, Smith JC (2008) Persistent Na+ and K+-dominated leak currents contribute to respiratory rhythm generation in the pre-Botzinger complex in vitro. J Neurosci 28(7):1773–1785. doi:10.1523/JNEUROSCI.3916-07.2008. 28/7/1773 [pii]

Kubli FW, Hon EH, Khazin AF, Takemura H (1969) Observations on heart rate and pH in the human fetus during labor. Am J Obstet Gynecol 104(8):1190–1206

Lalley PM, Bischoff AM, Richter DW (1994) 5-HT-1A receptor-mediated modulation of medullary expiratory neurones in the cat. J Physiol 476(1):117–130

Lauder JM, Bloom FE (1975) Ontogeny of monoamine neurons in the locus coeruleus, raphe nuclei and substantia nigra of the rat. II. Synaptogenesis. J Comp Neurol 163(3):251–264. doi:10.1002/cne.901630302

Li C, Horn JP (2008) Differential inhibition of Ca2+ channels by alpha2-adrenoceptors in three functional subclasses of rat sympathetic neurons. J Neurophysiol 100(6):3055–3063. doi:10.1152/jn.90590.2008. 90590.2008 [pii]

Li A, Nattie E (2006) Catecholamine neurones in rats modulate sleep, breathing, central chemoreception and breathing variability. J Physiol 570(Pt 2):385–396

Li Z, Morris KF, Baekey DM, Shannon R, Lindsey BG (1999) Responses of simultaneously recorded respiratory-related medullary neurons to stimulation of multiple sensory modalities. J Neurophysiol 82(1):176–187

Lu B, Su Y, Das S, Wang H, Wang Y, Liu J, Ren D (2009) Peptide neurotransmitters activate a cation channel complex of NALCN and UNC-80. Nature 457(7230):741–744. doi:10.1038/nature07579. nature07579 [pii]

Lu B, Zhang Q, Wang H, Wang Y, Nakayama M, Ren D (2010) Extracellular calcium controls background current and neuronal excitability via an UNC79-UNC80-NALCN cation channel complex. Neuron 68(3):488–499. doi:10.1016/j.neuron.2010.09.014. S0896-6273(10)00756-7 [pii]

Lumsden A (1999) Closing in on rhombomere boundaries. Nat Cell Biol 1:E83–E85

Manzke T, Guenther U, Ponimaskin EG, Haller M, Dutschmann M, Schwarzacher S, Richter DW (2003) 5-HT4(a) receptors avert opioid-induced breathing depression without loss of analgesia. Science 301(5630):226–229. doi:10.1126/science.1084674. 301/5630/226 [pii]

Manzke T, Preusse S, Hulsmann S, Richter DW (2008) Developmental changes of serotonin 4(a) receptor expression in the rat pre-Botzinger complex. J Comp Neurol 506(5):775–790. doi:10.1002/cne.21581

Meillerais A, Champagnat J, Morin-Surun MP (2010) Extracellular calcium induces quiescence of the low-frequency embryonic motor rhythm in the mouse isolated brainstem. J Neurosci Res 88(16):3555–3565. doi:10.1002/jnr.22518

Messier ML, Li A, Nattie EE (2002) Muscimol inhibition of medullary raphe neurons decreases the CO2 response and alters sleep in newborn piglets. Respir Physiol Neurobiol 133(3):197–214

Messier ML, Li A, Nattie EE (2004) Inhibition of medullary raphe serotonergic neurons has age-dependent effects on the CO2 response in newborn piglets. J Appl Physiol 96(5):1909–1919

Meyer LC, Fuller A, Mitchell D (2006) Zacopride and 8-OH-DPAT reverse opioid-induced respiratory depression and hypoxia but not catatonic immobilization in goats. Am J Physiol Regul Integr Comp Physiol 290(2):R405–R413. doi:10.1152/ajpregu.00440.2005. 00440.2005 [pii]

Morgado-Valle C, Baca SM, Feldman JL (2010) Glycinergic pacemaker neurons in pre-Botzinger complex of neonatal mouse. J Neurosci 30(10):3634–3639. doi:10.1523/JNEUROSCI.3040-09.2010. 30/10/3634 [pii]

Morin X, Cremer H, Hirsch MR, Kapur RP, Goridis C, Brunet JF (1997) Defects in sensory and autonomic ganglia and absence of locus coeruleus in mice deficient for the homeobox gene Phox2a. Neuron 18(3):411–423

Mulkey DK, Talley EM, Stornetta RL, Siegel AR, West GH, Chen X, Sen N, Mistry AM, Guyenet PG, Bayliss DA (2007) TASK channels determine pH sensitivity in select respiratory neurons but do not contribute to central respiratory chemosensitivity. J Neurosci 27(51):14049–14058. doi:10.1523/JNEUROSCI.4254-07.2007. 27/51/14049 [pii]

Nattie E (1999) CO2, brainstem chemoreceptors and breathing. Prog Neurobiol 59(4):299–331

Nattie EE, Li A (2001) CO2 dialysis in the medullary raphe of the rat increases ventilation in sleep. J Appl Physiol 90(4):1247–1257

Nattie E, Li A (2012) Central chemoreceptors: locations and functions. Compr Physiol 2(1):221–254. doi:10.1002/cphy.c100083

Nattie EE, Li A, Richerson GB, Lappi DA (2004) Medullary serotonergic neurones and adjacent neurones that express neurokinin-1 receptors are both involved in chemoreception in vivo. J Physiol 556(Pt 1):235–253. doi:10.1113/jphysiol.2003.059766

Niebert M, Vogelgesang S, Koch UR, Bischoff AM, Kron M, Bock N, Manzke T (2011) Expression and function of serotonin 2A and 2B receptors in the mammalian respiratory network. PLoS One 6(7):e21395. doi:10.1371/journal.pone.0021395. PONE-D-11-03965 [pii]

Olson L, Seiger A (1972) Early prenatal ontogeny of central monoamine neurons in the rat: fluorescence histochemical observations. Z Anat Entwicklungsgesch 137(3):301–316

Onimaru H, Homma I (2003) A novel functional neuron group for respiratory rhythm generation in the ventral medulla. J Neurosci 23(4):1478–1486. 23/4/1478 [pii]

Oyamada Y, Ballantyne D, Muckenhoff K, Scheid P (1998) Respiration-modulated membrane potential and chemosensitivity of locus coeruleus neurones in the in vitro brainstem-spinal cord of the neonatal rat. J Physiol 513(Pt 2):381–398

Pace RW, Mackay DD, Feldman JL, Del Negro CA (2007) Inspiratory bursts in the preBotzinger complex depend on a calcium-activated non-specific cation current linked to glutamate receptors in neonatal mice. J Physiol 582(Pt 1):113–125. doi:10.1113/jphysiol.2007.133660. jphysiol.2007.133660 [pii]

Pagliardini S, Ren J, Greer JJ (2003) Ontogeny of the pre-Botzinger complex in perinatal rats. J Neurosci 23(29):9575–9584. 23/29/9575 [pii]

Pagliardini S, Ren J, Gray PA, Vandunk C, Gross M, Goulding M, Greer JJ (2008) Central respiratory rhythmogenesis is abnormal in lbx1- deficient mice. J Neurosci 28(43):11030–11041. doi:10.1523/JNEUROSCI.1648-08.2008. 28/43/11030 [pii]

Pattyn A, Goridis C, Brunet JF (2000) Specification of the central noradrenergic phenotype by the homeobox gene Phox2b. Mol Cell Neurosci 15(3):235–243. doi:10.1006/mcne.1999.0826. S1044-7431(99)90826-6[pii]

Pattyn A, Vallstedt A, Dias JM, Samad OA, Krumlauf R, Rijli FM, Brunet JF, Ericson J (2003) Coordinated temporal and spatial control of motor neuron and serotonergic neuron generation from a common pool of CNS progenitors. Genes Dev 17(6):729–737. doi:10.1101/gad.255803

Pena F, Parkis MA, Tryba AK, Ramirez JM (2004) Differential contribution of pacemaker properties to the generation of respiratory rhythms during normoxia and hypoxia. Neuron 43(1):105–117. doi:10.1016/j.neuron.2004.06.023. S0896627304003927[pii]

Pilowsky PM, Jiang C, Lipski J (1990) An intracellular study of respiratory neurons in the rostral ventrolateral medulla of the rat and their relationship to catecholamine-containing neurons. J Comp Neurol 301(4):604–617. doi:10.1002/cne.903010409

Prinz AA, Bucher D, Marder E (2004) Similar network activity from disparate circuit parameters. Nat Neurosci 7(12):1345–1352. doi:10.1038/nn1352. nn1352 [pii]

Qian Y, Fritzsch B, Shirasawa S, Chen CL, Choi Y, Ma Q (2001) Formation of brainstem (nor) adrenergic centers and first-order relay visceral sensory neurons is dependent on homeodomain protein Rnx/Tlx3. Genes Dev 15(19):2533–2545. doi:10.1101/gad.921501

Ramanantsoa N, Hirsch MR, Thoby-Brisson M, Dubreuil V, Bouvier J, Ruffault PL, Matrot B, Fortin G, Brunet JF, Gallego J, Goridis C (2011) Breathing without CO(2) chemosensitivity in conditional Phox2b mutants. J Neurosci 31(36):12880–12888. doi:10.1523/JNEUROSCI.1721-11.2011. 31/36/12880 [pii]

Ramirez JM, Koch H, Garcia AJ 3rd, Doi A, Zanella S (2011) The role of spiking and bursting pacemakers in the neuronal control of breathing. J Biol Phys 37(3):241–261. doi:10.1007/s10867-011-9214-z. 9214 [pii]

Ramirez JM, Doi A, Garcia AJ 3rd, Elsen FP, Koch H, Wei AD (2012) The cellular building blocks of breathing. Compr Physiol 2(4):2683–2731. doi:10.1002/cphy.c110033

Ray RS, Corcoran AE, Brust RD, Kim JC, Richerson GB, Nattie E, Dymecki SM (2011) Impaired respiratory and body temperature control upon acute serotonergic neuron inhibition. Science 333(6042):637–642. doi:10.1126/science.1205295

Ray RS, Corcoran AE, Brust RD, Soriano LP, Nattie EE, Dymecki SM (2013) Egr2-neurons control the adult respiratory response to hypercapnia. Brain Res 1511:115–125. doi:10.1016/j.brainres.2012.12.017

Rekling JC, Feldman JL (1998) PreBotzinger complex and pacemaker neurons: hypothesized site and kernel for respiratory rhythm generation. Annu Rev Physiol 60:385–405. doi:10.1146/annurev.physiol.60.1.385

Rekling JC, Shao XM, Feldman JL (2000) Electrical coupling and excitatory synaptic transmission between rhythmogenic respiratory neurons in the preBotzinger complex. J Neurosci 20(23):RC113

Richardson L, Venkataraman S, Stevenson P, Yang Y, Moss J, Graham L, Burton N, Hill B, Rao J, Baldock RA, Armit C (2014) EMAGE mouse embryo spatial gene expression database: 2014 update. Nucleic Acids Res 42(Database issue):D835–D844. doi:10.1093/nar/gkt1155

Richerson GB (1995) Response to CO2 of neurons in the rostral ventral medulla in vitro. J Neurophysiol 73(3):933–944

Richerson GB (2004) Serotonergic neurons as carbon dioxide sensors that maintain pH homeostasis. Nat Rev Neurosci 5(6):449–461. doi:10.1038/nrn1409. nrn1409[pii]

Richter DW, Manzke T, Wilken B, Ponimaskin E (2003) Serotonin receptors: guardians of stable breathing. Trends Mol Med 9(12):542–548

Sahibzada N, Ferreira M, Wasserman AM, Taveira-DaSilva AM, Gillis RA (2000) Reversal of morphine-induced apnea in the anesthetized rat by drugs that activate 5-hydroxytryptamine(1A) receptors. J Pharmacol Exp Ther 292(2):704–713

Shideler KK, Yan J (2010) M1 muscarinic receptor for the development of auditory cortical function. Mol Brain 3:29. doi:10.1186/1756-6606-3-29. 1756-6606-3-29 [pii]

Shirasawa S, Arata A, Onimaru H, Roth KA, Brown GA, Horning S, Arata S, Okumura K, Sasazuki T, Korsmeyer SJ (2000) Rnx deficiency results in congenital central hypoventilation. Nat Genet 24(3):287–290. doi:10.1038/73516

Smith JC, Ellenberger HH, Ballanyi K, Richter DW, Feldman JL (1991) Pre-Botzinger complex: a brainstem region that may generate respiratory rhythm in mammals. Science 254(5032):726–729

Smith JC, Butera RJ, Koshiya N, Del Negro C, Wilson CG, Johnson SM (2000) Respiratory rhythm generation in neonatal and adult mammals: the hybrid pacemaker-network model. Respir Physiol 122(2–3):131–147

Solomon IC (2003) Focal CO2/H+ alters phrenic motor output response to chemical stimulation of cat pre-Botzinger complex in vivo. J Appl Physiol 94(6):2151–2157. doi:10.1152/japplphysiol.01192.2002. 01192.2002 [pii]

Soulier V, Peyronnet J, Pequignot JM, Cottet-Emard JM, Lagercrantz H, Dalmaz Y (1997) Long-term impairment in the neurochemical activity of the sympathoadrenal system after neonatal hypoxia in the rat. Pediatr Res 42(1):30–38. doi:10.1203/00006450-199707000-00006

Steindler DA, Trosko BK (1989) Two types of locus coeruleus neurons born on different embryonic days in the mouse. Anat Embryol (Berl) 179(5):423–434

St-John WM (2008) Eupnea of in situ rats persists following blockers of in vitro pacemaker burster activities. Respir Physiol Neurobiol 160(3):353–356. doi:10.1016/j.resp.2007.12.002. S1569-9048(07)00339-4 [pii]

Stornetta RL, Moreira TS, Takakura AC, Kang BJ, Chang DA, West GH, Brunet JF, Mulkey DK, Bayliss DA, Guyenet PG (2006) Expression of Phox2b by brainstem neurons involved in chemosensory integration in the adult rat. J Neurosci 26(40):10305–10314. doi:10.1523/JNEUROSCI.2917-06.2006. 26/40/10305 [pii]

Stunden CE, Filosa JA, Garcia AJ, Dean JB, Putnam RW (2001) Development of in vivo ventilatory and single chemosensitive neuron responses to hypercapnia in rats. Respir Physiol 127(2–3):135–155

Takakura AC, Moreira TS, Colombari E, West GH, Stornetta RL, Guyenet PG (2006) Peripheral chemoreceptor inputs to retrotrapezoid nucleus (RTN) CO2-sensitive neurons in rats. J Physiol 572(Pt 2):503–523. doi:10.1113/jphysiol.2005.103788. jphysiol.2005.103788 [pii]

Taylor NC, Li A, Nattie EE (2006) Ventilatory effects of muscimol microdialysis into the rostral medullary raphe region of conscious rats. Respir Physiol Neurobiol 153(3):203–216

Teppema LJ, Veening JG, Kranenburg A, Dahan A, Berkenbosch A, Olievier C (1997) Expression of c-fos in the rat brainstem after exposure to hypoxia and to normoxic and hyperoxic hypercapnia. J Comp Neurol 388(2):169–190

Thoby-Brisson M, Ramirez JM (2001) Identification of two types of inspiratory pacemaker neurons in the isolated respiratory neural network of mice. J Neurophysiol 86(1):104–112

Thoby-Brisson M, Trinh JB, Champagnat J, Fortin G (2005) Emergence of the pre-Botzinger respiratory rhythm generator in the mouse embryo. J Neurosci 25(17):4307–4318. doi:10.1523/JNEUROSCI.0551-05.2005. 25/17/4307 [pii]

Thoby-Brisson M, Karlen M, Wu N, Charnay P, Champagnat J, Fortin G (2009) Genetic identification of an embryonic parafacial oscillator coupling to the preBotzinger complex. Nat Neurosci 12(8):1028–1035. doi:10.1038/nn.2354. nn.2354 [pii]

Tryba AK, Pena F, Lieske SP, Viemari JC, Thoby-Brisson M, Ramirez JM (2008) Differential modulation of neural network and pacemaker activity underlying eupnea and sigh-breathing activities. J Neurophysiol 99(5):2114–2125. doi:10.1152/jn.01192.2007. 01192.2007 [pii]

Viemari JC, Ramirez JM (2006) Norepinephrine differentially modulates different types of respiratory pacemaker and nonpacemaker neurons. J Neurophysiol 95(4):2070–2082. doi:10.1152/jn.01308.2005. 01308.2005 [pii]

Viemari JC, Burnet H, Bevengut M, Hilaire G (2003) Perinatal maturation of the mouse respiratory rhythm-generator: in vivo and in vitro studies. Eur J Neurosci 17(6):1233–1244

Viemari JC, Bevengut M, Burnet H, Coulon P, Pequignot JM, Tiveron MC, Hilaire G (2004a) Phox2a gene, A6 neurons, and noradrenaline are essential for development of normal respiratory rhythm in mice. J Neurosci 24(4):928–937. doi:10.1523/JNEUROSCI.3065-03.2004. 24/4/928[pii]

Viemari JC, Bevengut M, Coulon P, Hilaire G (2004b) Nasal trigeminal inputs release the A5 inhibition received by the respiratory rhythm generator of the mouse neonate. J Neurophysiol 91(2):746–758. doi:10.1152/jn.01153.2002. 01153.2002[pii]

Viemari JC, Garcia AJ 3rd, Doi A, Ramirez JM (2011) Activation of alpha-2 noradrenergic receptors is critical for the generation of fictive eupnea and fictive gasping inspiratory activities in mammals in vitro. Eur J Neurosci 33(12):2228–2237. doi:10.1111/j.1460-9568.2011.07706.x

Wallen-Mackenzie A, Gezelius H, Thoby-Brisson M, Nygard A, Enjin A, Fujiyama F, Fortin G, Kullander K (2006) Vesicular glutamate transporter 2 is required for central respiratory rhythm generation but not for locomotor central pattern generation. J Neurosci 26(47):12294–12307. doi:10.1523/JNEUROSCI.3855-06.2006. 26/47/12294 [pii]

Wang W, Richerson GB (1999) Development of chemosensitivity of rat medullary raphe neurons. Neuroscience 90(3):1001–1011

Wang W, Pizzonia JH, Richerson GB (1998) Chemosensitivity of rat medullary raphe neurones in primary tissue culture. J Physiol 511(Pt 2):433–450

Wang W, Tiwari JK, Bradley SR, Zaykin RV, Richerson GB (2001) Acidosis-stimulated neurons of the medullary raphe are serotonergic. J Neurophysiol 85(5):2224–2235

Wang S, Xu DJ, Cai JB, Huang YZ, Zou JG, Cao KJ (2009) Rapid component I(Kr) of cardiac delayed rectifier potassium currents in guinea-pig is inhibited by alpha(1)-adrenoreceptor activation via protein kinase A and protein kinase C-dependent pathways. Eur J Pharmacol 608(1–3):1–6. doi:10.1016/j.ejphar.2009.02.017. S0014-2999(09)00156-3 [pii]

Wang CC, Borchert A, Ugun-Klusek A, Tang LY, Lui WT, Chu CY, Billett E, Kuhn H, Ufer C (2011a) Monoamine oxidase a expression is vital for embryonic brain development by modulating developmental apoptosis. J Biol Chem 286(32):28322–28330. doi:10.1074/jbc.M111.241422. M111.241422 [pii]

Wang ZQ, Chen K, Ying QL, Li P, Shih JC (2011b) Monoamine oxidase A regulates neural differentiation of murine embryonic stem cells. J Neural Transm 118(7):997–1001. doi:10.1007/s00702-011-0655-0

Wenker IC, Kreneisz O, Nishiyama A, Mulkey DK (2010) Astrocytes in the retrotrapezoid nucleus sense H+ by inhibition of a Kir4.1-Kir5.1-like current and may contribute to chemoreception by a purinergic mechanism. J Neurophysiol 104(6):3042–3052. doi:10.1152/jn.00544.2010. jn.00544.2010 [pii]

Wu C, Dias P, Kumar S, Lauder JM, Singh S (1999) Differential expression of serotonin 5-HT2 receptors during rat embryogenesis. Dev Neurosci 21(1):22–28. 17362 [pii]

Yamamoto S, Kanno T, Yamada K, Yasuda Y, Nishizaki T (2009) Dual regulation of heat-activated K+ channel in rat DRG neurons via alpha(1) and beta adrenergic receptors. Life Sci 85(3–4):167–171. doi:10.1016/j.lfs.2009.05.009. S0024-3205(09)00227-6 [pii]

Yew DT, Chan WY (1999) Early appearance of acetylcholinergic, serotoninergic, and peptidergic neurons and fibers in the developing human central nervous system. Microsc Res Tech 45(6):389–400. doi:10.1002/(SICI)1097-0029(19990615)45:6<389::AID-JEMT6>3.0.CO;2-Z. [pii]10.1002/(SICI)1097-0029(19990615)45:6<389::AID-JEMT6>3.0.CO;2-Z

Zanella S, Roux JC, Viemari JC, Hilaire G (2006) Possible modulation of the mouse respiratory rhythm generator by A1/C1 neurones. Respir Physiol Neurobiol 153(2):126–138. doi:10.1016/j. resp.2005.09.009. S1569-9048(05)00257-0 [pii]

Zavala-Tecuapetla C, Aguileta MA, Lopez-Guerrero JJ, Gonzalez-Marin MC, Pena F (2008) Calcium-activated potassium currents differentially modulate respiratory rhythm generation. Eur J Neurosci 27(11):2871–2884. doi:10.1111/j.1460-9568.2008.06214.x. EJN6214 [pii]

Zhao ZQ, Scott M, Chiechio S, Wang JS, Renner KJ, Gereau RW, Johnson RL, Deneris ES, Chen ZF (2006) Lmx1b is required for maintenance of central serotonergic neurons and mice lacking central serotonergic system exhibit normal locomotor activity. J Neurosci 26(49):12781–12788. doi:10.1523/JNEUROSCI.4143-06.2006. 26/49/12781 [pii]

Chapter 11
Neurodevelopmental Effects of Serotonin on the Brainstem Respiratory Network

Karina Bravo, Jaime Eugenín, and Isabel Llona

Abstract Serotonin has multiple roles during development of the nervous system. Human pathologies, mouse genetic models, and pharmacological experiments have demonstrated a role of serotonin in the development of neural networks. Here we summarize evidence showing that serotonin is important for the brainstem respiratory network. The available data highlight the role of serotonin as a developmental signal that previously has not been specifically considered for the respiratory network.

Keywords Chemoreception • Breathing • Fluoxetine • 5-HT • Development

Abbreviations

5-HT	5-hydroxytryptamine, serotonin
5-HTergic	Serotonergic
5-HTP	5-hydroxytryptophan
5-HTR	Serotonin receptors
BDNF	Brain-derived neurotrophic factor
E	Embryonic day
KO	Knock out
MAO-A	Monoamine oxidase A
NTS	Nucleus tractus solitarii
P	Postnatal day
PCPA	*Para*-chlorophenylalanine
pFRG	Parafacial respiratory group
preBötC	preBötzinger complex
RPG	Respiratory pattern generator

K . Bravo, PhD • J. Eugenín, MD, PhD • I. Llona, PhD (✉)
Laboratorio de Sistemas Neurales, Facultad de Química y Biología, Departamento de Biología, Universidad de Santiago de Chile, USACH, PO 9170022 Santiago, Chile
e-mail: isabel.llona@usach.cl

© Springer International Publishing AG 2017
R. von Bernhardi et al. (eds.), *The Plastic Brain*, Advances in Experimental Medicine and Biology 1015, DOI 10.1007/978-3-319-62817-2_11

RTN   Retrotrapezoid nucleus
SERT   Serotonin transporter
SIDS   Sudden infant death syndrome
SSRIs   Selective serotonin reuptake inhibitors
TPH   Tryptophan hydroxylase
Trp   Tryptophan
VMAT   Vesicular monoamine transporter
XII   Hypoglossal nucleus

Introduction

Serotonin (5-hydroxytryptamine, 5-HT) is a molecule with multiple functions as a neurotransmitter in the central nervous system as well as an autacoid in the periphery. Its synthesis is catalyzed by the enzyme tryptophan hydroxylase (TPH1 and TPH2) and its degradation by monoamine oxidase A (MAO-A). 5-HT is stored in vesicles by the vesicular monoamine transporter (VMAT2). Levels of 5-HT in the synaptic cleft are regulated by reuptake through the 5-HT transporter (SERT). Besides its role in synaptic signaling, 5-HT appears to be involved in a wide range of processes during mammalian development (Gaspar et al. 2003; Kepser and Homberg 2015). 5-HT modulates brain embryonic development by participating in cell proliferation (Klempin et al. 2013), apoptosis (Liu et al. 2013), cell and tissue regeneration, and cell migration (Liu et al. 2011), among others. 5-HT related molecules such as enzymes that are responsible for 5-HT synthesis and breakdown, 5-HT receptors, and SERT are expressed in the brain even before 5-HT neurons are born (Côté et al. 2007; Bonnin and Levitt 2011).

Developmental defects have been linked to disruption of the 5-HT system either by genetic or pharmacological manipulations (Narboux-Neme et al. 2013; Kepser and Homberg 2015). The roles of 5-HT and its signaling molecules during development are especially important to consider in light of recent findings of effects of 5-HT reuptake inhibitors (SSRIs) during pregnancy (Alwan and Friedman 2009; Oberlander et al. 2009; Homberg et al. 2010; Gentile and Galbally 2011; Velasquez et al. 2013; Bravo et al. 2016). SSRIs given to the pregnant mother to treat depression will increase extracellular 5-HT not only in the mother but also in the brain of the unborn child (Gentile and Galbally 2011). These children acquire an increased risk to develop reduced somatosensory responses (Oberlander et al. 2009) and/or psychomotor control, and appear to have a higher risk to develop autism-like symptoms (Gentile 2015).

As already described in Chaps. 8, 9 and 10, a neural network located in the brainstem generates respiration. It receives tonic input from such sensory modalities as central and peripheral chemoreceptors, thermoreceptors, and mechanoreceptors. Its output controls cranial- and spinal motoneurons that innervate the respiratory muscles. Serotonergic neurons originating in the raphe obscurus nucleus modulate respiration in normal and pathophysiologic conditions (Feldman et al. 2003; Paterson et al. 2006). Many genetic models where 5-HT is increased or decreased have significant consequences for the development of the respiratory network (Burnet et al. 2001; Erickson et al. 2007; Hodges et al. 2008; Li and Nattie 2008; Zanella et al. 2009). In the present chapter, we will summarize and discuss recent data on the developmental role of 5-HT in the brainstem respiratory network.

Ontogeny of Serotonergic Systems

The 5-HT phenotype is determined by sequential expression of several transcription factors. Transcription factors Nkx2.2 and Mash1 expressed in mice at embryonic day 8–10 (E8-E10) are required to generate precursors of the 5-HT phenotype in the raphe nucleus (Alenina et al. 2006). The expression of transcription factor Phox2b at E6.5 inhibits differentiation of the serotonergic neurons. However, Phox2b is switched off at E10.5 by the expression of Nkx2.2. Then, a second group of transcription factors is required between E10-E12 for differentiation of 5-HT neurons (Cordes 2005). In this second group, Lmx1b is a critical intermediate factor that couples Nkx2–2–mediated early specification with Pet1-mediated terminal differentiation (Ding et al. 2003). Pet-1 expression is restricted to the raphe nuclei and it is induced specifically in postmitotic 5-HT neuron precursors 0.5 days before brain 5-HT synthesis starts (Hendricks et al. 1999), making Pet-1 transcription an early marker of the serotonergic phenotype. Other members of this group of transcription factors, like Gata2 and Gata3, give specificity to 5-HT neurons (Alenina et al. 2006). Eng1 transcription factor expressed in serotonergic postmitotic neurons is needed for the maintenance of serotonergic phenotype (Fox and Deneris 2012). During development, Lmxb1b expression precedes Pet-1. Mutations of these transcription factors have been used to generate mice with different degrees of serotonergic neuron deficiency (Deneris 2011). The Lmx1b null mice are totally lacking in 5-HT neurons (Ding et al. 2003), while Pet-1 null mice lose 70–80% of 5-HT neurons (Hendricks et al. 1999).

As described in Chap. 10, the serotonergic neuronal somata are grouped in the raphe nuclei and project to practically all areas of the central nervous system (Jacobs and Azmitia 1992). The raphe nucleus is located in the midline of the brainstem and is divided into caudal and rostral raphe (Dahlstrom and Fuxe 1964). Raphe neurons also release neuropeptides like thyrotropin releasing hormone, somatostatin, and substance P (Covenas et al. 2001). The raphe nucleus receives inputs from the prelimbic and infralimbic cortex, dorsal hypothalamic area, preoptic area, midbrain periaqueductual nucleus, amygdala, and the pontine subcoeruleus area (Jacobs and Azmitia 1992). Raphe neurons have projections to various brain stem nuclei of the respiratory network, like the nucleus tractus solitarii (NTS), the nucleus ambiguous, the preBötzinger complex (preBötC), the hypoglossal nucleus (XII) (Manaker and Tischler 1993), and also to the phrenic nerve (Hosogai et al. 1998). The nucleus raphe obscurus modulates XII output, providing a tonic excitatory drive to hypoglossal motoneurons (Peever and Duffin 2001).

The serotonergic system appears early in embryonic development (Murrin et al. 2007). Neurons able to synthesize 5-HT can be observed between E10 and E13 in rats in the rostroventrolateral medulla (Aitken and Tork 1988) and serotonergic projections in the ventral portion of the brainstem are present from E13.5 in mice. SERT mRNA can be detected between E10 and E13 in the raphe nucleus (Hansson et al. 1999). This early onset of serotonergic systems during embryonic development coincides with the emergence of the respiratory rhythm (see Chap. 10). In mice, a respiratory rhythm that consists of firing of short duration can be recorded from cranial level C3-C4 in spinal cord beginning at E13 (Eugenin et al. 2003). At E18, neural networks controlling the generation of breathing are capable of maintaining survival ex utero. However, the rhythm is highly irregular (Di Pasquale et al. 1996).

5-HT and the Respiratory Network

The respiratory rhythm is generated by a neural network, the respiratory pattern generator (RPG), located in the ventral respiratory column of the brainstem (Feldman et al. 2013; Smith et al. 2013; Richter and Smith 2014). Actually, it is thought that the inspiratory activity of the preBötC oscillator coupled to the pre-inspiratory and post-inspiratory parafacial oscillator (pFRG) are the source of the respiratory rhythm (Feldman et al. 2013; Smith et al. 2013; Richter and Smith 2014). The RPG commands the synchronous and rhythmic discharge of spinal and cranial motoneurons innervating the respiratory muscles. Signal integration by the RPG enables respiration to match various physiological demands. There are many neurotransmitters and neuromodulators that interact at the RPG to generate and control breathing (Doi and Ramirez 2008). 5-HT is involved in normal functioning of the respiratory neural network as a neuromodulator (Hodges and Richerson 2008).

A subpopulation of the serotonergic neurons localized in the raphe nucleus is closely associated with medullary arteries. These serotonergic neurons likely are detectors of CO_2/H^+ in perivascular nervous tissue and cerebrospinal fluid and contribute to central chemoreception modulating respiratory rhythm generation (Nattie and Li 2010; Dreshaj et al. 1998; Teran et al. 2014). Acidified artificial cerebrospinal fluid applied in raphe nucleus increases the amplitude of phrenic nerve discharge in anesthetized cats and rats. In addition, hypercapnia (increase in PCO_2) increases the firing rate of neurons of the caudal raphe nucleus in rats (Taylor et al. 2004; Richerson 2004). Furthermore, the optogenetic stimulation of raphe neurons stimulates breathing (Depuy et al. 2011).

Also, when an acidic stimulus is applied to the locus coeruleus, NTS, ventrolateral bulbar region, and rostral ventral medulla, phrenic nerve discharge increases. Thus, central chemoreception is distributed across several brainstem nuclei and it modulates respiratory rhythmogenesis (Coates et al. 1993; Douglas et al. 2001; Richerson 2004; Nattie and Li 2010).

Likewise, in vivo studies have shown breathing changes, when serotonergic agents are applied (Holtman et al. 1986; Lalley 1986; Mitchell et al. 1992). Thereby, peripheral administration of a $5\text{-HT}_{2A/2C}$ receptor agonist in neonatal rats causes respiratory depression (Cayetanot et al. 2001). The inhibition of 5-HT synthesis with *para*-chlorophenylalanine (PCPA) in conscious unrestrained rats increases respiratory frequency, ventilation, and the ventilatory response to CO_2 (Olson et al. 1979); hiperventilation that can be reversed by administering 5-hydroxytryptophan (5-HTP), a precursor of 5-HT. Furthermore, 5-HTP depresses the ventilation, an effect that can be reverted partially by inhibition of amino acid decarboxylase (McCrimmon and Lalley 1982). These data suggest an inhibitory action of 5HT on the respiratory network.

Adult rats of the strain *Brown Norway*, which exhibit a modest 5-HT deficiency, have a deficit in the ventilatory CO_2 sensitivity. Systemic injection of fluoxetine (a 5-HT reuptake blocker) increases the ventilatory response to 7% CO_2 in these animals (Hodges et al. 2013), suggesting that increasing 5-HT may reverse the inherently low CO_2 sensitivity in this rat strain. Besides, fluoxetine delivered directly into the medullary raphe nucleus increases the ventilatory response to CO_2 in adult rats (Taylor et al. 2004).

The administration of 5-HT in vitro in the isolated spinal cord-brainstem ("en bloc") preparation increases both the frequency and amplitude of the fictive respiration recorded from phrenic roots and hypoglossal nucleus (Morin et al. 1990; Di Pasquale et al. 1996; Lindsay and Feldman 1993). In the brainstem slice preparation, local application of 5-HT in the raphe nucleus increases the firing rate of the hypoglossal nerve (Ptak et al. 2009). Also, 5-HT modulates excitability of motoneurons (Rekling et al. 2000). Most of the evidence points to a stimulatory effect on respiratory output when 5-HT neurons increase their firing rate (Hodges and Richerson 2008).

The 5-HT receptors involved in the respiratory brainstem neural network are the 5-HT_{1A}, 5-HT_{2A}, 5-HT_4, and 5-HT_7 receptors. The 5-HT_{1A} receptor is located somatodendritically in raphe neurons and in non-5-HT neurons that are innervated by 5-HT neurons. When 5-HT_{1A} receptors in raphe neurons are activated, the firing rate of the respiratory network is diminished (Lalley et al. 1997). By contrast, activation of this receptor in other brainstem respiratory nuclei (Liu and Wong-Riley 2010a) excites the respiratory network (Corcoran et al. 2014). $5\text{-HT}_{2A,B,C}$ receptor activation is associated with activation of phospholipase C, increasing neural excitability. In the brainstem slice preparation, superfusion with 5-HT_{2A} receptor agonists increases preBötC activity, which is reversed by the 5-HT_{2A} receptor antagonist kentanserin (Pena and Ramirez 2002). These receptors are important in respiratory rhythm generation (Tryba et al. 2006) as well as in modulating respiratory motoneurons (Brandes et al. 2006). Activation of 5-HT_4 receptors, which are highly expressed in preBötC neurons, excites the respiratory network, increasing respiration (Manzke et al. 2008). Systemic injection of BIMU-8, a 5-HT_{4A} receptor agonist (Manzke et al. 2003), increases cAMP production, which stimulates breathing and "protects" respiratory activity from the depression induced by μ-opioid agonists (Manzke et al. 2003). The expression of each 5-HT receptor subtype shows different patterns during the early postnatal maturation of the neural respiratory network, a critical period because a decrease in serotonergic drive to RPG could cause a decreased ventilatory response to homeostatic challenge (Liu and Wong-Riley 2010b).

Neural Respiratory Network in Models with Altered 5-HT

5-HT appears to be involved in a wide range of processes during mammalian development. In mice, 5-HT is involved in the development not only of brain serotonergic neurons but also of others. There are various conditions where 5-HT may be altered during brain development. In rodents, genetic manipulation of different genes either decreases or increases levels of 5-HT (Burnet et al. 2001; Ding et al. 2003). Levels also change in some human pathological conditions of genetic origin (Bervini and Herzog 2013). 5-HT can also be modified by pharmacological treatments aimed at various targets of the 5-HT system, such as blockade of MAO-A with clorgyline (Real et al. 2007) SERT blockade by fluoxetine (see next paragraph).

Models with Deficiencies in 5-HT

Mutations of transcription factors that determine the 5-HT phenotype have been used to generate mouse models with various degrees of deficiency in 5-HT system. Transcription factor Lmxb1b expression precedes that of Pet-1. The Lmx1b null mice are totally deficient in 5-HT neurons (Ding et al. 2003). Both Lmx1b$^{-/-}$ and Pet-1$^{-/-}$ mice have a deficit in 5-HT$_{1A}$ receptor binding in raphe neurons, none of which are serotonergic, but not in non-serotonergic neurons elsewhere in the nervous system (Massey et al. 2013).

Lmx1b null mice have less ventilatory response to CO_2 than wild type mice do, whereas the basal ventilation and hypoxic ventilatory response are normal. Intracerebroventricular infusion of 5-HT stimulates ventilation and restores the ventilatory response to hypercarbia (Hodges and Richerson 2008). Pet-1$^{-/-}$ mice lose between 70% and 80% of 5-HT neurons and exhibit a high variability in breathing pattern during the neonatal period in comparison with wild type mice (Hendricks et al. 2003; Erickson et al. 2007), with lower breathing frequency, prolonged apnea and reduced 5-HT$_{1A}$ binding in the raphe nucleus (Massey et al. 2013). In both Lmx1b$^{-/-}$ and Pet$^{-/-}$mice the respiratory abnormalities spontaneously reverse and return to normal between the second and third postnatal week. In Lmx1b$^{-/-}$ mutants can be reversed by 5-HT$_{2A}$ agonists before the abnormalities spontaneously reverse (Hodges et al. 2008; Erickson et al. 2007). These results suggest that alteration in the serotonergic system causes a delayed maturation of central respiratory chemosensitivity. Because chemosensitivity is a distributed property (Nattie and Li 2010), it is tempting to postulate that in the absence of 5-HT neurons or with fewer of them, such other nuclei as the RTN and locus coeruleus assume greater importance in regulating breathing in response to CO_2 changes.

Phox2b is a transcription factor essential for development of cardio-vascular, respiratory and digestive functions of the nervous system (Goridis and Brunet 2010). Phox2b expression is a marker of preinspiratory neurons in the RTN/pFRG. Lesion of RTN neurons with Substance P conjugated with saporin decreased the resting ventilation and increased the response to hypercapnia (7% CO_2) in rats with or without peripheral chemoreceptors (Takakura et al. 2014). Dissociated Phox2b positive neurons in culture are CO_2 and pH sensitive (Wang et al. 2013). Congenital central hypoventilation syndrome is related to a mutation in the Phox2b gene. The respiratory abnormalities include a decreased tidal volume and alveolar hypoventilation accentuated principally during sleep (Weese-Mayer et al. 2009). In Phox2b mutant mice the RTN neurons do not develop, leading to impaired central chemosensitivity and apnea in neonates, with partially recovery in adults (Ramanantsoa et al. 2011).

5-HT synthesis from tryptophan (Trp) is catalyzed by TPH. In vertebrate two isoenzymes, TPH1 and TPH2, that are encoded by distinct genes, are expressed. TPH1 is expressed in the gut, generating 95% of the peripheral 5-HT, and TPH2 is the isoenzyme present in central nervous system, particularly in the raphe nucleus (Walther and Bader 2003; Gutknecht et al. 2008). In mouse brain, TPH2 expression is first detected at E11.25; it increases exponentially until reaching its highest level around 3 weeks of age. TPH2 KO ($TPH2^{-/-}$) mice show no clear evidence of alterations in serotonergic neuron morphology and fiber distribution, determined by SERT immunoreactivity, suggesting that 5-HT is not required for serotonergic neuron formation (Gutknecht et al. 2008). However, $TPH2^{-/-}$ mice present growth retardation, decreased nocturnal blood pressure and heart rate, and altered body temperature control. Respiration frequency is reduced, but circadian variations are normal (Savelieva et al. 2008) in accord with other models of 5-HT depletion (Alenina et al. 2009; Gutknecht et al. 2009). TPH1 expression remains practically undetectable in the brain during prenatal and postnatal periods (Gutknecht et al. 2009). Interestingly Côté et al. (2007) provided evidence from TPH1 mutant mice, that the maternal 5-HT level is crucial for development.

The vesicular monoamine transporter VMAT has two isoforms; VMAT1 is principally expressed in chromaffin and enterochromaffin cells, and VMAT2 is primarily expressed in presynaptic monoaminergic neurons and has a higher affinity for catecholamines than VMAT1 (Erickson et al. 1996; Lebrand et al. 1996). The conditional deletion of vesicular monoamine transporter 2 gene VMAT2$^{sert/cre}$ mouse line has been characterized as having a uniform reduction of 5-HT in all brain regions (−95% reduction) in adults and no alteration of respiration has been reported. Also 5-HT$_{1A}$ receptor is increased in the raphe nucleus of VMAT2$^{sert/cre}$ mice in response to the decreased levels of 5-HT (Narboux-Neme et al. 2011).

As introduced in Chap. 8, *Sudden infant death syndrome* (SIDS) is defined as the sleep-related death of a child less than 1 year of age that remains unexplained after exhaustive study of the circumstances of death including pathological investigation on autopsy (Kinney and Thach 2009). It is a leading cause of death in infants under 1 year old in developed countries. The 5-HT system is associated with about 70% of deaths from SIDS (Paterson et al. 2006; Kinney 2009; Duncan et al. 2010). A triple risk hypothesis has been proposed as a framework involving the interaction of three main factors in the pathogenesis of SIDS: (1) underlying vulnerability in the infant, (2) a critical developmental period involving the first 6 months, and (3) an exogenous stressor that may induce asphyxia (hypoxia and hypercapnia), such as a prone sleeping position (Kinney et al. 2001; Kinney 2009). SIDS infants have more 5-HT neurons, fewer 5-HT_{1A} and 5-HT_{2A} receptors, less 5-HT in the brain and high levels of TRP, HVA and HIAA reflecting an altered 5-HT turnover (Duncan et al. 2010). Also, decreased levels of TPH2 and 5-HT have been found in SIDS cases (Duncan et al. 2010), and increased 5-hydroxyindoleacetic acid, a 5-HT metabolite, and tryptophan have been detected in cerebrospinal fluid (Cann-Moisan et al. 1999).

Individuals affected by Rett syndrome, a neurodevelopmental disorder usually caused by a mutation in the gene encoding MeCP2, present a progresive respiratory dysfunction characterized by apnea and decreased respiratory rhythm stability (Ramirez et al. 2013). Many neurochemical signaling systems participate in the altered respiratory network in Rett syndrome, including a deficit in BDNF, noradrenaline and 5-HT probably affecting the respiratory motoneuron output (Katz et al. 2009). It has been reported that there are low levels of monoamines in the brains of Rett syndome patients (Riederer et al. 1985, 1986) with altered 5-HT innervation in dorsal motor nucleus of the vagus (Paterson et al. 2005). MeCP2 null mice, a model of Rett syndrome, have decreased levels of NA and 5-HT in the forebrain, pons and medulla (Viemari et al. 2005) and altered breathing, with hyperventilation, apnea and increased variability in respiratory frequency, with an impaired response to CO_2, but the mechanism and similarity to human the condition remains unclear. Breathing disturbances are more pronounced during wakefulness, but irregular breathing also occurs during sleep (Ramirez et al. 2013).

Another way to decrease 5-HT content is to restrict intake of tryptophan (Trp), the precursor in 5-HT synthesis. Rat dams with a Trp-free diet gave birth to neonates of reduced body weight, fewer serotonergic neurons in the raphe nucleus (Flores-Cruz and Escobar 2012), and abnormal patterning of retinotectal projections (Gonzalez et al. 2008). Additionally, Trp intake reduces adult sleep apneas and increases REM sleep in infants (Maurizi 1985).

In sum, studies have shown that decreased 5-HT levels during fetal development either with genetic and environmental models in rodents or with human pathologies are associated with failures in the generation and control of the respiratory rhythm, with more apnea and impaired sensory responses to hypercapnia (see Fig. 11.1 and Table 11.1). However, respiratory abnormalities are overcome in affected animals and individuals allowed to reach adulthood. These results indicate that 5-HT is needed for normal development of the respiratory neural network, but compensatory mechanisms may arise, allowing an almost normal functioning of the network in adults.

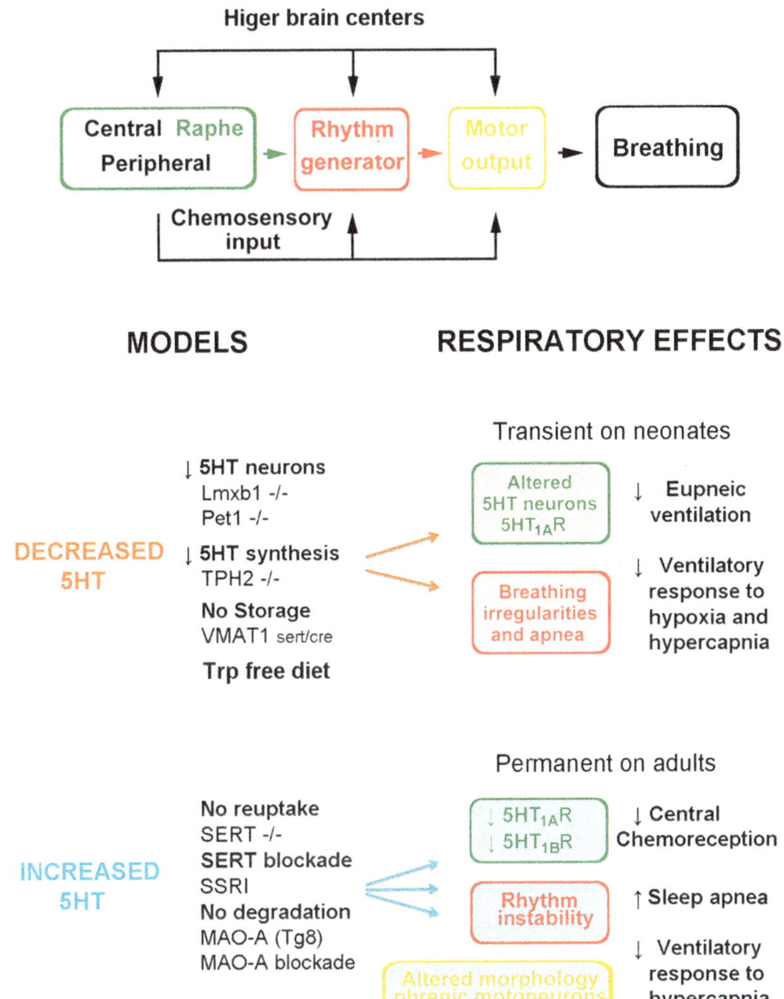

Fig. 11.1 Overview of respiratory consequences in the models with altered 5HT. Summary of the neurodevelompental effects of 5HT on respiratory network, described in the text. In *upper panel* a scheme shows the respiratory network. Rhythm generator (*red box*) projects to motor neurons (*yellow box*) that control respiratory muscles. Rhythm generator receives sensorial input by central and peripheral chemoreceptors. Raphe nucleus, a central chemosentive nuclei, is shown because it is a relevant target for the 5HT modifications (*green box*). For clarity, other inputs to the respiratory network are shown collectively as higher brain centers. *Arrows* show the direction of flux information. On the *left* hand, are despicted the models where 5HT is decreased (*blue*) or increased (*orange*). On the *right side* the summary of the respiratory effects are shown. The *boxes* show specific modifications at the raphe nucleus (*green outlined box*), the rhythm generator (*red outlined box*), and motor output (*yellow outlined box*). When 5HT is low during development, eupneic ventilation and ventilatory response to hypercarbia and hypoxia are diminished in early postnatal life. These impairments do not persist in adults. On the contrary, when 5HT is increased during development respiratory rhythm is instable and frequent apnea specially during sleep are seen. Ventilatory response to hypercarbia is diminished and morphology of phrenic motor neurons is altered. Respiratory changes persist in adults

Table 11.1 Models with 5HT deficiencies

Effects of decreased 5HT level on

	Models	Respiratory neural network	Non respiratory neural network	References
Genetic alterations in rodents	**Lmx1b$^{-/-}$**	Neonates before P10 present hypoventilation and long apnea. Irregular basal rhythm at P2–P14. 50% decreased ventilator/ response to hypercapnia in adults. Intracerebroventricular infusion of 5HT restores baseline ventilation and ventilator to response to hypercapnia. Decreased 5HT$_{1A}$ binding in raphe nucleus at P3, P10 and P25	Impaired shivering in hypothermic mice. Decreased mechanical sensitivity and enhanced inflammatory pain sensitivity. No analgesic effect of antidepressant. Impaired spatial memory. Lethal phenotype at 2–3 weeks old	Hodges et al. (2008, 2009), Hodges and Richerson (2008), Zhao et al. (2007), Ding et al. (2003), Massey et al. (2013), Dai et al. (2008)
	Pet-1$^{-/-}$	At birth variable breathing pattern. From P0 to P4.5 low breathing frequency, spontaneous and prolonged apnea, increased hypoxia induced apnea. Breathing is normalized at P9.5	Agressive behavior in male. Locomotor hypoactivity	Massey et al. (2013), Hendricks et al. (2003), Schaefer et al. (2009), Erickson et al. (2007)
	VMAT2 SERT-CRE	No evident alterations in breathing have been reported. Increased 5HT$_{1A}$ receptors in raphe nucleus. No survival beyond P5 after reserpine treatment	High mortality at P6. Growth retardation and feeding abnormalities. Increased escape behavior and anxiolytic response in behavioral test. Altered development of thalamocortical pathway and increased celldeath in cerebral cortex	Narboux-Neme et al. (2011), Alvarez et al. (2002), Stankovski et al. (2007), Fon et al. (1997)
	Mecp2 null	Decreased ventilator to response to mild hypercapnia (3–5% CO_2) in null mice. Increased apnea in heterozigous mice	Catecholaminergic deficit in the locus coeruleus of Mecp2 null male. Altered hyppocampal 5HTergic innervation during postnatal development. Abnormal BDNF expresion in nodose ganglia. Altered somatosensory barrel cortex refinement in the early postnatal age	Wang et al. (2006), Roux et al. (2010), Isoda et al. (2010), Zhang et al. (2011)

	TPH2$^{-/-}$, TPH1 and TPH2 double KO	Reduced respiratory frequency		Increased letality during first month of age. Altered control of body temperature. Altered sleep patterns and decreased heart rate and blood pressure. Adult male mice present pro-depresive and anxious behavior. Low body weight in male mice	Alenina et al. (2009), Savelieva et al. (2008)
Pharmacological or diet modification in rodents	Tryptophan-free diet	Reduced expression of TPOH in caudal raphe and increased TPOH immunopositive cells in rostral raphe			Flores-Cruz and Escobar (2012)
Humans pathologies	SIDS	Decreased ventilator to response to hypoxia and hypercapnia. In raphe nucleus low expression of TPH2, augmented 5HT neurons, and decreased expression of 5HT$_{1A}$ and 5HT$_{2A}$ receptors. Reduction in brainstem 5HT$_{1A}$ binding sites	In brainstem increased pro-inflammatory cytokine IL-1, and decreased BDNF and Trkb		Duncan et al. (2010), Brady et al. (1985), Shannon et al. (1977), Paterson et al. (2009), Tang et al. (2012), Kadhim et al. (2003), Massey et al. (2013)
	Rett syndrome	Increased respiratory frequency and rhythm irregularities	Seizures, autism, ataxia, and stereotyped hand movement. Increased heart rate		Weese-Mayer et al. (2008), Julu et al. (2001)

Summary of the neurodevelopment effects of decreased levels of serotonin in respiratory and non respiratory neural networks. The *left column* shows the models, and the *center columns* the main effects on different neural networks. The *right column* shows the references

Models of Increased 5-HT

Tg8 transgenic mice, as result of a disruption in the gene encoding MAO-A, have a four- to ten-fold increase in whole brain 5-HT content in fetuses and neonates, and twice the usual 5-HT content in adults (Viemari and Hilaire 2003), resulting in altered development of several neural networks and behavior (Popova et al. 2007; Vishnivetskaya et al. 2007). These mice have an abnormal development of primary somatosensory cortex that can be prevented by reducing 5-HT levels during early postnatal development (Cases et al. 1998). Additionally, these mutants show a transient delay of locomotor network maturation (Cazalets et al. 2000). These mice also have several breathing abnormalities, such as an unstable respiratory rhythm, an abnormal dendrite morphology of phrenic motoneurons (Bou-Flores et al. 2000), increased sleep apnea (Real et al. 2007), no response to 5-HT_{1A} receptor activation (Lanoir et al. 2006), a decreased sensitivity to hypoxia and lung inflation (Burnet et al. 2001), and an abnormal organization of phrenic motoneuron premotor network (Bras et al. 2008). Some functional abnormalities can be prevented with pharmacological treatments that modify serotonergic systems (Burnet et al. 2001). Pharmacological blockade of MAO-A with clorgyline increases the number of apneas during sleep in adult wild type mice, but not in MAO-A null mice; in agreement with MAO-A null mouse results (Real et al. 2007). By contrast, the inhibition of 5-HT synthesis with PCPA, decreased the apnea index in mutant but not in wild type mice (Real et al. 2007).

Another genetic model of elevated extracellular 5-HT level is the SERT-KO mice. These mutants do not express 5-HT transporter and have higher extracellular 5-HT levels than wild type animals (Mathews et al. 2004). The expression of 5-HT_{1A} receptors is diminished in the dorsal raphe nucleus (Fabre et al. 2000). Among others, these mutant mice show altered development of the retino-geniculate projections and patterning of thalamocortical projections (Persico et al. 2001) and the topographically organized whisker-barrel fields (Migliarini et al. 2013), as occurs with MAO-A KO. In adults, a pro-depressive abnormal behavior was reported (Lira et al. 2003). SERT-KO mice showed a decreased ventilatory response to CO_2 with male predominance (Li and Nattie 2008). The double SERT/MAO-A KO mice also show impaired development of somatosensory cortex (Salichon et al. 2001) due to changes in 5-HT levels during development.

Pathological conditions of genetic origin in humans also affect genes leading to increased levels of 5-HT. Prader Willi syndrome is a neurodevelopmental disorder of low prevalence (Zanella et al. 2009) characterized by short stature, small hands and feet, dysmorphic facial features, hypogonadism and intellectual disability (Angulo et al. 2015). Patients show hypoventilation, respiratory rhythm instability and altered ventilatory response to hypoxia and hypercapnia (Zanella et al. 2009). Necdin deficient mice (Ndn-KO), a model of Prader Willi syndrome, display irregular breathing and severe apnea as well as abnormally high levels of 5-HT in the medulla (Zanella et al. 2009; Bervini and Herzog 2013).

The administration of SERT inhibitors, like fluoxetine, leads to an increase of 5-HT level. We have evaluated whether fluoxetine administered to mice during pregnancy may alter central respiratory chemoreception in offspring. In control animals, fictive respiratory activity recorded from the medullary slice preparation from P8 neonates is increased in response to exogenously added 5-HT. However, in fluoxetine exposed animals the dose-response curve for exogenous 5-HT showed lower reactivity with respect to controls, as shown in Fig. 11.2. These results suggest that high levels of 5-HT during development alter 5-HT receptors. Because serotonergic neurons from the raphe nucleus play an important role in chemosensitivity, we expect that fluoxetine exposed animals may have impaired chemosensitivity, in agreement with the previous results in the other models of increased 5-HT.

In summary, all circumstances where 5-HT is increased during the development of neural networks results in a respiratory dysfunction that persists in adulthood (see Fig. 11.1 and Table 11.2). In many of the models discussed a reduced ventilatory response to hypercapnia and/or apnea is observed. These data point to a role of 5-HT in the development of chemoreception. However, no data have been provided on perturbations of other chemosensitive-specific nuclei besides the raphe nucleus as a consequence of increasing 5-HT during development.

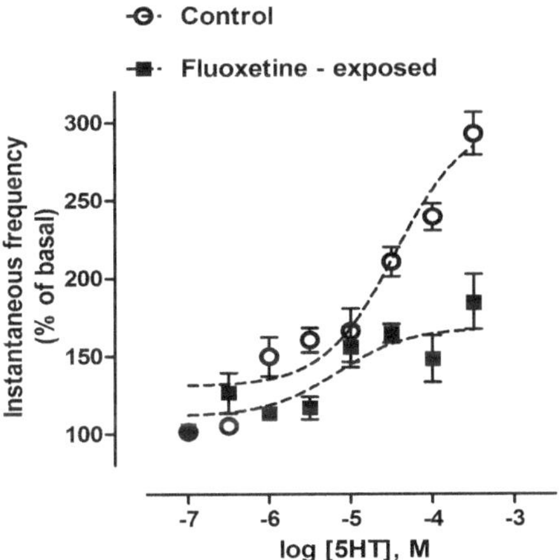

Fig. 11.2 Perinatal exposure to fluoxetine decreased reactivity to 5HT of fictive respiration. Dose response curve of exogenous 5HT on fictive respiration. Osmotic mini pump were implanted in pregnant CF1 mice at E5–7 to deliver 7 mg kg^{-1} day^{-1} fluoxetine. Fluoxetine exposure lasted from E5–7 until weaning at P12. Fictive respiration was recorded with suction electrodes from ventral respiratory column from medullary slices prepared from P8 neonates (Coddou et al. 2009). Results (mean ± SEM) are expressed as % of basal frequency. Basal instantaneous respiratory frequency was similar in control (0.21 ± 0.014 Hz) and fluoxetine exposed animals (0.23 ± 0.010 Hz). Five slices coming from control (o) and fluoxetine exposed animals (■) were averaged for each dosis. Animal from control and fluoxetine exposed groups come from 7 l. F test show different curves for control vs fluoxetine exposed (p < 0.0001)

Table 11.2 Models of increased 5HT

Effects of increased 5HT level on				
Models		Respiratory neural network	Non respiratory neural network	References
Genetic alterations, in rodents	**MAO-A deficient (Tg8) MAO-A KO**	Instable respiratory rhythm. Abnormal morphology of motoneurons and organization of the premotor network of phrenic motoneurons. Increased sleep apnea. Decreased sensitivity to hypoxia and lung inflation. Lack of response to $5HT_{1A}$ activation. Decreased density of $5HT_{1A}$ receptors in raphe nucleus	Abnormal development of the primary somatosensory cortex. Aggresive behavior. Altered dopamine metabolism. Altered vasopresin expression in hypothalamus. Delayed locomotor maturation. Altered segregation of retinal proyections	Bou-Flores et al. (2000), Bras et al. (2008), Burnet et al. (2001), Cases et al. (1998), Cazalets et al. (2000), Vishnivetskaya et al. (2007), Popova et al. (2007), Vacher et al. (2004), Upton et al. (1999, 2002), Mejia et al. (2002)
	SERT KO	Decreased ventilatory response to hypercapnia at P15, and adult male mice. Low expression of $5HT_{1A}$ and $5HT_{1B}$ receptors in dorsal raphe nucleus	Altered pattering of somatosensory neurons similar to MAO-A/SERT double K.O. Altered segregation of retinogeniculate and retinotectal afferents. Depressive-like behavior and increased anxiety in adult mice. Decreased functional $5HT_{2B}$ causing altered development of cardiovascular system	Cases et al. (1998), Persico et al. (2001), Salichon et al. (2001), Lira et al. (2003), Penatti et al. (2011), Ansorge et al. (2004), Migliarini (2013), Fabre et al. (2000)

Pharmacological or diet modification in rodents	Blockade of SERT	Fluoxetine administration (ip) increases ventilator response to 7% CO_2 in Brown Norway but not in Sprague Dawly rats. Increased ventilatory response to CO_2 when focal microdialysis of fluoxetine was applied into the medullary raphe of the rat. Recordings at ventral respiratory column from P8 neonates perinatally exposed to fluoxetine show decreased reactivity to 5HT	Diminished weight gain in Sprague Dawley rats after fluoxetine treatment. 5HIAA and 5HIAA/5HT ratios reduced in pons, hypothalamus and forebrain by fluoxetine treatment in Brown Norway rats	Hodges et al. (2013), Taylor et al. (2004)
	Injection of L-tryptophan	High levels of 5HT in brain. Transient apnea		Hilaire et al. (1993)
Humans pathologies	Prader Willi	Rhyhtm instability, sleep apnea and blunted response to hypoxia and hypercapnia	Overeating, overweight, perinatal hypotonia, hypofunction of hypotalamus, hypogonadism and short stature. Learning disorder	Angulo et al. (2015), Ramirez et al. (2013), Zanella et al. (2009)

Summary of the neurodevelopment effects of increased levels of serotonin in respiratory and non respiratory neural networks. The *left column* shows the models, and the *center columns* the main effects on different neural networks. The *right column* shows the references

Concluding Remarks

5-HT plays an important role in the early development of neural networks. In the particular case of the brainstem respiratory network, growing data from humans and rodents show that increases *and* decreases of 5-HT alter the respiratory network. In general terms, the generation of respiratory rhythm, the motor output and the che-mosensory inputs that control respiration are all disturbed (Fig. 11.1). Physiological changes are more apparent in neonates than in adulthood in animals in which 5-HT is decreased by various means. In contrast, changes typically persist in adults in response to manipulations or mutations that increase 5-HT. SSRIs are among the most prescribed drugs to treat psychiatric disorders, such as depression. The data discussed show that increasing extracellular 5-HT levels during embryonic develop-ment has consequences, suggesting that caution be taken in the use of SSRIs during pregnancy. The work presented here highlights the role of 5-HT as a developmental signal that has not previously been considered for the respiratory network.

Acknowledgment Grants Fondo Nacional de Desarrollo Científico y Tecnológico (FONDECYT) # 1171434 (JE) and DICYT-USACH. K. Bravo is a PhD candidate at PhD Program in Neuroscience, Universidad de Santiago de Chile.

References

Aitken AR, Tork I (1988) Early development of serotonin-containing neurons and pathways as seen in wholemount preparations of the fetal rat brain. J Comp Neurol 274(1):32–47. doi:10.1002/cne.902740105

Alenina N, Bashammakh S, Bader M (2006) Specification and differentiation of serotonergic neu-rons. Stem Cell Rev 2(1):5–10. doi:10.1007/s12015-006-0002-2

Alenina N, Kikic D, Todiras M, Mosienko V, Qadri F, Plehm R, Boye P, Vilianovitch L, Sohr R, Tenner K, Hortnagl H, Bader M (2009) Growth retardation and altered autonomic control in mice lacking brain serotonin. Proc Natl Acad Sci U S A 106(25):10332–10337. doi:10.1073/pnas.0810793106. 0810793106

Alvarez C, Vitalis T, Fon EA, Hanoun N, Hamon M, Seif I, Edwards R, Gaspar P, Cases O (2002) Effects of genetic depletion of monoamines on somatosensory cortical development. Neuroscience 115(3):753–764

Alwan S, Friedman JM (2009) Safety of selective serotonin reuptake inhibitors in pregnancy. CNS Drugs 23(6):493–509. doi:10.2165/00023210-200923060-00004

Angulo MA, Butler MG, Cataletto ME (2015) Prader-Willi syndrome: a review of clinical, genetic, and endocrine findings. J Endocrinol Investig 38(12):1249–1263. doi:10.1007/s40618-015-0312-9. 10.1007/s40618-015-0312-9

Ansorge MS, Zhou M, Lira A, Hen R, Gingrich JA (2004) Early-life blockade of the 5-HT trans-porter alters emotional behavior in adult mice. Science 306(5697):879–881. doi:10.1126/science.1101678

Bervini S, Herzog H (2013) Mouse models of Prader-Willi syndrome: a systematic review. Front Neuroendocrinol 34(2):107–119. doi:10.1016/j.yfrne.2013.01.002

Bonnin A, Levitt P (2011) Fetal, maternal, and placental sources of serotonin and new impli-cations for developmental programming of the brain. Neuroscience 197:1–7. doi:10.1016/j.neuroscience.2011.10.005

Bou-Flores C, Lajard AM, Monteau R, De Maeyer E, Seif I, Lanoir J, Hilaire G (2000) Abnormal phrenic motoneuron activity and morphology in neonatal monoamine oxidase A-deficient transgenic mice: possible role of a serotonin excess. J Neurosci 20(12):4646–4656

Brady JP, McCann EM (1985) Control of ventilation in subsequent siblings of victims of sudden infant death syndrome. J Pediatr 106(2):212–217

Brandes IF, Zuperku EJ, Stucke AG, Jakovcevic D, Hopp FA, Stuth EA (2006) Serotonergic modulation of inspiratory hypoglossal motoneurons in decerebrate dogs. J Neurophysiol 95(6):3449–3459. doi:10.1152/jn.00823.2005

Bras H, Gaytan SP, Portalier P, Zanella S, Pasaro R, Coulon P, Hilaire G (2008) Prenatal activation of 5-HT2A receptor induces expression of 5-HT1B receptor in phrenic motoneurons and alters the organization of their premotor network in newborn mice. Eur J Neurosci 28(6):1097–1107. doi:10.1111/j.1460-9568.2008.06407.x. EJN6407 [pii]

Bravo K, Eugenín J, Llona I (2016) Perinatal fluoxetine exposure impairs CO_2 chemoreflex: implications for sudden infant death syndrome. Am J Respir Cell Mol Biol 55(3):368–376

Burnet H, Bevengut M, Chakri F, Bou-Flores C, Coulon P, Gaytan S, Pasaro R, Hilaire G (2001) Altered respiratory activity and respiratory regulations in adult monoamine oxidase A-deficient mice. J Neurosci 21(14):5212–5221. 21/14/5212 [pii]

Cann-Moisan C, Girin E, Giroux JD, Le Bras P, Caroff J (1999) Changes in cerebrospinal fluid monoamine metabolites, tryptophan, and gamma-aminobutyric acid during the 1st year of life in normal infants. Comparison with victims of sudden infant death syndrome. Biol Neonate 75(3):152–159. 14091 [pii]

Cases O, Lebrand C, Giros B, Vitalis T, De Maeyer E, Caron MG, Price DJ, Gaspar P, Seif I (1998) Plasma membrane transporters of serotonin, dopamine, and norepinephrine mediate serotonin accumulation in atypical locations in the developing brain of monoamine oxidase A knock-outs. J Neurosci 18(17):6914–6927

Cayetanot F, Gros F, Larnicol N (2001) 5-HT(2A/2C) receptor-mediated hypopnea in the newborn rat: relationship to Fos immunoreactivity. Pediatr Res 50(5):596–603. doi:10.1203/00006450-200111000-00011

Cazalets JR, Gardette M, Hilaire G (2000) Locomotor network maturation is transiently delayed in the MAOA-deficient mouse. J Neurophysiol 83(4):2468–2470

Coates EL, Li A, Nattie EE (1993) Widespread sites of brain stem ventilatory chemoreceptors. J Appl Physiol (1985) 75(1):5–14

Coddou C, Bravo E, Eugenin J (2009) Alterations in cholinergic sensitivity of respiratory neurons induced by pre-natal nicotine: a mechanism for respiratory dysfunction in neonatal mice. Philos Trans R Soc Lond B Biol Sci 364(1529):2527–2535. doi:10.1098/rstb.2009.0078. 364/1529/2527 [pii]

Corcoran AE, Commons KG, Wu Y, Smith JC, Harris MB, Richerson GB (2014) Dual effects of 5-HT(1a) receptor activation on breathing in neonatal mice. J Neurosci 34(1):51–59. doi:10.1523/JNEUROSCI.0864-13.2014. 34/1/51 [pii]

Cordes SP (2005) Molecular genetics of the early development of hindbrain serotonergic neurons. Clin Genet 68(6):487–494. doi:10.1111/j.1399-0004.2005.00534.x. CGE534 [pii]

Côté F, Fligny C, Bayard E, Launay JM, Gershon MD, Mallet J, Vodjdani G (2007) Maternal serotonin is crucial for murine embryonic development. Proc Natl Acad Sci U S A 104(1):329–334. doi:10.1073/pnas.0606722104

Covenas R, Marcos P, Belda M, de Leon M, Narvaez JA, Aguirre JA, Gonzalez-Baron S (2001) Neuropeptides in the raphe nuclei: an immunocytochemical study. Rev Neurol 33(2):131–137

Dai JX, Hu ZL, Shi M, Guo C, Ding YQ (2008) Postnatal ontogeny of the transcription factor Lmx1b in the mouse central nervous system. J Comp Neurol 509(4):341–355. doi:10.1002/cne.21759

Dahlstrom A, Fuxe K (1964) Localization of monoamines in the lower brain stem. Experientia 20(7):398–399

Deneris ES (2011) Molecular genetics of mouse serotonin neurons across the lifespan. Neuroscience 197:17–27. doi:10.1016/j.neuroscience.2011.08.061

Depuy SD, Kanbar R, Coates MB, Stornetta RL, Guyenet PG (2011) Control of breathing by raphe obscurus serotonergic neurons in mice. J Neurosci 31(6):1981–1990. doi:10.1523/JNEUROSCI.4639-10.2011

Di Pasquale E, Tell F, Monteau R, Hilaire G (1996) Perinatal developmental changes in respiratory activity of medullary and spinal neurons: an in vitro study on fetal and newborn rats. Brain Res Dev Brain Res 91(1):121–130

Ding YQ, Marklund U, Yuan W, Yin J, Wegman L, Ericson J, Deneris E, Johnson RL, Chen ZF (2003) Lmx1b is essential for the development of serotonergic neurons. Nat Neurosci 6(9):933–938. doi:10.1038/nn1104

Doi A, Ramirez JM (2008) Neuromodulation and the orchestration of the respiratory rhythm. Respir Physiol Neurobiol 164(1–2):96–104. doi:10.1016/j.resp.2008.06.007

Douglas RM, Trouth CO, James SD, Sexcius LM, Kc P, Dehkordi O, Valladares ER, McKenzie JC (2001) Decreased CSF pH at ventral brain stem induces widespread c-Fos immunoreactivity in rat brain neurons. J Appl Physiol (1985) 90(2):475–485

Dreshaj IA, Haxhiu MA, Martin RJ (1998) Role of the medullary raphe nuclei in the respiratory response to CO2. Respir Physiol 111(1):15–23

Duncan JR, Paterson DS, Hoffman JM, Mokler DJ, Borenstein NS, Belliveau RA, Krous HF, Haas EA, Stanley C, Nattie EE, Trachtenberg FL, Kinney HC (2010) Brainstem serotonergic deficiency in sudden infant death syndrome. JAMA 303(5):430–437. doi:10.1001/jama.2010.45

Erickson JD, Schafer MK, Bonner TI, Eiden LE, Weihe E (1996) Distinct pharmacological properties and distribution in neurons and endocrine cells of two isoforms of the human vesicular monoamine transporter. Proc Natl Acad Sci U S A 93(10):5166–5171

Erickson JT, Shafer G, Rossetti MD, Wilson CG, Deneris ES (2007) Arrest of 5HT neuron differentiation delays respiratory maturation and impairs neonatal homeostatic responses to environmental challenges. Respir Physiol Neurobiol 159(1):85–101. doi:10.1016/j.resp.2007.06.002

Eugenin J, Ampuero E, Infante CD, Silva E, Llona I (2003) pH sensitivity of spinal cord rhythm in fetal mice in vitro. Adv Exp Med Biol 536:535–539

Fabre V, Beaufour C, Evrard A, Rioux A, Hanoun N, Lesch KP, Murphy DL, Lanfumey L, Hamon M, Martres MP (2000) Altered expression and functions of serotonin 5-HT1A and 5-HT1B receptors in knock-out mice lacking the 5-HT transporter. Eur J Neurosci 12(7):2299–2310

Feldman JL, Mitchell GS, Nattie EE (2003) Breathing: rhythmicity, plasticity, chemosensitivity. Annu Rev Neurosci 26:239–266. doi:10.1146/annurev.neuro.26.041002.131103

Feldman JL, Del Negro CA, Gray PA (2013) Understanding the rhythm of breathing: so near, yet so far. Annu Rev Physiol 75:423–452. doi:10.1146/annurev-physiol-040510-130049

Flores-Cruz GM, Escobar A (2012) Reduction of serotonergic neurons in the dorsal raphe due to chronic prenatal administration of a tryptophan-free diet. Int J Dev Neurosci 30(2):63–67. doi:10.1016/j.ijdevneu.2012.01.002

Fon EA, Pothos EN, Sun BC, Killeen N, Sulzer D, Edwards RH (1997) Vesicular transport regulates monoamine storage and release but is not essential for amphetamine action. Neuron 19(6):1271–1283

Fox SR, Deneris ES (2012) Engrailed is required in maturing serotonin neurons to regulate the cytoarchitecture and survival of the dorsal raphe nucleus. J Neurosci 32(23):7832–7842. doi:10.1523/JNEUROSCI.5829-11.2012

Gaspar P, Cases O, Maroteaux L (2003) The developmental role of serotonin: news from mouse molecular genetics. Nat Rev Neurosci 4(12):1002–1012. doi:10.1038/nrn1256

Gentile S (2015) Prenatal antidepressant exposure and the risk of autism spectrum disorders in children. Are we looking at the fall of gods? J Affect Disord 182:132–137. doi:10.1016/j.jad.2015.04.048

Gentile S, Galbally M (2011) Prenatal exposure to antidepressant medications and neurodevelopmental outcomes: a systematic review. J Affect Disord 128(1–2):1–9. doi:10.1016/j.jad.2010.02.125

Gonzalez EM, Penedo LA, Oliveira-Silva P, Campello-Costa P, Guedes RC, Serfaty CA (2008) Neonatal tryptophan dietary restriction alters development of retinotectal projections in rats. Exp Neurol 211(2):441–448. doi:10.1016/j.expneurol.2008.02.009

Goridis C, Brunet JF (2010) Central chemoreception: lessons from mouse and human genetics. Respir Physiol Neurobiol 173(3):312–321. doi:10.1016/j.resp.2010.03.014

Gutknecht L, Waider J, Kraft S, Kriegebaum C, Holtmann B, Reif A, Schmitt A, Lesch KP (2008) Deficiency of brain 5-HT synthesis but serotonergic neuron formation in Tph2 knockout mice. J Neural Transm (Vienna) 115(8):1127–1132. doi:10.1007/s00702-008-0096-6

Gutknecht L, Kriegebaum C, Waider J, Schmitt A, Lesch KP (2009) Spatio-temporal expression of tryptophan hydroxylase isoforms in murine and human brain: convergent data from Tph2 knockout mice. Eur Neuropsychopharmacol 19(4):266–282. doi:10.1016/j.euroneuro.2008.12.005

Hansson SR, Mezey E, Hoffman BJ (1999) Serotonin transporter messenger RNA expression in neural crest-derived structures and sensory pathways of the developing rat embryo. Neuroscience 89(1):243–265

Hendricks T, Francis N, Fyodorov D, Deneris ES (1999) The ETS domain factor Pet-1 is an early and precise marker of central serotonin neurons and interacts with a conserved element in serotonergic genes. J Neurosci 19(23):10348–10356

Hendricks TJ, Fyodorov DV, Wegman LJ, Lelutiu NB, Pehek EA, Yamamoto B, Silver J, Weeber EJ, Sweatt JD, Deneris ES (2003) Pet-1 ETS gene plays a critical role in 5-HT neuron development and is required for normal anxiety-like and aggressive behavior. Neuron 37(2):233–247

Hilaire G, Morin D, Lajard AM, Monteau R (1993) Changes in serotonin metabolism may elicit obstructive apnoea in the newborn rat. J Physiol 466:367–381

Hodges MR, Richerson GB (2008) Contributions of 5-HT neurons to respiratory control: neuromodulatory and trophic effects. Respir Physiol Neurobiol 164(1–2):222–232. doi:10.1016/j.resp.2008.05.014

Hodges MR, Tattersall GJ, Harris MB, McEvoy SD, Richerson DN, Deneris ES, Johnson RL, Chen ZF, Richerson GB (2008) Defects in breathing and thermoregulation in mice with near-complete absence of central serotonin neurons. J Neurosci 28(10):2495–2505. doi:10.1523/JNEUROSCI.4729-07.2008

Hodges MR, Echert AE, Puissant MM, Mouradian GC Jr (2013) Fluoxetine augments ventilatory CO_2 sensitivity in Brown Norway but not Sprague Dawley rats. Respir Physiol Neurobiol 186(2):221–228. doi:10.1016/j.resp.2013.02.020

Hodges MR, Wehner M, Aungst J, Smith JC, Richerson GB (2009) Transgenic mice lacking serotonin neurons have severe apnea and high mortality during development. J Neurosci 29(33):10341–10349. doi:10.1523/JNEUROSCI.1963-09.2009

Holtman JR Jr, Dick TE, Berger AJ (1986) Involvement of serotonin in the excitation of phrenic motoneurons evoked by stimulation of the raphe obscurus. J Neurosci 6(4):1185–1193

Homberg JR, Schubert D, Gaspar P (2010) New perspectives on the neurodevelopmental effects of SSRIs. Trends Pharmacol Sci 31(2):60–65. doi:10.1016/j.tips.2009.11.003

Hosogai M, Matsuo S, Sibahara T, Kawai Y (1998) Projection of respiratory neurons in rat medullary raphe nuclei to the phrenic nucleus. Respir Physiol 112(1):37–50

Isoda K, Morimoto M, Matsui F, Hasegawa T, Tozawa T, Morioka S, Chiyonobu T, Nishimura A, Yoshimoto K, Hosoi H (2010) Postnatal changes in serotonergic innervation to the hippocampus of methyl-CpG-binding protein 2-null mice. Neuroscience 165(4):1254–1260. doi:10.1016/j.neuroscience.2009.11.036

Jacobs BL, Azmitia EC (1992) Structure and function of the brain serotonin system. Physiol Rev 72(1):165–229

Julu PO, Kerr AM, Apartopoulos F, Al-Rawas S, Engerstrom IW, Engerstrom L, Jamal GA, Hansen S (2001) Characterisation of breathing and associated central autonomic dysfunction in the Rett disorder. Arch Dis Child 85(1):29–37

Kadhim H, Kahn A, Sebire G (2003) Distinct cytokine profile in SIDS brain: a common denominator in a multifactorial syndrome? Neurology 61(9):1256–1259

Katz DM, Dutschmann M, Ramirez JM, Hilaire G (2009) Breathing disorders in Rett syndrome: progressive neurochemical dysfunction in the respiratory network after birth. Respir Physiol Neurobiol 168(1–2):101–108. doi:10.1016/j.resp.2009.04.017

Kepser LJ, Homberg JR (2015) The neurodevelopmental effects of serotonin: a behavioural perspective. Behav Brain Res 277:3–13. doi:10.1016/j.bbr.2014.05.022

Kinney HC (2009) Brainstem mechanisms underlying the sudden infant death syndrome: evidence from human pathologic studies. Dev Psychobiol 51(3):223–233. doi:10.1002/dev.20367

Kinney HC, Thach BT (2009) The sudden infant death syndrome. N Engl J Med 361(8):795–805. doi:10.1056/NEJMra0803836

Kinney HC, Filiano JJ, White WF (2001) Medullary serotonergic network deficiency in the sudden infant death syndrome: review of a 15-year study of a single dataset. J Neuropathol Exp Neurol 60(3):228–247

Klempin F, Beis D, Mosienko V, Kempermann G, Bader M, Alenina N (2013) Serotonin is required for exercise-induced adult hippocampal neurogenesis. J Neurosci 33(19):8270–8275. doi:10.1523/JNEUROSCI.5855-12.2013

Lalley PM (1986) Serotoninergic and non-serotoninergic responses of phrenic motoneurones to raphe stimulation in the cat. J Physiol 380:373–385

Lalley PM, Benacka R, Bischoff AM, Richter DW (1997) Nucleus raphe obscurus evokes 5-HT-1A receptor-mediated modulation of respiratory neurons. Brain Res 747(1):156–159

Lanoir J, Hilaire G, Seif I (2006) Reduced density of functional 5-HT1A receptors in the brain, medulla and spinal cord of monoamine oxidase-A knockout mouse neonates. J Comp Neurol 495(5):607–623. doi:10.1002/cne.20916

Lebrand C, Cases O, Adelbrecht C, Doye A, Alvarez C, El Mestikawy S, Seif I, Gaspar P (1996) Transient uptake and storage of serotonin in developing thalamic neurons. Neuron 17(5):823–835

Li A, Nattie E (2008) Serotonin transporter knockout mice have a reduced ventilatory response to hypercapnia (predominantly in males) but not to hypoxia. J Physiol 586(9):2321–2329. doi:10.1113/jphysiol.2008.152231

Lindsay AD, Feldman JL (1993) Modulation of respiratory activity of neonatal rat phrenic motoneurones by serotonin. J Physiol 461:213–233

Lira A, Zhou M, Castanon N, Ansorge MS, Gordon JA, Francis JH, Bradley-Moore M, Lira J, Underwood MD, Arango V, Kung HF, Hofer MA, Hen R, Gingrich JA (2003) Altered depression-related behaviors and functional changes in the dorsal raphe nucleus of serotonin transporter-deficient mice. Biol Psychiatry 54(10):960–971

Liu Q, Wong-Riley MT (2010a) Postnatal changes in the expressions of serotonin 1A, 1B, and 2A receptors in ten brain stem nuclei of the rat: implication for a sensitive period. Neuroscience 165(1):61–78. doi:10.1016/j.neuroscience.2009.09.078

Liu Q, Wong-Riley MT (2010b) Postnatal changes in tryptophan hydroxylase and serotonin transporter immunoreactivity in multiple brainstem nuclei of the rat: implications for a sensitive period. J Comp Neurol 518(7):1082–1097. doi:10.1002/cne.22265

Liu Y, Wei L, Laskin DL, Fanburg BL (2011) Role of protein transamidation in serotonin-induced proliferation and migration of pulmonary artery smooth muscle cells. Am J Respir Cell Mol Biol 44(4):548–555. doi:10.1165/rcmb.2010-0078OC

Liu Y, Tian H, Yan X, Fan F, Wang W, Han J (2013) Serotonin inhibits apoptosis of pulmonary artery smooth muscle cells through 5-HT2A receptors involved in the pulmonary artery remodeling of pulmonary artery hypertension. Exp Lung Res 39(2):70–79. doi:10.3109/01902148.2012.758191

Manaker S, Tischler LJ (1993) Origin of serotoninergic afferents to the hypoglossal nucleus in the rat. J Comp Neurol 334(3):466–476. doi:10.1002/cne.903340310

Manzke T, Guenther U, Ponimaskin EG, Haller M, Dutschmann M, Schwarzacher S, Richter DW (2003) 5-HT4(a) receptors avert opioid-induced breathing depression without loss of analgesia. Science 301(5630):226–229. doi:10.1126/science.1084674

Manzke H, Lehmann K, Klopocki E, Caliebe A (2008) Catel-Manzke syndrome: two new patients and a critical review of the literature. Eur J Med Genet 51(5):452–465. doi:10.1016/j.ejmg.2008.03.005

Massey CA, Kim G, Corcoran AE, Haynes RL, Paterson DS, Cummings KJ, Dymecki SM, Richerson GB, Nattie EE, Kinney HC, Commons KG (2013) Development of brainstem 5-HT1A receptor-binding sites in serotonin-deficient mice. J Neurochem 126(6):749–757. doi:10.1111/jnc.12311

Mathews TA, Fedele DE, Coppelli FM, Avila AM, Murphy DL, Andrews AM (2004) Gene dose-dependent alterations in extraneuronal serotonin but not dopamine in mice with reduced serotonin transporter expression. J Neurosci Methods 140(1–2):169–181. doi:10.1016/j.jneumeth.2004.05.017

Maurizi CP (1985) Could supplementary dietary tryptophan and taurine prevent epileptic seizures? Med Hypotheses 18(4):411–415

McCrimmon DR, Lalley PM (1982) Inhibition of respiratory neural discharges by clonidine and 5-hydroxytryptophan. J Pharmacol Exp Ther 222(3):771–777

Mejia JM, Ervin FR, Baker GB, Palmour RM (2002) Monoamine oxidase inhibition during brain development induces pathological aggressive behavior in mice. Biol Psychiatry 52(8):811–821

Migliarini S, Pacini G, Pelosi B, Lunardi G, Pasqualetti M (2013) Lack of brain serotonin affects postnatal development and serotonergic neuronal circuitry formation. Mol Psychiatry 18(10):1106–1118. doi:10.1038/mp.2012.128

Mitchell JB, Betito K, Rowe W, Boksa P, Meaney MJ (1992) Serotonergic regulation of type II corticosteroid receptor binding in hippocampal cell cultures: evidence for the importance of serotonin-induced changes in cAMP levels. Neuroscience 48(3):631–639

Morin D, Hennequin S, Monteau R, Hilaire G (1990) Serotonergic influences on central respiratory activity: an in vitro study in the newborn rat. Brain Res 535(2):281–287

Murrin LC, Sanders JD, Bylund DB (2007) Comparison of the maturation of the adrenergic and serotonergic neurotransmitter systems in the brain: implications for differential drug effects on juveniles and adults. Biochem Pharmacol 73(8):1225–1236. doi:10.1016/j.bcp.2007.01.028

Narboux-Neme N, Sagne C, Doly S, Diaz SL, Martin CB, Angenard G, Martres MP, Giros B, Hamon M, Lanfumey L, Gaspar P, Mongeau R (2011) Severe serotonin depletion after conditional deletion of the vesicular monoamine transporter 2 gene in serotonin neurons: neural and behavioral consequences. Neuropsychopharmacology 36(12):2538–2550. doi:10.1038/npp.2011.142

Narboux-Neme N, Angenard G, Mosienko V, Klempin F, Pitychoutis PM, Deneris E, Bader M, Giros B, Alenina N, Gaspar P (2013) Postnatal growth defects in mice with constitutive depletion of central serotonin. ACS Chem Neurosci 4(1):171–181. doi:10.1021/cn300165x

Nattie E, Li A (2010) Central chemoreception in wakefulness and sleep: evidence for a distributed network and a role for orexin. J Appl Physiol (1985) 108(5):1417–1424. doi:10.1152/japplphysiol.01261.2009

Oberlander TF, Gingrich JA, Ansorge MS (2009) Sustained neurobehavioral effects of exposure to SSRI antidepressants during development: molecular to clinical evidence. Clin Pharmacol Ther 86(6):672–677. doi:10.1038/clpt.2009.201

Olson EB Jr, Dempsey JA, McCrimmon DR (1979) Serotonin and the control of ventilation in awake rats. J Clin Invest 64(2):689–693. doi:10.1172/JCI109510

Paterson DS, Hilaire G, Weese-Mayer DE (2009) Medullary serotonin defects and respiratory dysfunction in sudden infant death syndrome. Respir Physiol Neurobiol 168(1–2):133–143. doi:10.1016/j.resp.2009.05.010

Paterson DS, Thompson EG, Belliveau RA, Antalffy BA, Trachtenberg FL, Armstrong DD, Kinney HC (2005) Serotonin transporter abnormality in the dorsal motor nucleus of the vagus in Rett syndrome: potential implications for clinical autonomic dysfunction. J Neuropathol Exp Neurol 64(11):1018–1027

Paterson DS, Trachtenberg FL, Thompson EG, Belliveau RA, Beggs AH, Darnall R, Chadwick AE, Krous HF, Kinney HC (2006) Multiple serotonergic brainstem abnormalities in sudden infant death syndrome. JAMA 296(17):2124–2132. doi:10.1001/jama.296.17.2124

Peever JH, Duffin J (2001) Respiratory control of hypoglossal motoneurons. Adv Exp Med Biol 499:101–106

Pena F, Ramirez JM (2002) Endogenous activation of serotonin-2A receptors is required for respiratory rhythm generation in vitro. J Neurosci 22(24):11055–11064

Penatti E, Barina A, Schram K, Li A, Nattie E (2011) Serotonin transporter null male mouse pups have lower ventilation in air and 5% CO_2 at postnatal ages P15 and P25. Respir Physiol Neurobiol 177(1):61–65. doi:10.1016/j.resp.2011.02.006

Persico AM, Mengual E, Moessner R, Hall FS, Revay RS, Sora I, Arellano J, DeFelipe J, Gimenez-Amaya JM, Conciatori M, Marino R, Baldi A, Cabib S, Pascucci T, Uhl GR, Murphy DL, Lesch KP, Keller F (2001) Barrel pattern formation requires serotonin uptake by thalamocortical afferents, and not vesicular monoamine release. J Neurosci 21(17):6862–6873

Popova NK, Naumenko VS, Plyusnina IZ (2007) Involvement of brain serotonin 5-HT1A receptors in genetic predisposition to aggressive behavior. Neurosci Behav Physiol 37(6):631–635. doi:10.1007/s11055-007-0062-z

Ptak K, Yamanishi T, Aungst J, Milescu LS, Zhang R, Richerson GB, Smith JC (2009) Raphe neurons stimulate respiratory circuit activity by multiple mechanisms via endogenously released serotonin and substance P. J Neurosci 29(12):3720–3737. doi:10.1523/JNEUROSCI.5271-08.2009

Ramanantsoa N, Hirsch MR, Thoby-Brisson M, Dubreuil V, Bouvier J, Ruffault PL, Matrot B, Fortin G, Brunet JF, Gallego J, Goridis C (2011) Breathing without CO(2) chemosensitivity in conditional Phox2b mutants. J Neurosci 31(36):12880–12888. doi:10.1523/JNEUROSCI.1721-11.2011

Ramirez JM, Ward CS, Neul JL (2013) Breathing challenges in Rett syndrome: lessons learned from humans and animal models. Respir Physiol Neurobiol 189(2):280–287. doi:10.1016/j.resp.2013.06.022

Real C, Popa D, Seif I, Callebert J, Launay JM, Adrien J, Escourrou P (2007) Sleep apneas are increased in mice lacking monoamine oxidase A. Sleep 30(10):1295–1302

Rekling JC, Funk GD, Bayliss DA, Dong XW, Feldman JL (2000) Synaptic control of motoneuronal excitability. Physiol Rev 80(2):767–852

Richerson GB (2004) Serotonergic neurons as carbon dioxide sensors that maintain pH homeostasis. Nat Rev Neurosci 5(6):449–461. doi:10.1038/nrn1409

Richter DW, Smith JC (2014) Respiratory rhythm generation in vivo. Physiology (Bethesda) 29(1):58–71. doi:10.1152/physiol.00035.2013

Riederer P, Brucke T, Sofic E, Kienzl E, Schnecker K, Schay V, Kruzik P, Killian W, Rett A (1985) Neurochemical aspects of the Rett syndrome. Brain and Development 7(3):351–360

Riederer P, Weiser M, Wichart I, Schmidt B, Killian W, Rett A (1986) Preliminary brain autopsy findings in progredient Rett syndrome. Am J Med Genet Suppl 1:305–315

Roux JC, Panayotis N, Dura E, Villard L (2010) Progressive noradrenergic deficits in the locus coeruleus of Mecp2 deficient mice. J Neurosci Res 88(7):1500–1509. doi:10.1002/jnr.22312

Salichon N, Gaspar P, Upton AL, Picaud S, Hanoun N, Hamon M, De Maeyer E, Murphy DL, Mossner R, Lesch KP, Hen R, Seif I (2001) Excessive activation of serotonin (5-HT) 1B receptors disrupts the formation of sensory maps in monoamine oxidase a and 5-ht transporter knock-out mice. J Neurosci 21(3):884–896

Savelieva KV, Zhao S, Pogorelov VM, Rajan I, Yang Q, Cullinan E, Lanthorn TH (2008) Genetic disruption of both tryptophan hydroxylase genes dramatically reduces serotonin and affects behavior in models sensitive to antidepressants. PLoS One 3(10):e3301. doi:10.1371/journal.pone.0003301

Schaefer TL, Vorhees CV, Williams MT (2009) Mouse plasmacytoma-expressed transcript 1 knock out induced 5-HT disruption results in a lack of cognitive deficits and an anxiety phenotype complicated by hypoactivity and defensiveness. Neuroscience 164(4):1431–1443. doi:10.1016/j.neuroscience.2009.09.059

Shannon DC, Kelly DH, O'Connell K (1977) Abnormal regulation of ventilation in infants at risk for sudden-infantdeath syndrome. N Engl J Med 297(14):747–750. doi:10.1056/NEJM197710062971403

Smith JC, Abdala AP, Borgmann A, Rybak IA, Paton JF (2013) Brainstem respiratory networks: building blocks and microcircuits. Trends Neurosci 36(3):152–162. doi:10.1016/j.tins.2012.11.004

Stankovski L, Alvarez C, Ouimet T, Vitalis T, El-Hachimi KH, Price D, Deneris E, Gaspar P, Cases O (2007) Developmental cell death is enhanced in the cerebral cortex of mice lacking the brain vesicular monoamine transporter. J Neurosci 27(6):1315–1324. doi:10.1523/JNEUROSCI.4395-06.2007

Takakura AC, Barna BF, Cruz JC, Colombari E, Moreira TS (2014) Phox2b-expressing retrotrapezoid neurons and the integration of central and peripheral chemosensory control of breathing in conscious rats. Exp Physiol 99(3):571–585. doi:10.1113/expphysiol.2013.076752

Tang S, Machaalani R, Waters KA (2012) Expression of brain-derived neurotrophic factor and TrkB receptor in the sudden infant death syndrome brainstem. Respir Physiol Neurobiol 180(1):25–33. doi:10.1016/j.resp.2011.10.004

Taylor NC, Li A, Green A, Kinney HC, Nattie EE (2004) Chronic fluoxetine microdialysis into the medullary raphe nuclei of the rat, but not systemic administration, increases the ventilatory response to CO2. J Appl Physiol (1985) 97(5):1763–1773. doi:10.1152/japplphysiol.00496.2004

Teran FA, Massey CA, Richerson GB (2014) Serotonin neurons and central respiratory chemoreception: where are we now? Prog Brain Res 209:207–233. doi:10.1016/B978-0-444-63274-6.00011-4

Tryba AK, Pena F, Ramirez JM (2006) Gasping activity in vitro: a rhythm dependent on 5-HT2A receptors. J Neurosci 26(10):2623–2634. doi:10.1523/JNEUROSCI.4186-05.2006

Upton AL, Salichon N, Lebrand C, Ravary A, Blakely R, Seif I, Gaspar P (1999) Excess of serotonin (5-HT) alters the segregation of ispilateral and contralateral retinal projections in monoamine oxidase A knock-out mice: possible role of 5-HT uptake in retinal ganglion cells during development. J Neurosci 19(16):7007–7024

Upton AL, Ravary A, Salichon N, Moessner R, Lesch KP, Hen R, Seif I, Gaspar P (2002) Lack of 5-HT(1B) receptor and of serotonin transporter have different effects on the segregation of retinal axons in the lateral geniculate nucleus compared to the superior colliculus. Neuroscience 111:597–610

Vacher CM, Calas A, Maltonti F, Hardin-Pouzet H (2004) Postnatal regulation by monoamines of vasopressin expression in the neuroendocrine hypothalamus of MAO-A-deficient mice. Eur J Neurosci 19(4):1110–1114

Velasquez JC, Goeden N, Bonnin A (2013) Placental serotonin: implications for the developmental effects of SSRIs and maternal depression. Front Cell Neurosci 7:47. doi:10.3389/fncel.2013.00047

Viemari JC, Hilaire G (2003) Monoamine oxidase A-deficiency and noradrenergic respiratory regulations in neonatal mice. Neurosci Lett 340(3):221–224

Viemari JC, Roux JC, Tryba AK, Saywell V, Burnet H, Pena F, Zanella S, Bevengut M, Barthelemy-Requin M, Herzing LB, Moncla A, Mancini J, Ramirez JM, Villard L, Hilaire G (2005) Mecp2 deficiency disrupts norepinephrine and respiratory systems in mice. J Neurosci 25(50):11521–11530. doi:10.1523/JNEUROSCI.4373-05.2005

Vishnivetskaya GB, Skrinskaya JA, Seif I, Popova NK (2007) Effect of MAO A deficiency on different kinds of aggression and social investigation in mice. Aggress Behav 33(1):1–6. doi:10.1002/ab.20161

Walther DJ, Bader M (2003) A unique central tryptophan hydroxylase isoform. Biochem Pharmacol 66(9):1673–1680

Wang H, Chan SA, Ogier M, Hellard D, Wang Q, Smith C, Katz DM (2006) Dysregulation of brain-derived neurotrophic factor expression and neurosecretory function in Mecp2 null mice. J Neurosci 26(42):10911–10915. doi:10.1523/JNEUROSCI.1810-06.2006

Wang S, Shi Y, Shu S, Guyenet PG, Bayliss DA (2013) Phox2b-expressing retrotrapezoid neurons are intrinsically responsive to H+ and CO2. J Neurosci 33(18):7756–7761. doi:10.1523/JNEUROSCI.5550-12.2013

Weese-Mayer DE, Lieske SP, Boothby CM, Kenny AS, Bennett HL, Ramirez JM (2008) Autonomic dysregulation in young girls with Rett Syndrome during nighttime in-home recordings. Pediatr Pulmonol 43(11):1045–1060. doi:10.1002/ppul.20866

Weese-Mayer DE, Rand CM, Berry-Kravis EM, Jennings LJ, Loghmanee DA, Patwari PP, Ceccherini I (2009) Congenital central hypoventilation syndrome from past to future: model for translational and transitional autonomic medicine. Pediatr Pulmonol 44(6):521–535. doi:10.1002/ppul.21045

Zanella S, Tauber M, Muscatelli F (2009) Breathing deficits of the Prader-Willi syndrome. Respir Physiol Neurobiol 168(1–2):119–124. doi:10.1016/j.resp.2009.03.010

Zhang X, Su J, Cui N, Gai H, Wu Z, Jiang C (2011) The disruption of central CO_2 chemosensitivity in a mouse model of Rett syndrome. Am J Physiol Cell Physiol 301(3):C729–738. doi:10.1152/ajpcell.00334.2010

Zhao ZQ, Chiechio S, Sun YG, Zhang KH, Zhao CS, Scott M, Johnson RL, Deneris ES, Renner KJ, Gereau RWt, Chen ZF (2007) Mice lacking central serotonergic neurons show enhanced inflammatory pain and an impaired analgesic response to antidepressant drugs. J Neurosci 27(22):6045–6053. doi:10.1523/JNEUROSCI.1623-07.2007

Chapter 12
Neural Network Reconfigurations: Changes of the Respiratory Network by Hypoxia as an Example

Fernando Peña-Ortega

Abstract Neural networks, including the respiratory network, can undergo a reconfiguration process by just changing the number, the connectivity or the activity of their elements. Those elements can be either brain regions or neurons, which constitute the building blocks of macrocircuits and microcircuits, respectively. The reconfiguration processes can also involve changes in the number of connections and/or the strength between the elements of the network. These changes allow neural networks to acquire different topologies to perform a variety of functions or change their responses as a consequence of physiological or pathological conditions. Thus, neural networks are not hardwired entities, but they constitute flexible circuits that can be constantly reconfigured in response to a variety of stimuli. Here, we are going to review several examples of these processes with special emphasis on the reconfiguration of the respiratory rhythm generator in response to different patterns of hypoxia, which can lead to changes in respiratory patterns or lasting changes in frequency and/or amplitude.

Keywords Network topology • Intrinsic properties • Synaptic properties • Gasping • Long term facilitation

Abbreviations

BDNF   Brain-derived neurotrophic factor
CNS   Central nervous system
D1/D2   Dopamine receptors
EEG   Electroencephalography

F. Peña-Ortega (✉)
Departamento de Neurobiología del Desarrollo y Neurofisiología, Instituto de Neurobiología,
Universidad Nacional Autónoma de México, UNAM-Campus Juriquilla,
Boulevard Juriquilla 3001, Querétaro 76230, Mexico
e-mail: jfpena@unam.mx

© Springer International Publishing AG 2017
R. von Bernhardi et al. (eds.), *The Plastic Brain*, Advances in Experimental
Medicine and Biology 1015, DOI 10.1007/978-3-319-62817-2_12

E	Expiratory
FFA	Flufenamic acid
fMRI	Functional Magnetic resonance imaging
I	Inspiratory
I_{CAN}	Ca^{2+}-activated cationic current
I_{NaP}	Persistent Na^+ current
LTF	Long-term facilitation
MEA	Multielectrode array
NTS	Nucleus tractus solitarius
PI	Postinspiratory
preBötC	Pre-Bötzinger complex
RVLM	Rostral ventrolateral medulla
STP	Short-term potentiation

Introduction

The term neural network has been broadly used in neuroscience to identify groups of neurons, neuronal assemblies, or brain regions that are interconnected in specific configurations or that are simultaneously active during a particular process (Lindsey et al. 2000; Morris et al. 2000; Greicius et al. 2003; Ramirez et al. 2004, 2007; Carrillo-Reid et al. 2008; Segers et al. 2008; Galán et al. 2010; Jáidar et al. 2010; Ott et al. 2011; Peña-Ortega 2013; Zhang et al. 2016; Schultz and Cole 2016). Thus, the "nodes" of these networks can be either neurons (forming microcircuits) or brain regions (forming macrocircuits) that are functionally or anatomically interconnected (Greicius et al. 2003; Nieto-Posadas et al. 2014; Cohen and D'Esposito 2016; Zhang et al. 2016; Schultz and Cole 2016; Qin et al. 2017; Flores-Martínez and Peña-Ortega, 2017). The nature of the connections between the nodes can be established from structural (e.g., synaptic contacts or tractography) (Zavala-Tecuapetla et al. 2014; Vargas-Barroso et al. 2016; Qin et al. 2017) or dynamic observations (e.g., signal correlations) (Hartelt et al. 2007; Nieto-Posadas et al. 2014; Cohen and D'Esposito 2016; Becker et al. 2016; Zhang et al. 2016).

Regardless of the organizational level (from microcircuits to macrocircuits), the configurations of neural networks can be defined by the group of nodes or units, representing different regions or neurons, respectively, and the connections among them, representing their communication pathways and strengths (Carrillo-Reid et al. 2008; Segers et al. 2008; Galán et al. 2010; Jáidar et al. 2010; Ott et al. 2011; Nieto-Posadas et al. 2014; De Vico Fallani et al. 2014; Lv et al. 2016; Guedj et al. 2016). Of course, the functioning of these networks cannot be explained by studying only one component of the circuit; it requires consideration of the emergent properties produced by the dynamic interactions among the elements of the entire network (Hooper and Marder 1987; Hooper and Moulins 1990; De Vico Fallani et al. 2014).

It is necessary to consider that neural networks are not hard-wired, but are flexible systems that may acquire different configurations to perform different neural processes (Hooper and Moulins 1990; Marder et al. 1996; Parker et al. 1998; Thoby-Brisson and Simmers 1998; Peña et al. 2004; Ramirez et al. 2004; Nieto-Posadas et al. 2014; Gandolfi et al. 2015; Brovelli et al. 2017), making this scenario a bit more complex. Furthermore, the changes in network configuration may persist for minutes to hours or even days to weeks (Marder et al. 1996; Parker et al. 1998; Thoby-Brisson and Simmers 1998; Gandolfi et al. 2015; Lv et al. 2016; Guedj et al. 2016; Becker et al. 2016). Thus, neuronal networks are highly plastic and can reconfigure in a state-dependent manner to produce a variety of network configurations with their respective outputs (Peña 2009; Nieto-Posadas et al. 2014; Cohen and D'Esposito 2016; Schultz and Cole 2016; Brovelli et al. 2017). The reconfiguration process allows neuronal networks to become efficient multifunctional entities capable of producing more than one output (Marder 1994; Peña 2009) by having flexible neural network boundaries; thus, it is possible to create the composition of different neural assemblies as required (Marder 1994; Peña 2009; Carrillo-Reid et al. 2008, 2011).

The fact that neuronal networks can assume different configurations to produce different forms of activities was first established by studying the neuronal networks of invertebrates (Nagy and Dickinson 1983; Flamm and Harris-Warrick 1986; Nusbaum and Marder 1989; Harris-Warrick and Marder 1991; Marder and Calabrese 1996; Ramirez 1998). The hypothesis that individual neuronal networks could reconfigure to produce multiple outputs was first advanced by Getting (1983) based on studies of the sea slug *Tritonia,* and today this phenomenon has also been observed in the mammalian nervous system (including humans), both at the microcircuit (Peña 2009; Nieto-Posadas et al. 2014; Gandolfi et al. 2015) and at the macrocircuit levels (Cao and Slobounov 2010; Hermans et al. 2011; Choi et al. 2012; Zhang et al. 2016; Brovelli et al. 2017).

Study of neuronal network configuration has recently benefited from the development of methodological approaches that can simultaneously evaluate the activity of dozens of nodes/units at the microcircuit level; for instance, voltage-sensitive dye imaging (Koshiya et al. 2014; Gandolfi et al. 2015), multielectrode recordings (Nieto-Posadas et al. 2014; Lorea-Hernández et al. 2016), and multineuronal, single-cell calcium imaging techniques (Sasaki et al. 2007; Carrillo-Reid et al. 2008; Peña et al. 2010; Flores-Martínez and Peña-Ortega, 2017) or neuroimaging techniques that can visualize the activity of the whole brain, such as functional magnetic resonance imaging (fMRI) (Hermans et al. 2011; Choi et al. 2012; Cohen and D'Esposito 2016; Becker et al. 2016; Zhang et al. 2016; Schultz and Cole 2016; Guedj et al. 2016) or electroencephalography (EEG) (Ponten et al. 2007; Cao and Slobounov 2010; Zhang et al. 2016; Brovelli et al. 2017). Moreover, the development of a variety of analytical tools has allowed not only the precise determination of network connectivity with unprecedented detail, but also the detection of dynamic changes in such connectivity under different physiological and pathological conditions (Carter et al. 2012; Zhang et al. 2016). Graph metrics, such as the clustering coefficient, path length, and efficiency measures, are often used to characterize the topology of brain networks (Bassett et al. 2006; Cohen and D'Esposito 2016), whereas centrality metrics, such as degree, betweenness, closeness, and eigenvector centrality, are used to identify crucial areas within the network (Rubinov and Sporns 2010). Community structure analysis, which detects those groups of regions more densely connected to each other, is also essential for understanding brain network organization and topology (Rubinov and Sporns 2010). Thus, the combination of these techniques and analytical approaches is allowing us to unravel the precise network configurations, from micro- to macroscopic levels, involved in the generation of different behaviors or brain computations (Choi et al. 2012; Nieto-Posadas et al. 2014), and it is also providing us with a view of the specific changes in the configuration within these neural networks that are involved in generating different network outputs under physiological (Peña et al. 2004; Nieto-Posadas et al. 2014; Schultz and Cole 2016; Guedj et al. 2016) or pathological conditions (Choi et al. 2012; Nieto-Posadas et al. 2014; Zhang et al. 2016; Qin et al. 2017; Flores-Martínez and Peña-Ortega, 2017). Next, I will provide some examples of reconfiguration processes in physiological and pathological conditions using various methodological approaches at the micro- and macrocircuit levels.

At the macrocircuit level (which involves brain regions), it has been shown by EEG that mild traumatic brain injury produces a significant decrease in long-distance connectivity and a significant increase in short-distance connectivity, which together reduce the brain's "small-world" properties (Cao and Slobounov 2010). A network exhibiting small-world properties is more clustered (high clustering coefficient) than a random network, but has a similar path length (distance between nodes) than the one found in a random network, which can be achieved by the presence of hub-nodes (Watts and Strogatz 1998; Rubinov and Sporns 2010). In contrast to mild traumatic brain injury, during seizures in the epileptic brain (ictal periods), neuronal networks move toward a more ordered configuration, with high clustering among brain areas compared to a more randomly organized interictal network configuration (Ponten et al. 2007). By using fMRI, it has been shown that seizures are also related to an inhibitory effect on the default mode of brain function, which gradually disappears after the seizure (Liao et al. 2014). It has also been shown with this technique that during exposure to a fear-related acute stressor, responsiveness and interconnectivity within the brain increases (Hermans et al. 2011), and that during a delirium episode, the functional connectivities of the intralaminar thalamic nuclei and of the caudate nuclei with other subcortical regions were reversibly reduced (Choi et al. 2012). Recently, it has been shown that a fast reconfiguration of connections across brain regions occurs during different cognitive tasks (Cohen et al. 2014; Schultz and Cole 2016; Brovelli et al. 2017) and that those changes in connectivity closely correlate with the performance of the subjects in those tasks (Cohen et al. 2014; Schultz and Cole 2016; Brovelli et al. 2017).

At the microcircuit level (which involves individual neurons), multielectrode arrays show that a brief glutamate exposure produces hypersynchrony in primary cultures of hippocampal neurons and a loss of its small-world network topology

(Srinivas et al. 2007). A different scenario is seen with multielectrode arrays and hippocampal slices (Gong et al. 2014), where the interactions among the neural groups appear to be random in control conditions; but the network becomes more regular and connections increase during epileptiform activity (Gong et al. 2014). Furthermore, in contrast to primary cell cultures (Srinivas et al. 2007), the epileptic hippocampal network exhibits small-world properties, whereas during control activity it does not (Gong et al. 2014). As for more permanent reconfiguration at the microcircuit level assessed by array recordings in vivo, it has been found that as learning progresses, some CA1 interneurons are connected to new pyramidal assemblies, while some others dissociate from them (Dupret et al. 2013).

Another experimental approach to study neural network configuration at the microcircuit level is multineuron calcium imaging (Sasaki et al. 2007; Peña et al. 2010; Jáidar et al. 2012; Carrillo-Reid et al. 2008, 2011; Flores-Martínez and Peña-Ortega, 2017). This approach has shown that the activation of dopamine receptors in striatal slices increases neuronal synchronization and induces different topological configurations depending on whether the D1 or the D2 receptor is activated (Carrillo-Reid et al. 2011). This same technique reveals that even in control conditions, the CA3 area of hippocampal slices spontaneously produces different "network states" that were defined by specific active cell ensembles (Sasaki et al. 2007). Interestingly, a particular network state can be active for a long period of time (tens of seconds), but eventually switches (state-transition) to different, long-lasting network states (different active cell ensembles) (Sasaki et al. 2007). This observation corroborates a similar finding using multielectrode arrays in cortical slice cultures (Beggs and Plenz 2004). These network states were stabilized by synaptic activity and maintained despite external perturbations (Sasaki et al. 2007). Later, it was established that such states are generated by a network with scale-free structure, which engaged varying sets of densely interwired (thus highly synchronized) neuron groups (Takahashi et al. 2010).

Besides the configuration changes that occur spontaneously in the hippocampal and cortical networks, neuromodulators can change network configurations both in vivo (Becker et al. 2016; Guedj et al. 2016; Lv et al. 2016) and in vitro (Peña et al. 2010; Peña-Ortega 2013; Isla et al. 2016; Flores-Martínez and Peña-Ortega 2017). For instance, in vitro cholinergic modulation in the cortex reduces weak, pairwise relationships and excludes neurons that during circuit activity are already unreliable (Runfeldt et al. 2014); and dopamine reorganizes the hippocampal network by expanding the diversity of active cell assemblies (Miyawaki et al. 2014). Changes in network configuration at the microcircuit level and related to a specific behavioral state have been reported by using multineuron calcium imaging in the auditory cortex of lactating rodents (Rothschild et al. 2013), which show an increase in pairwise and higher-order correlations among neurons of the auditory cortex in lactating mothers (Rothschild et al. 2013). A long-lasting reconfiguration has also been revealed through calcium imaging in hippocampal slices during associative synaptic plasticity, wherein neural ensembles change by incorporating neurons belonging to various ensembles (Yuan et al. 2011). We have recently observed that voluntary exercise produces a long-lasting change in the hippocampal network activity (Isla et al. 2016). In vivo calcium imaging in the behaving rat has shown that correlations between pairs of neurons increased throughout learning in the motor cortex (Komiyama et al. 2010).

In addition to all the examples of network reconfiguration discussed so far, among the best examples of network flexibility are the reconfiguration processes that the respiratory network undergoes under different hypoxic conditions (Lieske et al. 2000; Morris et al. 2000; Galán et al. 2010; Peña-Ortega 2012). It is known that during prolonged hypoxic conditions, the respiratory network changes its motor output from normal breathing (eupnea) to gasping (Lieske et al. 2000; Nieto-Posadas et al. 2014; Rivera-Angulo and Peña-Ortega 2014), and that a short-term potentiation of the respiratory network output can be observed after hypoxic exposure (Blitz and Ramirez 2002; Galán et al. 2010; Lee et al. 2015; Sandhu et al. 2015) (for further discussion on short-term plasticity, see Chap. 3). Interestingly, in all cases the changes in respiratory activity are accompanied by changes in network configuration (Galán et al. 2010; Nieto-Posadas et al. 2014). Additionally, the exposure to intermittent hypoxia produces a long-term increase in respiratory motor output that seems to also involve a change in respiratory network configuration (Morris et al. 2000; Blitz and Ramirez 2002).

Changes in the Respiratory Network Under Hypoxia

The Reconfiguration Leading to Gasping Generation

Hypoxia can be caused by a reduction in partial pressure of oxygen in the environment, inadequate oxygen transport, or the inability of tissues or organs to use oxygen (Peña and Ramirez 2005; Ramirez et al. 2007; Peña 2009). Hypoxia, as well as hypercapnia, are also associated with diverse pathological conditions, such as those occurring during a compromise in blood supply or during respiratory dysfunction (e.g., obstructive sleep apnea) (Peña and Ramirez 2005; Peña 2009). The respiratory network responds to severe hypoxia (normally associated with hypercapnia) by generating gasping, which is considered to be the 'last-resort' respiratory effort to auto-resuscitate and sustain life. Indeed, failure to respond in this way can result in death, contributing to Sudden Infant Death Syndrome (SIDS) (Hunt 1992; Poets et al. 1999), as previously mentioned in Chaps. 8 and 11.

Breathing is commanded and regulated by the respiratory circuits of the brainstem (Smith et al. 1991; Peña and García 2006; Peña-Ortega 2012). Within the respiratory network, the central respiratory pattern generator is located in the pre-Bötzinger complex (preBötC; Smith et al. 1991) (see Chaps. 8, 9 and 10). The pre-BötC contains several types of neurons, including expiratory, inspiratory, and postinspiratory neurons (Peña and Ramirez 2002, 2004; Zavala-Tecuapetla et al. 2014) (Fig. 12.1) that interact through fast synaptic, and often reciprocal, connections that can be either inhibitory (GABA and glycine) (Fig. 12.1) or excitatory

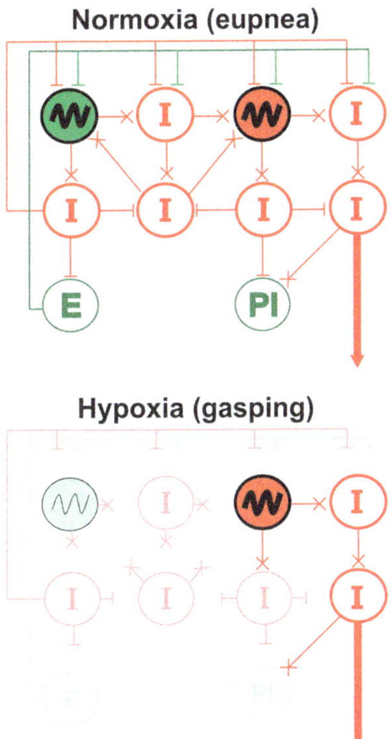

Fig. 12.1 Schematic representation of the reconfiguration of the preBötzinger complex network in hypoxia generated by cell-focused studies. The schematic diagram illustrates the proposed reconfiguration of the respiratory network from normoxia (eupnea) to hypoxia (gasping) based on cell-focused evidence. In normoxia, the network contains expiratory (*E*), postinspiratory (*PI*), and inspiratory (*I*) neurons reciprocally connected by inhibitory (represented by the *T-bars*) and excitatory synapses (represented by the *X-bars*). I_{CAN}-dependent pacemaker neurons (*green circles*) coexist with inspiratory I_{NaP}-dependent pacemaker neurons (*red circles*). Both pacemakers participate in eupnea generation, whereas in hypoxia most neurons shut down, including I_{CAN}-dependent pacemaker neurons, and a reduction in synaptic transmission is observed (mainly in inhibitory synapses). Thus, the gasping respiratory network relies on I_{NaP}-dependent pacemaker neurons (Modified from Peña 2009)

(Peña and Ramirez 2002, 2004; Nieto-Posadas et al. 2014). There are also intrinsic bursting respiratory pacemaker neurons within the preBötC (Del Negro et al. 2002, 2005; Peña and Ramirez 2002, 2004; Peña 2008). There are at least two types of intrinsic bursting respiratory pacemaker neurons (Peña et al. 2004; Del Negro et al. 2005; Peña 2008; Ramírez-Jarquín et al. 2012) (Fig. 12.1), which can be distinguished by their electrophysiological properties (Peña et al. 2004; Del Negro et al. 2005; Peña 2008) and their sensitivity to ion channel blockers (Peña et al. 2004; Del Negro et al. 2005; Peña 2008). One type of pacemaker neuron relies on the Ca^{2+}-activated cationic current (I_{CAN}) to generate intrinsic bursting (see Chaps. 9 and 10), which is selectively blocked by the I_{CAN} blocker flufenamic acid (FFA) (Peña et al. 2004; Del Negro et al. 2005; Peña and Ordaz 2008). In contrast, the other type of pacemaker relies on the persistent Na^+ current (I_{NaP}) to generate intrinsic bursting, which is selectively blocked by the I_{NaP} blocker riluzole (Del Negro et al. 2005; Peña et al. 2004; Peña 2009). It has been shown that under normoxic conditions, the blockade of both pacemaker neuron populations is required to abolish eupnea generation both in vitro and in vivo (Peña et al. 2004; Peña and Aguileta 2007). In contrast, the blockade of only one of these populations is not enough to stop rhythm generation in normoxia (Peña et al. 2004; Peña and Aguileta 2007).

The hypoxic respiratory response in vitro and in vivo is biphasic and leads, after a long period of hypoxia, to a change in the respiratory pattern (Lieske et al. 2000; Peña et al. 2004; Zavala-Tecuapetla et al. 2008, Peña et al. 2008; Armstrong et al. 2010). Initially, the frequency and amplitude of the respiratory activity increases and, if hypoxic conditions are maintained, a depression of the respiratory activity leads to gasping generation (Hunt 1992; Poets et al. 1999; Peña and Aguileta 2007; Lorea-Hernández et al. 2016). Gasping, in contrast to eupnea, is characterized by rapidly rising inspiratory activity without subsequent expiratory activity, both in vivo and in vitro (Lieske et al. 2000; Peña et al. 2008; Peña and Aguileta 2007; Ramírez-Jarquín et al. 2012; Nieto-Posadas et al. 2014). Gasping is also characterized by shorter inspiratory durations (Lieske et al. 2000; Peña et al. 2008; Peña and Aguileta 2007; Ramírez-Jarquín et al. 2012; Nieto-Posadas et al. 2014). Interestingly, Lieske et al. (2000) showed that both eupnea and gasping patterns are generated within the preBötC and proposed that a reconfiguration process is responsible for the reversible change in respiratory patterns depending on oxygen availability.

Cell-focused studies have shown that hypoxia, which can certainly be associated with secondary acidosis (Roberts et al. 1998), dramatically changes the intrinsic and synaptic properties of the respiratory network (Peña and Ramirez 2005; Peña 2009) (Fig. 12.1). The effects of hypoxia on the intrinsic excitability of the preBötC respiratory neurons are heterogeneous: whereas some cell-focused studies reported that hypoxia depressed the firing of large subsets of respiratory neurons (St John and Bianchi 1985; Richter et al. 1991; Ballanyi et al. 1994; England et al. 1995; St John 1999; Thoby-Brisson and Ramirez 2000), others showed that hypoxia increased (Richter et al. 1991; Lovering et al. 2006) or did not change the firing of neurons within the respiratory network (Lieske et al. 2000; Thoby-Brisson and Ramirez 2000; Lovering et al. 2006). Likewise, one subtype of pacemaker neurons maintains its bursting properties during prolonged hypoxia without an apparent change in membrane potential (Thoby-Brisson and Ramirez 2000; Peña et al. 2004) (Fig. 12.1), whereas another type of pacemaker loses its ability to maintain intrinsic bursting activity in hypoxia (Thoby-Brisson and Ramirez 2000; Peña et al. 2004) (Fig. 12.1). Despite the one study using multi-neuron calcium imaging in vitro that did not find a reduction in the number of active respiratory neurons during hypoxia (Barnes et al. 2007), our cell-focused studies allowed us to propose a model of gasping generation (Fig. 12.1), in which a considerable proportion of respiratory neurons shut down in hypoxia (Ramirez et al. 2007; Peña 2009).

Along with the changes in the intrinsic properties of the respiratory network, it is well known that hypoxia, which can certainly be associated with secondary acidosis (Roberts et al. 1998), induces changes in synaptic transmission in the respiratory network (Peña and Ramirez 2005; Peña 2009). The most consistent effect of hypoxia on the preBötC is a suppression of synaptic inhibition (Fig. 12.1), which has been observed both in vivo and in vitro (Richter et al. 1991; Ballanyi et al. 1994; England et al. 1995; Schmidt et al. 1995; Ramirez et al. 1998; Lieske et al. 2000; Thoby-Brisson and Ramirez 2000). This depression of synaptic inhibition leads expiratory neurons to cease discharging rhythmically (Lieske et al. 2000; Thoby-Brisson and Ramirez 2000), and it leads postinspiratory neurons to fire in phase with inspiration (Lieske et al. 2000). The effect of hypoxia on glutamatergic transmission is highly heterogeneous (Ballanyi et al. 1994, 1999; Ramirez et al. 1998); there is a component of glutamatergic transmission that is depressed in hypoxia, and it has been detected as a reduction in the phasic excitation of some inspiratory neurons (Ballanyi et al. 1994, 1999; Ramirez et al. 1998). Another component of glutamatergic transmission is resistant to hypoxia and maintains the synchronization of the respiratory neurons in the absence of oxygen (Ballanyi et al. 1994, 1999; Ramirez et al. 1998). Changes in electrical transmission is not involved in gasping generation because blockers of gap junctions fail to disrupt the hypoxic response of the respiratory network (Rodman et al. 2006).

As mentioned, pacemaker neurons are differentially modulated by hypoxia (Peña et al. 2004; Peña 2008) (Fig. 12.1). We have found that the application of prolonged hypoxic conditions in brainstem slices, which can certainly be associated with secondary acidosis (Roberts et al. 1998), dramatically reduced (almost abolished) the intrinsic bursting of I_{CAN}-dependent pacemakers; whereas I_{NaP}-dependent pacemakers under the same conditions continued bursting (Peña et al. 2004; Peña 2008) (Fig. 12.1). We proposed, based on these cell-focused results, that I_{NaP}-dependent pacemakers are responsible for gasping generation (Fig. 12.1). In fact, we found that blocking I_{NaP}-dependent pacemaker activity with riluzole abolishes gasping generation not just in vitro (Peña et al. 2004; Peña 2008) but also in vivo (Peña and Aguileta 2007); this has been confirmed using the in situ preparation as well (Paton et al. 2006). In summary, we proposed, based on cell-focused studies, that gasping generation relies on hypoxia-resistant and I_{NaP}-dependent pacemaker neurons (Peña et al. 2004; Peña 2008).

Despite the great insights about the respiratory network configurations obtained from cell-focused studies (St John and Bianchi 1985; Richter et al. 1991; Ballanyi et al. 1994; England et al. 1995; Ramirez et al. 1998; St John 1999; Lieske et al. 2000; Thoby-Brisson and Ramirez 2000; Peña et al. 2004, 2006; Zavala-Tecuapetla et al. 2008; Ramírez-Jarquín et al. 2012), a more detailed description of respiratory circuit configurations is emerging from structural imaging (Hartelt et al. 2008; Mironov et al. 2009) as well as from the evaluation of cell assemblies, while maintaining single-cell resolution using multineuron calcium imaging (Okada et al.

2012; Gourévitch and Mellen 2014) or multielectrode arrays (Morris et al. 2000; Segers et al. 2008; Galán et al. 2010; Ott et al. 2011; Carroll and Ramirez 2013; Carroll et al. 2013). These techniques are allowing functional network analysis of the respiratory network and may reveal distinct configurations in control conditions or under various physiological and/or pathological conditions (Ramirez et al. 2004; Carrillo-Reid et al. 2008; Mironov et al. 2009; Peña et al. 2010; Lorea-Hernández et al. 2016). For instance, the network-focused studies of the respiratory network have started to reveal that the preBötC is constituted of dense clusters of respiratory cells with occasional connections between them (Hartelt et al. 2008; Mironov et al. 2009; Gaiteri and Rubin 2011). Moreover, it looks like these respiratory ensembles can be reconfigured in a cycle-by-cycle manner (Carroll et al. 2013; Carroll and Ramirez 2013; Koshiya et al. 2014). Modeling based on this evidence indicates that preBötC activity is highly dependent on its circuit configurations, the intrinsic dynamics of neurons at central network positions, and the strength of synaptic connections between neurons (Gaiteri and Rubin 2011). With the help of multi-electrode array recordings (Lindsey et al. 2000; Segers et al. 2008; Galán et al. 2010; Ott et al. 2011; Carroll and Ramirez 2013; Carroll et al. 2013), we have recently established that, contrary to previous suggestions from cell-focused studies (Peña 2009; Fig. 12.1), the number of active respiratory elements and the number of their functional links do not change dramatically in hypoxia during the transition to fictive-gasping generation (Fig. 12.2). The main

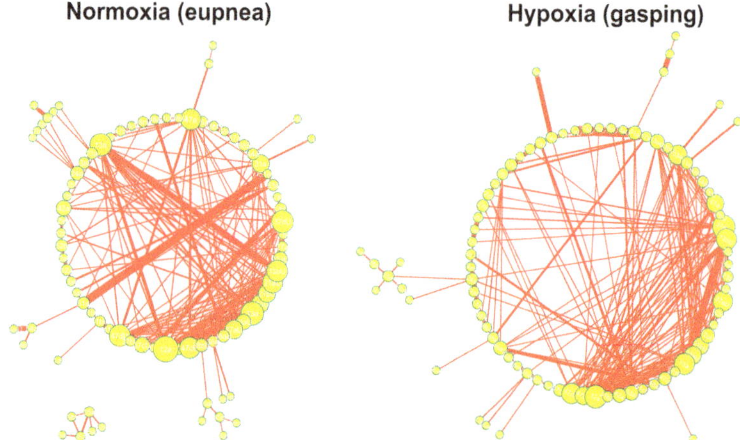

Fig. 12.2 Graph representation generated by network studies of the reconfiguration of the pre-Bötzinger complex network in hypoxia. One diagram represents the respiratory network in normoxia (eupnea) and the other in hypoxia (gasping). In both diagrams, each respiratory element is represented as a *yellow circle*, and each connection is represented as a *red line*. The diameter of the circles is proportional to the number of connections that each respiratory element has with other elements in the network. The width of the line is proportional to the strength of the connection. Note that the reconfiguration of the respiratory network in hypoxia is not due to a loss in the number of respiratory elements or of their connections, but to a reduction in the strength of such interactions (Modified from Nieto-Posadas et al. 2014)

change in the reconfiguration of the respiratory network during hypoxia involves a complex modification in the potency of the correlated activity between the elements of the circuit, suggesting a global reduction in the strength of network interactions (Nieto-Posadas et al. 2014; Fig. 12.2), which is accompanied by a complex modification of the firing frequency of the respiratory elements that, on average, leads to a global reduction in firing frequency (with just an exceptional cell shut down) (Nieto-Posadas et al. 2014). Thus, by using a network approach we have found that the reconfiguration of the preBötC under hypoxia, which can certainly be associated with secondary acidosis (Roberts et al. 1998), is a bit more complex (Fig. 12.2) than was thought using cell-focused approaches (Ramirez et al. 2007; Peña 2009; Fig. 12.1). In agreement with cell-focused studies that found that the effects of hypoxia on synaptic excitation were diverse (Ballanyi et al. 1994; Ramirez et al. 1998), we also found the heterogeneity of the effects of hypoxia on preBötC synaptic excitation using a network approach (Nieto-Posadas et al. 2014) (Fig. 12.2). Additionally, our observation of the heterogeneous effect of hypoxia on the firing frequency of the respiratory elements (Nieto-Posadas et al. 2014) confirmed the diversity of hypoxia-induced changes in firing frequency observed in cell-focused experiments (St John and Bianchi 1985; Richter et al. 1991; Ballanyi et al. 1994; England et al. 1995; St John 1999; Lieske et al. 2000; Thoby-Brisson and Ramirez 2000; Lovering et al. 2006). However, it is important to note that we did not observe the switching off of many respiratory neurons in hypoxia using a network approach (Nieto-Posadas et al. 2014) (Fig. 12.2) as suspected from the cell-focused approach (St John and Bianchi 1985; Richter et al. 1991; Ballanyi et al. 1994; England et al. 1995; St John 1999; Thoby-Brisson and Ramirez 2000) (Fig. 12.1) and as anticipated by using multineuron calcium imaging (Barnes et al. 2007). We also observed, using functional connectivity as is done in systems neurophysiology (Lindsey et al. 2000; Segers et al. 2008; Ott et al. 2011), the predicted heterogeneity in the changes of connections and interactions during hypoxic conditions (Ballanyi et al. 1994; Ramirez et al. 1998) (Fig. 12.2), and we provided quantitative evidence that these changes in functional coupling, and not in the number of respiratory elements or functional links among them, seem to be a major component of the respiratory network reconfiguration in hypoxia (Fig. 12.2). The observed re-arrangements of network interactions among the respiratory elements (Fig. 12.2) is consistent with a previous report that a brief application of cyanide (chemical hypoxia) induced the retraction of neuronal processes of respiratory neurons, which was interpreted as a reduction in connectivity among respiratory neurons (Mironov et al. 2009). Moreover, this change in the strength of interactions within the respiratory network in hypoxia can be related to the uncoupling of preBötC activity from one of its motor outputs (the hypoglossal nucleus) during fictive-gasping generation (Ramirez et al. 1998; Peña et al. 2008). In any case, the network analysis of the respiratory circuit under hypoxic conditions reveals that the network can become more vulnerable under these conditions (Peña 2009).

Other Possible Respiratory Reconfigurations Leading to Hypoxia-Induced Plasticity

It is well known that after a hypoxic period, which can certainly be associated with secondary acidosis (Roberts et al. 1998), the respiratory network undergoes a plastic change that leads to a short-term potentiation (STP) (Powell et al. 1998; Mitchell et al. 2001a, b; Blitz and Ramirez 2002; Hayashi et al. 2003; Sandhu et al. 2015; Lee et al. 2009a, b, 2014). STP of respiratory motor output has been described in humans (Fregosi 1991) and animal models (Powell et al. 1998; Mitchell et al. 2001a, b; Blitz and Ramirez 2002; Hayashi et al. 2003; Sandhu et al. 2015; Lee et al. 2009a, b, 2014), and it is manifested as a brief increase in respiratory activity after oxygen levels are restored (Fregosi 1991; Powell et al. 1998; Mitchell et al. 2001a, b; Blitz and Ramirez 2002; Hayashi et al. 2003; Sandhu et al. 2015; Lee et al. 2009a, b, 2014). It has been shown, by using multielectrode arrays, that during STP almost half of the respiratory interneurons in the C3–4 spinal cord increase their firing frequency (Sandhu et al. 2015), and it was found, by using bootstrap analysis of the synchrony between spike trains of respiratory neurons recorded with arrays in the brainstem, that during STP most pairs of respiratory elements were less synchronized, although some were more, suggesting that the respiratory network was transiently reconfigured after the hypoxic period (Galán et al. 2010).

Intermittent hypoxic episodes are typically a consequence of immature respiratory control (Mitchell et al. 2001a, b; Martin et al. 2012) and are a characteristic feature of sleep apnea (Gozal et al. 2001). It has been shown that chronic intermittent hypoxia causes neurocognitive deficits such as spatial learning impairments with increased cell death and structural changes to hippocampal and cortical regions (Gozal et al. 2001; Row et al. 2002, 2003; Klein et al. 2003; Cai et al. 2010; Nair et al. 2011; Xie and Yung 2012). There is also evidence that these effects are correlated with impairments in synaptic plasticity (Payne et al. 2004; Xie et al. 2010; Xie and Yung 2012). However, intermittent hypoxia can also have beneficial effects. It has been used as a method for training mountaineers and athletes (Beidleman et al. 2003; Millet et al. 2010), and it has even been applied as treatment and prevention for hypertension (Serebrovskaya et al. 2008) and ischemic coronary artery diseases (Zhu et al. 2006). In recent years, intermittent hypoxia has been found to exert many positive effects on the CNS in animal and human studies, such as enhancement of ventilation in normal subjects (Morris and Gozal 2004; Griffin et al. 2012) or in subjects with spinal cord injury (Trumbower et al. 2012; Hayes et al. 2014; Tester et al. 2014), increase in spatial learning and memory (Zhang et al. 2005; Lu et al. 2009), production of antidepressant-like effects (Zhu et al. 2010), and treatment of Parkinson's disease (Belikova et al. 2012). At the cellular level, it has been found that intermittent hypoxia-treated animals recovered synaptic transmission more rapidly following re-oxygenation in both the CA1 and the dentate gyrus (Wall et al. 2014) and that the maturation of hippocampal neurons is enhanced by postischemia intermittent hypoxia intervention (Tsai et al. 2011, 2013), which is reflected in the alleviation of ischemia-induced long-term memory impairment via brain-derived neurotrophic factor (BDNF) expression (Tsai et al. 2011, 2013).

As mentioned, intermittent hypoxia has been used to treat ventilator dysfunction in humans (Trumbower et al. 2012; Hayes et al. 2014; Tester et al. 2014), and it is very likely that this beneficial effect is mediated by a plastic change called long-term facilitation (LTF), that the respiratory network undergoes after intermittent hypoxia (Bach and Mitchell 1996; Fregosi and Mitchell 1994; Hayashi et al. 1993; Turner and Mitchell 1997; Mitchell et al. 2001a). It is known that brief (3–5 min) repetitive episodes of hypoxia, known as acute intermittent hypoxia, elicit increases both in respiratory frequency and in the amplitude of the respiratory motor output in vivo (Bach and Mitchell 1996; Fregosi and Mitchell 1994; Hayashi et al. 1993; Turner and Mitchell 1997; Mitchell et al. 2001a, b), as well as in in vitro preparations such as the working heart brainstem preparation (Tadjalli et al. 2007) or the rhythmogenic brainstem slice (Blitz and Ramirez 2002). Long-term facilitation has been reported in humans (Lee et al. 2009a, b), dogs (Cao et al. 1992), goats (Turner and Mitchell 1997), ducks (Mitchell et al. 2001b), rabbits (Sokolowska and Pokorski 2006), rats (Bach and Mitchell 1996), and mice (Terada et al. 2008).

There is some evidence that the respiratory network is reconfigured by acute intermittent hypoxia (Morris et al. 2000, Kline et al. 2007; Griffioen et al. 2007; Moraes et al. 2013; Almado et al. 2014; Zanella et al. 2014). One example is the recent finding that intermittent hypoxia leads to a change in the neuromodulatory response of the preBötC. Zanella et al. (2014) found that norepinephrine, which normally regularizes respiratory activity (see Chap. 10), renders respiratory activity irregular after intermittent hypoxia. Moreover, they observed that intermittent hypoxia increases synaptic inhibition within the preBötC; thus, norepinephrine-induced rhythm irregularity is prevented by blocking synaptic inhibition before acute intermittent hypoxia (Zanella et al. 2014). More direct evidence of the respiratory network reconfiguration after intermittent hypoxia is found in the nucleus tractus solitarius (NTS) (Kline et al. 2007), in which the augmentation in afferent input and enhancement in spontaneous synaptic discharge are counterbalanced by a reduction in evoked synaptic transmission between sensory afferents and second-order NTS cells (Kline et al. 2007). It has also been shown that the reconfiguration of the respiratory network leading to long-term facilitation recruits an inspiratory-evoked excitatory neurotransmission to cardioinhibitory vagal neurons that is dependent on the generation of reactive oxygen species (Griffioen et al. 2007); furthermore, a subpopulation of non-C1 respiratory-modulated rostral ventrolateral medulla (RVLM) presympathetic neurons presented enhanced excitatory synaptic inputs from the respiratory network after intermittent hypoxia (Moraes et al. 2013). Interestingly, these presympathetic neurons, as well as phrenic nucleus-projecting RVLM neurons, do not change their intrinsic electrophysiological properties after chronic intermittent hypoxia (Almado et al. 2014). Finally, it has been found by using MEAs in vivo that long-term facilitation produces a very complex reconfiguration of the brainstem respiratory network that includes heterogeneous changes in firing frequency, effective connectivity and synchrony (Morris et al. 2000). However, the actual configuration of the brainstem underlying long-term facilitation remains unknown, and the reconfiguration of the preBötC during long-term facilitation needs to be determined in order to understand the change in respiratory frequency that occurs after intermittent hypoxia.

In summary, the evidence reviewed here shows that neural networks in general, and the respiratory network in particular, are able to change their network configuration by changing the number or the activity of their elements (either brain regions or neurons), by changing the number of their connections, or by just changing the strength of those connections. This clearly indicates that neural networks are not simple hardwired collections of brain regions or neurons that produce simple behaviors but, on the contrary, they constitute flexible circuits that can be reconfigured in response to the environmental, behavioral, metabolic, or pathologic states of the animal (Hooper and Moulins 1990; Marder et al. 1996; Parker et al. 1998; Thoby-Brisson and Simmers 1998; Peña et al. 2004; Ramirez et al. 2004; Nieto-Posadas et al. 2014).

Acknowledgments We thank Dr. Dorothy Pless and Jessica González for editorial comments. This study was supported by Fundación Marcos Moshinsky, CONACyT Grants 117, 151261, 235789, 24688 and 181323; and by DGAPA-UNAM Grants IN200715.

References

Almado CE, Leão RM, Machado BH (2014) Intrinsic properties of rostral ventrolateral medulla presympathetic and bulbospinal respiratory neurons of juvenile rats are not affected by chronic intermittent hypoxia. Exp Physiol 99(7):937–950

Armstrong GA, López-Guerrero JJ, Dawson-Scully K, Peña F, Robertson RM (2010) Inhibition of protein kinase G activity protects neonatal mouse respiratory network from hyperthermic and hypoxic stress. Brain Res 1311:64–72

Bach KB, Mitchell GS (1996) Hypoxia-induced long-term facilitation of respiratory activity is serotonin dependent. Respir Physiol 104(2–3):251–260

Ballanyi K, Völker A, Richter DW (1994) Anoxia induced functional inactivation of neonatal respiratory neurones in vitro. Neuroreport 6(1):165–168

Ballanyi K, Onimaru H, Homma I (1999) Respiratory network function in the isolated brainstem-spinal cord of newborn rats. Prog Neurobiol 59(6):583–634

Barnes BJ, Tuong CM, Mellen NM (2007) Functional imaging reveals respiratory network activity during hypoxic and opioid challenge in the neonate rat tilted sagittal slab preparation. J Neurophysiol 97(3):2283–2292

Bassett DS, Meyer-Lindenberg A, Achard S, Duke T, Bullmore E (2006) Adaptive reconfiguration of fractal small-world human brain functional networks. Proc Natl Acad Sci U S A 103(51):19518–19523

Becker R, Braun U, Schwarz AJ, Gass N, Schweiger JI, Weber-Fahr W, Schenker E, Spedding M, Clemm von Hohenberg C, Risterucci C, Zang Z, Grimm O, Tost H, Sartorius A, Meyer-Lindenberg A (2016) Species-conserved reconfigurations of brain network topology induced by ketamine. Transl Psychiatry 6:e786

Beggs JM, Plenz D (2004) Neuronal avalanches are diverse and precise activity patterns that are stable for many hours in cortical slice cultures. J Neurosci 24(22):5216–5229

Beidleman BA, Muza SR, Fulco CS, Cymerman A, Ditzler DT, Stulz D, Staab JE, Robinson SR, Skrinar GS, Lewis SF, Sawka MN (2003) Intermittent altitude exposures improve muscular performance at 4,300 m. J Appl Physiol (1985) 95(5):1824–1832

Belikova MV, Kolesnikova EE, Serebrovskaya TV (2012) Intermittent hypoxia and experimental Parkinson's disease. In: Xi L, Serebrovskaya TV (eds) Intermittent hypoxia and human diseases. Springer, New York, pp 147–153

Blitz DM, Ramirez JM (2002) Long-term modulation of respiratory network activity following anoxia in vitro. J Neurophysiol 87(6):2964–2971

Brovelli A, Badier JM, Bonini F, Bartolomei F, Coulon O, Auzias G (2017) Dynamic reconfiguration of visuomotor-related functional connectivity networks. J Neurosci 37(4):839–853

Cai XH, Zhou YH, Zhang CX, Hu LG, Fan XF, Li CC, Zheng GQ, Gong YS (2010) Chronic intermittent hypoxia exposure induces memory impairment in growing rats. Acta Neurobiol Exp (Wars) 70(3):279–287

Cao C, Slobounov S (2010) Alteration of cortical functional connectivity as a result of traumatic brain injury revealed by graph theory, ICA, and sLORETA analyses of EEG signals. IEEE Trans Neural Syst Rehabil Eng 18(1):11–19

Cao KY, Zwillich CW, Berthon-Jones M, Sullivan CE (1992) Increased normoxic ventilation induced by repetitive hypoxia in conscious dogs. J Appl Physiol (1985) 73(5):2083–2088

Carrillo-Reid L, Tecuapetla F, Tapia D, Hernández-Cruz A, Galarraga E, Drucker-Colin R, Bargas J (2008) Encoding network states by striatal cell assemblies. J Neurophysiol 99(3):1435–1450

Carrillo-Reid L, Hernández-López S, Tapia D, Galarraga E, Bargas J (2011) Dopaminergic modulation of the striatal microcircuit: receptor-specific configuration of cell assemblies. J Neurosci 31(42):14972–14983

Carroll MS, Ramirez JM (2013) Cycle-by-cycle assembly of respiratory network activity is dynamic and stochastic. J Neurophysiol 109(2):296–305

Carroll MS, Viemari JC, Ramirez JM (2013) Patterns of inspiratory phase-dependent activity in the in vitro respiratory network. J Neurophysiol 109(2):285–295

Carter AR, Shulman GL, Corbetta M (2012) Why use a connectivity-based approach to study stroke and recovery of function? NeuroImage 62(4):2271–2280

Choi SH, Lee H, Chung TS, Park KM, Jung YC, Kim SI, Kim JJ (2012) Neural network functional connectivity during and after an episode of delirium. Am J Psychiatry 169(5):498–507

Cohen JR, D'Esposito M (2016) The segregation and integration of distinct brain networks and their relationship to cognition. J Neurosci 36(48):12083–12094

Cohen JR, Gallen CL, Jacobs EG, Lee TG, D'Esposito M (2014) Quantifying the reconfiguration of intrinsic networks during working memory. PLoS One 9(9):e106636

De Vico Fallani F, Richiardi J, Chavez M, Achard S (2014) Graph analysis of functional brain networks: practical issues in translational neuroscience. Philos Trans R Soc Lond B Biol Sci 369(1653):pii: 20130521

Del Negro CA, Morgado-Valle C, Feldman JL (2002) Respiratory rhythm: an emergent network property? Neuron 34(5):821–830

Del Negro CA, Morgado-Valle C, Hayes JA, Mackay DD, Pace RW, Crowder EA, Feldman JL (2005) Sodium and calcium current-mediated pacemaker neurons and respiratory rhythm generation. J Neurosci 25(2):446–453

Dupret D, O'Neill J, Csicsvari J (2013) Dynamic reconfiguration of hippocampal interneuron circuits during spatial learning. Neuron 78(1):166–180

England SJ, Melton JE, Douse MA, Duffin J (1995) Activity of respiratory neurons during hypoxia in the chemodenervated cat. J Appl Physiol (1985) 78(3):856–861

Flamm RE, Harris-Warrick RM (1986) Aminergic modulation in lobster stomatogastric ganglion. I. Effects on motor pattern and activity of neurons within the pyloric circuit. J Neurophysiol 55(5):847–865

Flores-Martínez E, Peña-Ortega F (2017) Amyloid β peptide-induced changes in prefrontal cortex activity and its response to hippocampal input. Int J Pept 2017:7386809

Fregosi RF (1991) Short-term potentiation of breathing in humans. J Appl Physiol (1985) 71(3):892–899

Fregosi RF, Mitchell GS (1994) Long-term facilitation of inspiratory intercostal nerve activity following carotid sinus nerve stimulation in cats. J Physiol 477(Pt 3):469–479

Gaiteri C, Rubin JE (2011) The interaction of intrinsic dynamics and network topology in determining network burst synchrony. Front Comput Neurosci 5:10

Galán RF, Dick TE, Baekey DM (2010) Analysis and modeling of ensemble recordings from respiratory pre-motor neurons indicate changes in functional network architecture after acute hypoxia. Front Comput Neurosci 4:pii: 131

Gandolfi D, Mapelli J, D'Angelo E (2015) Long-term spatiotemporal reconfiguration of neuronal activity revealed by voltage-sensitive dye imaging in the cerebellar granular layer. Neural Plast 2015:284986

Getting PA (1983) Neural control of swimming in Tritonia. Symp Soc Exp Biol 37:89–128

Gong XW, Li JB, Lu QC, Liang PJ, Zhang PM (2014) Effective connectivity of hippocampal neural network and its alteration in Mg2+-free epilepsy model. PLoS One 9(3):e92961

Gourévitch B, Mellen N (2014) The preBötzinger complex as a hub for network activity along the ventral respiratory column in the neonate rat. NeuroImage 98:460–474

Gozal E, Row BW, Schurr A, Gozal D (2001) Developmental differences in cortical and hippocampal vulnerability to intermittent hypoxia in the rat. Neurosci Lett 305(3):197–201

Greicius MD, Krasnow B, Reiss AL, Menon V (2003) Functional connectivity in the resting brain: a network analysis of the default mode hypothesis. Proc Natl Acad Sci U S A 100(1):253–258

Griffin HS, Pugh K, Kumar P, Balanos GM (2012) Long-term facilitation of ventilation following acute continuous hypoxia in awake humans during sustained hypercapnia. J Physiol 590(Pt 20):5151–5165

Griffioen KJ, Kamendi HW, Gorini CJ, Bouairi E, Mendelowitz D (2007) Reactive oxygen species mediate central cardiorespiratory network responses to acute intermittent hypoxia. J Neurophysiol 97(3):2059–2066

Guedj C, Monfardini E, Reynaud AJ, Farnè A, Meunier M, Hadj-Bouziane F (2016) Boosting norepinephrine transmission triggers flexible reconfiguration of brain networks at rest. Cereb Cortex (In press)

Harris-Warrick RM, Marder E (1991) Modulation of neural networks for behavior. Annu Rev Neurosci 14:39–57

Hartelt N, Skorova E, Manzke T, Suhr M, Mironova L, Kügler S, Mironov SL (2008) Imaging of respiratory network topology in living brainstem slices. Mol Cell Neurosci 37(3):425–431

Hayashi F, Coles SK, Bach KB, Mitchell GS, McCrimmon DR (1993) Time-dependent phrenic nerve responses to carotid afferent activation: intact vs. decerebellate rats. Am J Phys 265(4 Pt 2):R811–R819

Hayashi F, Hinrichsen CF, McCrimmon DR (2003) Short-term plasticity of descending synaptic input to phrenic motoneurons in rats. J Appl Physiol (1985) 94(4):1421–1430

Hayes HB, Jayaraman A, Herrmann M, Mitchell GS, Rymer WZ, Trumbower RD (2014) Daily intermittent hypoxia enhances walking after chronic spinal cord injury: a randomized trial. Neurology 82(2):104–113

Hermans EJ, van Marle HJ, Ossewaarde L, Henckens MJ, Qin S, van Kesteren MT, Schoots VC, Cousijn H, Rijpkema M, Oostenveld R, Fernández G (2011) Stress-related noradrenergic activity prompts large-scale neural network reconfiguration. Science 334(6059):1151–1153

Hooper SL, Marder E (1987) Modulation of the lobster pyloric rhythm by the peptide proctolin. J Neurosci 7(7):2097–2112

Hooper SL, Moulins M (1990) Cellular and synaptic mechanisms responsible for a long-lasting restructuring of the lobster pyloric network. J Neurophysiol 64(5):1574–1589

Hunt CE (1992) The cardiorespiratory control hypothesis for sudden infant death syndrome. Clin Perinatol 19(4):757–771

Isla AG, Vázquez-Cuevas FG, Peña-Ortega F (2016) Exercise prevents amyloid-β-induced hippocampal network disruption by inhibiting GSK3β activation. J Alzheimers Dis 52(1):333–343

Jáidar O, Carrillo-Reid L, Hernández A, Drucker-Colín R, Bargas J, Hernández-Cruz A (2010) Dynamics of the Parkinsonian striatal microcircuit: entrainment into a dominant network state. J Neurosci 30(34):11326–11336

Klein JB, Gozal D, Pierce WM, Thongboonkerd V, Scherzer JA, Sachleben LR, Guo SZ, Cai J, Gozal E (2003) Proteomic identification of a novel protein regulated in CA1 and CA3 hippocampal regions during intermittent hypoxia. Respir Physiol Neurobiol 136(2–3):91–103

Kline DD, Ramirez-Navarro A, Kunze DL (2007) Adaptive depression in synaptic transmission in the nucleus of the solitary tract after in vivo chronic intermittent hypoxia: evidence for homeostatic plasticity. J Neurosci 27(17):4663–4673

Komiyama T, Sato TR, O'Connor DH, Zhang YX, Huber D, Hooks BM, Gabitto M, Svoboda K (2010) Learning-related fine-scale specificity imaged in motor cortex circuits of behaving mice. Nature 464(7292):1182–1186

Koshiya N, Oku Y, Yokota S, Oyamada Y, Yasui Y, Okada Y (2014) Anatomical and functional pathways of rhythmogenic inspiratory premotor information flow originating in the pre-Bötzinger complex in the rat medulla. Neuroscience 268:194–211

Lee DS, Badr MS, Mateika JH (2009a) Progressive augmentation and ventilator long-term facilitation are enhanced in sleep apnoea patients and are mitigated by antioxidant administration. J Physiol 587(Pt 22):5451–5467

Lee KZ, Reier PJ, Fuller DD (2009b) Phrenic motoneuron discharge patterns during hypoxia-induced short-term potentiation in rats. J Neurophysiol 102(4):2184–2193

Lee KZ, Sandhu MS, Dougherty BJ, Reier PJ, Fuller DD (2015) Hypoxia triggers short term potentiation of phrenic motoneuron discharge after chronic cervical spinal cord injury. Exp Neurol 263:314–324

Lieske SP, Thoby-Brisson M, Telgkamp P, Ramirez JM (2000) Reconfiguration of the neural network controlling multiple breathing patterns: eupnea, sighs and gasps [see comment]. Nat Neurosci 3(6):600–607

Lindsey BG, Morris KF, Segers LS, Shannon R (2000) Respiratory neuronal assemblies. Respir Physiol 122(2–3):183–196

Liao W, Zhang Z, Mantini D, Xu Q, Ji GJ, Zhang H, Wang J, Wang Z, Chen G, Tian L, Jiao Q, Zang YF, Lu G (2014) Dynamical intrinsic functional architecture of the brain during absence seizures. Brain Struct Funct 219(6):2001–2015

Lorea-Hernández JJ, Morales T, Rivera-Angulo AJ, Alcantara-Gonzalez D, Peña-Ortega F (2016) Microglia modulate respiratory rhythm generation and autoresuscitation. Glia 64(4):603–619

Lovering AT, Fraigne JJ, Dunin-Barkowski WL, Vidruk EH, Orem JM (2006) Medullary respiratory neural activity during hypoxia in NREM and REM sleep in the cat. J Neurophysiol 95(2):803–810

Lu XJ, Chen XQ, Weng J, Zhang HY, Pak DT, Luo JH, Du JZ (2009) Hippocampal spine-associated rap-specific GTPase-activating protein induces enhancement of learning and memory in post-natally hypoxia-exposed mice. Neuroscience 162(2):404–414

Lv Q, Yang L, Li G, Wang Z, Shen Z, Yu W, Jiang Q, Hou B, Pu J, Hu H, Wang Z (2016) Large-scale persistent network reconfiguration induced by ketamine in anesthetized monkeys: relevance to mood disorders. Biol Psychiatry 79(9):765–775

Marder E (1994) Invertebrate neurobiology. Polymorphic neural networks. Curr Biol 4(8):752–754

Marder E, Calabrese RL (1996) Principles of rhythmic motor pattern generation. Physiol Rev 76(3):687–717

Marder E, Abbott LF, Turrigiano GG, Liu Z, Golowasch J (1996) Memory from the dynamics of intrinsic membrane currents. Proc Natl Acad Sci U S A 93(24):13481–13486

Martin RJ, Di Fiore JM, Macfarlane PM, Wilson CG (2012) Physiologic basis for intermittent hypoxic episodes in preterm infants. Adv Exp Med Biol 758:351–358

Millet GP, Roels B, Schmitt L, Woorons X, Richalet JP (2010) Combining hypoxic methods for peak performance. Sports Med 40(1):1–25

Mironov SL, Skorova E, Hartelt N, Mironova LA, Hasan MT, Kügler S (2009) Remodelling of the respiratory network in a mouse model of Rett syndrome depends on brain-derived neurotrophic factor regulated slow calcium buffering. J Physiol 587(Pt 11):2473–2485

Mitchell GS, Baker TL, Nanda SA, Fuller DD, Zabka AG, Hodgeman BA, Bavis RW, Mack KJ, Olson EB Jr (2001a) Invited review: intermittent hypoxia and respiratory plasticity. J Appl Physiol (1985) 90(6):2466–2475

Mitchell GS, Powell FL, Hopkins SR, Milsom WK (2001b) Time domains of the hypoxic ventilatory response in awake ducks: episodic and continuous hypoxia. Respir Physiol 124(2):117–128

Miyawaki T, Norimoto H, Ishikawa T, Watanabe Y, Matsuki N, Ikegaya Y (2014) Dopamine receptor activation reorganizes neuronal ensembles during hippocampal sharp waves in vitro. PLoS One 9(8):e104438

Moraes DJ, da Silva MP, Bonagamba LG, Mecawi AS, Zoccal DB, Antunes-Rodrigues J, Varanda WA, Machado BH (2013) Electrophysiological properties of rostral ventrolateral medulla presympathetic neurons modulated by the respiratory network in rats. J Neurosci 33(49):19223–19237

Morris KF, Gozal D (2004) Persistent respiratory changes following intermittent hypoxic stimulation in cats and human beings. Respir Physiol Neurobiol 140(1):1–8

Morris KF, Baekey DM, Shannon R, Lindsey BG (2000) Respiratory neural activity during long-term facilitation. Respir Physiol 121(2–3):119–133

Nagy F, Dickinson PS (1983) Control of a central pattern generator by an identified modulatory interneurone in crustacea. I. Modulation of the pyloric motor output. J Exp Biol 105:33–58

Nair D, Dayyat EA, Zhang SX, Wang Y, Gozal D (2011) Intermittent hypoxia-induced cognitive deficits are mediated by NADPH oxidase activity in a murine model of sleep apnea. PLoS One 6(5):e19847

Nieto-Posadas A, Flores-Martínez E, Lorea-Hernández JJ, Rivera-Angulo AJ, Pérez-Ortega JE, Bargas J, Peña-Ortega F (2014) Change in network connectivity during fictive-gasping generation in hypoxia: prevention by a metabolic intermediate. Front Physiol 5:265

Nusbaum MP, Marder E (1989) A modulatory proctolin-containing neuron (MPN). II. State-dependent modulation of rhythmic motor activity. J Neurosci 9(5):1600–1607

Okada Y, Sasaki T, Oku Y, Takahashi N, Seki M, Ujita S, Tanaka KF, Matsuki N, Ikegaya Y (2012) Preinspiratory calcium rise in putative pre-Botzinger complex astrocytes. J Physiol 590(Pt 19):4933–4944

Ott MM, Nuding SC, Segers LS, Lindsey BG, Morris KF (2011) Ventrolateral medullary functional connectivity and the respiratory and central chemoreceptor-evoked modulation of retrotrapezoid-parafacial neurons. J Neurophysiol 105(6):2960–2975

Parker D, Zhang W, Grillner S (1998) Substance P modulates NMDA responses and causes long-term protein synthesis-dependent modulation of the lamprey locomotor network. J Neurosci 18(12):4800–4813

Paton JF, Abdala AP, Koizumi H, Smith JC, St-John WM (2006) Respiratory rhythm generation during gasping depends on persistent sodium current. Nat Neurosci 9(3):311–313

Payne RS, Goldbart A, Gozal D, Schurr A (2004) Effect of intermittent hypoxia on long-term potentiation in rat hippocampal slices. Brain Res 1029(2):195–199

Peña F (2008) Contribution of pacemaker neurons to respiratory rhythms generation in vitro. Adv Exp Med Biol 605:114–118

Peña F (2009) Neuronal network properties underlying the generation of gasping. Clin Exp Pharmacol Physiol 36(12):1218–1228

Peña F, Aguileta MA (2007) Effects of riluzole and flufenamic acid on eupnea and gasping of neonatal mice in vivo. Neurosci Lett 415(3):288–293

Peña F, García O (2006) Breathing generation and potential pharmacotherapeutic approaches to central respiratory disorders. Curr Med Chem 13(22):2681–2693

Peña F, Parkis MA, Tryba AK, Ramirez JM (2004) Differential contribution of pacemaker properties to the generation of respiratory rhythms during normoxia and hypoxia. Neuron 43(1):105–117

Peña F, Ramirez JM (2002) Endogenous activation of serotonin-2A receptors is required for respiratory rhythm generation in vitro. J Neurosci 22(24):11055–11064

Peña F, Ramirez JM (2004) Substance P-mediated modulation of pacemaker properties in the mammalian respiratory network. J Neurosci 24(34):7549–7556

Peña F, Ramirez JM (2005) Hypoxia-induced changes in neuronal network properties. Mol Neurobiol 32(3):251–283

Peña F, Ordaz B (2008) Non-selective cation channel blockers: potential use in nervous system basic research and therapeutics. Mini Rev Med Chem 8(8):812–819

Peña F, Meza-Andrade R, Páez-Zayas V, González-Marín MC (2008) Gasping generation in developing Swiss-Webster mice in vitro and in vivo. Neurochem Res 33(8):1492–1500

Peña F, Ordaz B, Balleza-Tapia H, Bernal-Pedraza R, Márquez-Ramos A, Carmona-Aparicio L, Giordano M (2010) Beta-amyloid protein (25-35) disrupts hippocampal network activity: role of Fyn-kinase. Hippocampus 20(1):78–96

Peña-Ortega F (2012) Tonic neuromodulation of the inspiratory rhythm generator. Front Physiol 3:253

Peña-Ortega F (2013) Amyloid beta-protein and neural network dysfunction. J Neurodegener Dis 2013:657470

Poets CF, Meny RG, Chobanian MR, Bonofiglo RE (1999) Gasping and other cardiorespiratory patterns during sudden infant deaths. Pediatr Res 45(3):350–354

Ponten SC, Bartolomei F, Stam CJ (2007) Small-world networks and epilepsy: graph theoretical analysis of intracerebrally recorded mesial temporal lobe seizures. Clin Neurophysiol 118(4):918–927

Powell FL, Milsom WK, Mitchell GS (1998) Time domains of the hypoxic ventilator response. Respir Physiol 112(2):123–134

Qin J, Liu H, Wei M, Zhao K, Chen J, Zhu J, Shen X, Yan R, Yao Z, Lu Q (2017) Reconfiguration of hub-level community structure in depressions: a follow-up study via diffusion tensor imaging. J Affect Disord 207:305–312

Ramirez JM (1998) Reconfiguration of the respiratory network at the onset of locust flight. J Neurophysiol 80(6):3137–3147

Ramirez JM, Quellmalz UJ, Wilken B, Richter DW (1998) The hypoxic response of neurones within the in vitro mammalian respiratory network. J Physiol 507(Pt 2):571–582

Ramirez JM, Tryba AK, Peña F (2004) Pacemaker neurons and neuronal networks: an integrative view. Curr Opin Neurobiol 14(6):665–674

Ramirez JM, Folkow LP, Blix AS (2007) Hypoxia tolerance in mammals and birds: from the wilderness to the clinic. Annu Rev Physiol 69:113–143

Ramírez-Jarquín JO, Lara-Hernández S, López-Guerrero JJ, Aguileta MA, Rivera-Angulo AJ, Sampieri A, Vaca L, Ordaz B, Peña-Ortega F (2012) Somatostatin modulates generation of inspiratory rhythms and determines asphyxia survival. Peptides 34(2):360–372

Richter DW, Bischoff A, Anders K, Bellingham M, Windhorst U (1991) Response of the medullary respiratory network of the cat to hypoxia. J Physiol 443:231–256

Rivera-Angulo AJ, Peña-Ortega F (2014) Isocitrate supplementation promotes breathing generation, gasping, and autoresuscitation in neonatal mice. J Neurosci Res 92(3):375–388

Roberts EL Jr, He J, Chih CP (1998) The influence of glucose on intracellular and extracellular pH in rat hippocampal slices during and after anoxia. Brain Res 783(1):44–50

Rodman JR, Harris MB, Rudkin AH, St-John WM, Leiter JC (2006) Gap junction blockade does not alter eupnea or gasping in the juvenile rat. Respir Physiol Neurobiol 152(1):51–60

Rothschild G, Cohen L, Mizrahi A, Nelken I (2013) Elevated correlations in neuronal ensembles of mouse auditory cortex following parturition. J Neurosci 33(31):12851–12861

Row BW, Kheirandish L, Neville JJ, Gozal D (2002) Impaired spatial learning and hyperactivity in developing rats exposed to intermittent hypoxia. Pediatr Res 52(3):449–453

Row BW, Liu R, Xu W, Kheirandish L, Gozal D (2003) Intermittent hypoxia is associated with oxidative stress and spatial learning deficits in the rat. Am J Respir Crit Care Med 167(11):1548–1553

Rubinov M, Sporns O (2010) Complex network measures of brain connectivity: uses and interpretations. NeuroImage 52(3):1059–1069

Runfeldt MJ, Sadovsky AJ, MacLean JN (2014) Acetylcholine functionally reorganizes neocortical microcircuits. J Neurophysiol 112(5):1205–1216

Sandhu MS, Baekey D, Mailing NG, Sanchez J, Reier PJ, Fuller DD (2015) Mid-cervical neuronal discharge patterns during and following hypoxia. J Neurophysiol 113(7):2091–2101

Sasaki T, Matsuki N, Ikegaya Y (2007) Metastability of active CA3 networks. J Neurosci 27(3):517–528

Schmidt C, Bellingham MC, Richter DW (1995) Adenosinergic modulation of respiratory neurones and hypoxic responses in the anaesthetized cat. J Physiol 483(Pt 3):769–781

Schultz DH, Cole MW (2016) Higher intelligence is associated with less task-related brain network reconfiguration. J Neurosci 36(33):8551–8561

Segers LS, Nuding SC, Dick TE, Shannon R, Baekey DM, Solomon IC, Morris KF, Lindsey BG (2008) Functional connectivity in the pontomedullary respiratory network. J Neurophysiol 100(4):1749–1769

Serebrovskaya TV, Manukhina EB, Smith ML, Downey HF, Mallet RT (2008) Intermittent hypoxia: cause of or therapy for systemic hypertension? Exp Biol Med (Maywood) 233(6):627–650

Smith JC, Ellenberger HH, Ballanyi K, Richter DW, Feldman JL (1991) Pre-Bötzinger complex: a brainstem region that may generate respiratory rhythm in mammals. Science 254(5032):726–729

Sokołowska B, Pokorski M (2006) Ventilatory augmentation by acute intermittent hypoxia in the rabbit. J Physiol Pharmacol 57(Suppl 4):341–347

Srinivas KV, Jain R, Saurav S, Sikdar SK (2007) Small-world network topology of hippocampal neuronal network is lost, in an in vitro glutamate injury model of epilepsy. Eur J Neurosci 25(11):3276–3286

St John WM (1999) Rostral medullary respiratory neuronal activities of decerebrate cats in eupnea, apneusis and gasping. Respir Physiol 116(1):47–65

St John WM, Bianchi AL (1985) Responses of bulbospinal and laryngeal respiratory neurons to hypercapnia and hypoxia. J Appl Physiol (1985) 59(4):1201–1207

Tadjalli A, Duffin J, Li YM, Hong H, Peever J (2007) Inspiratory activation is not required for episodic hypoxia-induced respiratory long-term facilitation in postnatal rats. J Physiol 585(Pt 2):593–606

Takahashi N, Sasaki T, Matsumoto W, Matsuki N, Ikegaya Y (2010) Circuit topology for synchronizing neurons in spontaneously active networks. Proc Natl Acad Sci U S A 107(22):10244–10249

Terada J, Nakamura A, Zhang W, Yanagisawa M, Kuriyama T, Fukuda Y, Kuwaki T (2008) Ventilatory long-term facilitation in mice can be observed during both sleep and wake periods and depends on orexin. J Appl Physiol (1985) 104(2):499–507

Tester NJ, Fuller DD, Fromm JS, Spiess MR, Behrman AL, Mateika JH (2014) Long-term facilitation of ventilation in humans with chronic spinal cord injury. Am J Respir Crit Care Med 189(1):57–65

Thoby-Brisson M, Simmers J (1998) Neuromodulatory inputs maintain expression of a lobster motor pattern-generating network in a modulation-dependent state: evidence from long-term decentralization in vitro. J Neurosci 18(6):2212–2225

Thoby-Brisson M, Ramirez JM (2000) Role of inspiratory pacemaker neurons in mediating the hypoxic response of the respiratory network in vitro. J Neurosci 20(15):5858–5866

Trumbower RD, Jayaraman A, Mitchell GS, Rymer WZ (2012) Exposure to acute intermittent hypoxia augments somatic motor function in humans with incomplete spinal cord injury. Neurorehabil Neural Repair 26(2):163–172

Tsai YW, Yang YR, Wang PS, Wang RY (2011) Intermittent hypoxia after transient focal ischemia induces hippocampal neurogenesis and c-Fos expression and reverses spatial memory deficits in rats. PLoS One 6(8):e24001

Tsai YW, Yang YR, Sun SH, Liang KC, Wang RY (2013) Post ischemia intermittent hypoxia induces hippocampal neurogenesis and synaptic alterations and alleviates long-term memory impairment. J Cereb Blood Flow Metab 33(5):764–773

Turner DL, Mitchell GS (1997) Long-term facilitation of ventilation following repeated hypoxic episodes in awake goats. J Physiol 499(Pt 2):543–550

Vargas-Barroso V, Ordaz-Sánchez B, Peña-Ortega F, Larriva-Sahd JA (2016) Electrophysiological Evidence for a Direct Link between the Main and Accessory Olfactory Bulbs in the Adult Rat. Front Neurosci 9:518

Wall AM, Corcoran AE, O'Halloran KD, O'Connor JJ (2014) Effects of prolyl-hydroxylase inhibition and chronic intermittent hypoxia on synaptic transmission and plasticity in the rat CA1 and dentate gyrus. Neurobiol Dis 62:8–17

Watts DJ, Strogatz SH (1998) Collective dynamics of 'small-world' networks. Nature 393(6684):440–442

Xie H, Yung WH (2012) Chronic intermittent hypoxia-induced deficits in synaptic plasticity and neurocognitive functions: a role for brain-derived neurotrophic factor. Acta Pharmacol Sin 33(1):5–10

Xie H, Leung KL, Chen L, Chan YS, Ng PC, Fok TF, Wing YK, Ke Y, Li AM, Yung WH (2010) Brain-derived neurotrophic factor rescues and prevents chronic intermittent hypoxia-induced impairment of hippocampal long-term synaptic plasticity. Neurobiol Dis 40(1):155–162

Yuan Q, Isaacson JS, Scanziani M (2011) Linking neuronal ensembles by associative synaptic plasticity. PLoS One 6(6):e20486

Zanella S, Doi A, Garcia AJ 3rd, Elsen F, Kirsch S, Wei AD, Ramirez JM (2014) When norepinephrine becomes a driver of breathing irregularities: how intermittent hypoxia fundamentally alters the modulatory response of the respiratory network. J Neurosci 34(1):36–50

Zavala-Tecuapetla C, Aguileta MA, Lopez-Guerrero JJ, González-Marín MC, Peña F (2008) Calcium-activated potassium currents differentially modulate respiratory rhythm generation. Eur J Neurosci 27(11):2871–2884

Zavala-Tecuapetla C, Tapia D, Rivera-Angulo AJ, Galarraga E, Peña-Ortega F (2014) Morphological characterization of respiratory neurons in the pre-Bötzinger complex. Prog Brain Res 209:39–56

Zhang JX, Chen XQ, Du JZ, Chen QM, Zhu CY (2005) Neonatal exposure to intermittent hypoxia enhances mice performance in water maze and 8-arm radial maze tasks. J Neurobiol 65(1):72–84

Zhang J, Cheng W, Liu Z, Zhang K, Lei X, Yao Y, Becker B, Liu Y, Kendrick KM, Lu G, Feng J (2016) Neural, electrophysiological and anatomical basis of brain-network variability and its characteristic changes in mental disorders. Brain 139(Pt 8):2307–2321

Zhu WZ, Xie Y, Chen L, Yang HT, Zhou ZN (2006) Intermittent high altitude hypoxia inhibits opening of mitochondrial permeability transition pores against reperfusion injury. J Mol Cell Cardiol 40(1):96–106

Zhu XH, Yan HC, Zhang J, Qu HD, Qiu XS, Chen L, Li SJ, Cao X, Bean JC, Chen LH, Qin XH, Liu JH, Bai XC, Mei L, Gao TM (2010) Intermittent hypoxia promotes hippocampal neurogenesis and produces antidepressant-like effects in adult rats. J Neurosci 30(38):12653–12663

Part IV
Damage-Triggered Neural Plasticity

Chapter 13
Progenitors in the Ependyma of the Spinal Cord: A Potential Resource for Self-Repair After Injury

Nicolás Marichal, Cecilia Reali, María Inés Rehermann, Omar Trujillo-Cenóz, and Raúl E. Russo

Abstract Traumatic injury of the spinal cord leads to devastating conditions that affect ~2.5 million people worldwide. This is because the mammalian spinal cord reacts to injury with only limited endogenous repair. Functional restoration requires the replacement of lost cells, the growth and navigation of regenerating axons on a permissive scaffold and axon re-myelination. The manipulation of endogenous spinal stem cells is regarded as a potential strategy to restore function. For this type of therapy it is necessary to determine the molecular and functional mechanisms regulating the proliferation, migration and differentiation of adult spinal progenitors. The spinal cord of animal models in which self-repair normally occurs may provide some clues. Salamanders, some fish and turtles regenerate their spinal cord after massive injury, achieving substantial functional recovery. This regeneration is orchestrated by progenitors that line the central canal (CC). Although mammals have lost the ability for self-repair, some cells in the CC react to injury by proliferating and migrating toward the lesion, where most become astrocytes in the core of the scar. Thus, CC-contacting progenitors in mammals have "latent" programs for endogenous repair of the spinal cord. Progenitor-like cells in the CC are functionally organized in lateral and midline domains, with heterogeneous molecular and membrane properties that represent targets for modulation. Understanding the mechanisms by which CC-can be manipulated will give valuable clues for endogenous spinal cord repair leading to successful functional recovery.

Keywords Spinal cord injury • Regeneration • Neural stem cells • Ependyma • Neural plasticity • Spinal progenitors

N. Marichal • C. Reali • M.I. Rehermann • O. Trujillo-Cenóz • R.E. Russo (⊠)
Neurofisiología Celular y Molecular, Instituto de Investigaciones Biológicas Clemente Estable, Avenida Italia 3318, CP 11600 Montevideo, Uruguay
e-mail: rrusso@iibce.edu.uy

Abbreviations

BLBP	Brain lipid binding protein
CC	Central canal
CSPG	Chondroitin sulphate proteoglycans
Cx43	Connexin 43
CXCR4	CXC chemokine receptor 4
DCX	Doublecortin
EGF	Epidermal growth factor
FGF2	Fribroblast growth factor 2
GFAP	Glial fibrillary acidic protein
I_{KA}	A-type currents
I_{KD}	Outward-rectifying K^+ currents
PSA-NCAM:	Poly-sialylated neural cell adhesion molecule
PTPσ	Protein tyrosine phosphatase σ
RG	Radial glia
SD1-α	Stromal cell-derived factor 1α
SCI	Spinal cord injury
SVZ	Subventricular zone

Introduction: The Problem of Spinal Cord Injury

Traumatic injury of the spinal cord leads to devastating conditions affecting ~2.5 million people worldwide (Rossignol et al. 2007). The quality of life of people suffering from spinal cord injury (SCI) is compromised not only by the loss of mobility below the lesion (quadriplegia or paraplegia) but also because of autonomic dysfunction with lack of bladder/bowel control and loss of sexual function (Thuret et al. 2006; Hou and Rabchevsky 2014). The first medical description of a spinal cord lesion is found in the Smith's papyrus and attributed to the Egyptian physician Imothep (XXVI B.C) (van Middendorp et al. 2010). In this ancient document, Imothep describes the various signs and symptoms of SCI stating that is "a medical condition that cannot be healed" (van Middendorp et al. 2010). Despite the enormous advances of medicine in many fields, this harsh statement still holds true.

SCI disrupts spinal circuits by interruption of axonal pathways and demyelination of surviving axons, the death of neurons and oligodendrocytes and glial scarring with cavitation (Thuret et al. 2006). Functional recovery would then be ideally achieved by the replacement of lost elements, navigation of regenerating axons towards proper targets on a permissive scaffold and axonal re-myelination for efficient conduction of nerve impulses. The manipulation of endogenous spinal stem cells is regarded as a potential strategy to bring about this kind of scenario (Rossignol et al. 2007; Thuret et al. 2006; Horner and Gage 2000). Unfortunately, as mentioned in Chap. 1, the normal reaction to injury of endogenous spinal progenitors in adult mammals is far from adequate to repair spinal circuits. Despite the fact that severed axons have the ability to re-grow, they cannot cross the lesion because of an unfavorable molecular environment generated by reactive astrocytes and other glial cells within the glial scar (Ramón y Cajal 1914; Yiu and He 2006; Rolls et al. 2009). Myelin related proteins (e.g., myelin-associated glycoprotein, oligodendrocyte myelin glycoprotein) and in particular chondroitin sulphate proteoglycans (CSPG) generated by astrocytes are major molecular players for the inhibition of axonal regeneration (Davies et al. 1997; Yiu and He 2006; Rolls et al. 2009). However, the inhibitors of axonal growth are not the only limitation that prevents successful reconnection, as when these extrinsic factors are removed, axonal regeneration is misdirected and fail to form functional contacts, suggesting there are intrinsic limitations for long range axonal re-growth (Harel and Strittmatter 2006). Finally, SCI activates endogenous progenitor cells that are recruited at the injury site, but they differentiate towards the glial lineage, failing to replace the lost neurons (Meletis et al. 2008; Barnabé-Heider et al. 2010).

Unlike mammals, some low vertebrates are able to reach considerable functional recovery after complete transection of the cord (Dervan and Roberts 2003; Rehermann et al. 2009; Tanaka and Ferretti 2009). The outstanding self-repair of the spinal cord in these animals is orchestrated by cells that line the central canal (CC) (Tanaka and Ferretti 2009). Although mammals have lost the ability for self-repair, some cells in the CC still react to injury by proliferating and migrating towards the lesion (Beattie et al. 1997; Johansson et al. 1999; Mothe and Tator 2005), where most become astrocyte-like cells in the core of the scar with few cells differentiating into oligodendrocytes (Meletis et al. 2008; Barnabé-Heider et al. 2010). Whether there is generation of new neurons from CC-contacting progenitors in the spinal cord of mammals remains controversial, with some reports supporting neurogenesis (Danilov et al. 2006; Shechter et al. 2007) whereas most suggest that differentiation toward the neuronal lineage is forbidden (Meletis et al. 2008; Horner et al. 2000; Shihabuddin et al. 2000). Collectively, the work discussed above supports the idea that the CC is a latent stem cell niche (Göritz and Frisén 2012) whose cells may have retained part of the programs needed to reconstruct the spinal cord after a lesion. In this chapter, we focus on the properties of cells within this niche in different animal models. We first summarize current knowledge about the identity of progenitor cells and the CC as a stem cell niche and then discuss the endogenous reaction of the ependymal region to SCI in animal models with and without regeneration capability and the implications for self-repair in mammals.

Identity and Properties of Neural Progenitors

The first progenitors in the neural tube are neuroephitelial cells, which at the onset of neurogenesis become radial glia (RG), the founders of most neurogenic lineages during development (Kriegstein and Alvarez-Buylla 2009). Both neuroepithelial cells and RG have a pronounced polarity with an apical pole bearing a single primary cilium protruding into the ventricular lumen and a distal process in contact with the pia (Kriegstein and Alvarez-Buylla 2009). This polarity is critical for determining and regulating the phenotype of neural stem cells (Alvarez-Buylla et al. 2001; Pinto and Götz 2007). For example, the apical pole of RG contains components like the centrosome and various key proteins (e.g., prominin, PAR3) whose inheritance during cell division determines the fate of daughter cells. In the adult mammalian brain, it is currently accepted that progenitors are a subtype of astrocyte that retains key features of neuroepithelial cells and RG (Doetsch et al. 1997; Horner and Palmer 2003; Ming and Song 2005; Lledo et al. 2006; Lim et al. 2008). However, intermediate progenitors in the subventricular zone (SVZ) of the developing cortex have multipolar processes that do not contact the ventricle or pial surface (Kriegstein and Alvarez-Buylla 2009). In fact, progenitors are heterogeneous cells expressing different molecules affecting their lineage potential (Pinto and Götz 2007). For example, expression of brain lipid binding protein (BLBP) seems to determine RG as bi-potent or multipotent progenitors (Hartfuss et al. 2001; Anthony et al. 2007; Pinto and Götz 2007). The transcription factor Pax6 is a major regulator of the subpopulation of neurogenic RG (Götz et al. 1998; Heins et al. 2002; Bel-Vialar et al. 2007) and progenitors in the adult mammalian brain (Kohwi et al. 2005; Maekawa et al. 2005; Nacher et al. 2005). In the developing spinal cord Pax6 interacts in a combinatorial manner with other transcription factors such as Olig2, Nkx2.2 and Sox9 to control neurogenesis and gliogenesis (Lee and Pfaff 2001; Rowitch 2004; Bel-Vialar et al. 2007; Guillemot 2007; Sugimori et al. 2007). The importance of the combination of key transcription factors in determining the biology of progenitors is highlighted by the possibility of reprogramming somatic cells to pluripotent stem cells with just a handful of factors (Yamanaka 2012).

In addition to this complex molecular signaling, progenitors have functional properties that regulate their behavior. For example, neural progenitors express various voltage-gated K^+ channels (Noctor et al. 2002; Chittajallu et al. 2002; Bahrey and Moody 2003; Filippov et al. 2003; Wang et al. 2003; Liu et al. 2006; Pardo 2004; Schaarschmidt et al. 2009) thought to regulate proliferation (Chittajallu et al. 2002; Pardo 2004; Schaarschmidt et al. 2009). Delayed outward-rectifying K^+ currents (I_{KD}) are essential for the G1/S transition in oligodendrocyte progenitors (OPs) (Chittajallu et al. 2002), whereas A-type currents (I_{KA}) are critical for proliferation of multipotent human neural progenitors (Schaarschmidt et al. 2009). Another key feature among precursors is their functional clustering via gap junctions (Lo Turco and Kriegstein 1991; Bittman et al. 1997; Noctor et al. 2002; Bruzzone and Dermietzel 2006; Liu et al. 2006), which regulates both proliferation and migration during development (Bittman et al. 1997; Bruzzone and Dermietzel 2006; Elias et al. 2007). These functional properties are likely to be part of epigenetic mechanisms for the control of progenitors and thus represent a potential target for therapeutic intervention.

The CC as a Stem Cell Niche

Although ependymal cells in the brain do not fulfill the classical criteria for stem cells, they represent a potential reservoir of neurogenic progenitors recruited by injury (Carlén et al. 2009). Ependymal cells in the adult spinal cord have an evolutionarily conserved origin in the ventral neuroepithelium (Fu et al. 2003), some originating from Olig2 progenitors from the pMN domain, which also originates motoneurons and oligodendrocytes (Masahira et al. 2006). Although ependymal cells in the mouse spinal cord stop dividing about 9 weeks after birth (Sabourin et al. 2009), they resume their mitotic activity in response to injury (Mothe and Tator 2005; Meletis et al. 2008). Both under normal conditions and after injury, these cells differentiate toward the glial lineage according to most reports (Johansson et al. 1999; Meletis et al. 2008). However, there are some studies that reported the generation of new neurons in the normal condition (Shechter et al. 2007) or in the diseased spinal cord (Danilov et al. 2006). It has been recently reported that sensory stimulation can induce the generation of a transient population of immature GABAergic interneurons in the spinal cord of mice (Shechter et al. 2011). In any case, the lack of neurogenic potential in the spinal cord is not an intrinsic limitation of spinal progenitors since when transplanted to a "neurogenic" microenvironment such as the dentate gyrus, they generate new neurons (Shihabuddin et al. 2000). The identity of stem cells in the adult spinal cord is difficult to establish and remains controversial. Although Horner et al. (2000) described proliferating cells in the grey and white matter of the spinal cord, other studies using in vitro neural stem assays (e.g., neurospheres) have shown that the vast majority of stem cell potential resides within the ependyma (Meletis et al. 2008; Sabourin et al. 2009). The cells lining the CC in mice are morphologically heterogeneous, with some having a cuboidal shape and few cilia whereas others -located mostly on the lateral aspects of the ependyma and called radial ependymocytes or tanycytes have a long basal process that contacts the basal lamina of blood vessels (Hugnot and Franzen 2011). As in other stem cell niches, some of these cells are related to their neighbors by zonnula occludens and gap junctions in their apical poles. Finally, the dorsal and ventral aspects of the ependyma harbor cells with characteristics of RG whose cell bodies lie at ependymal and subependymal levels (Meletis et al. 2008; Hamilton et al. 2009). Several markers related to neural stem/progenitor cells, including Sox2, CD15, CD133, nestin, vimentin, BLBP and glial fibrillary acidic protein (GFAP), are expressed by subpopulations of cells lining the CC (Meletis et al. 2008; Hamilton et al. 2009; Hugnot and Franzen 2011). These features may be linked to the maintenance of stem cell signaling pathways such as Notch, epithelial mesenchymal transition and bone morphogenetic protein signaling (Hugnot 2012). It has been suggested that the true neural stem cells that are able to sustain several passages of neurospheres and generate astrocytes, oligodendrocytes and neurons are GFAP RG-like cells located on the dorsal pole of the ependyma (Sabourin et al. 2009).

The ependymal region also contains a sub-class of cells that express the early neuronal markers HuC/D, doublecortin (DCX) and poly-sialylated neural cell adhesion molecule (PSA-NCAM) but do not express the mature neuronal marker NeuN (Fig. 13.1a) (Marichal et al. 2009). These cells originate in the embryo and have electrophysiological properties ranging from cells that generate slow Ca^{2+} spikes to cells with the ability to fire Na^+ spikes repetitively (Fig. 13.1b–d), suggesting different maturational stages. Because PSA-NCAM and DCX are involved in migration, neurite outgrowth and regeneration (Couillard-Despres et al. 2005; Bonfanti 2006), CC-contacting neuroblasts may be a source of plasticity similar to those in neurogenic niches (Lledo et al. 2006). In a model of multiple sclerosis it has been suggested that inflammation may trigger neurogenesis from CC-contacting progenitors (Danilov et al. 2006). An alternative possibility to neurogenesis is that the ependyma of the rodent spinal cord may be a reservoir of immature neurons in "standby mode" that under some circumstances may migrate and resume their differentiation to incorporate within damaged spinal circuits (Marichal et al. 2009).

Progenitors in the Central Canal Are Functionally Segregated in Spatial Domains

Central canal-contacting progenitors in the turtle Neural progenitors in the developing and adult brain are heterogeneous and regionally specified in terms of lineage potential (Graf and Stadtfeld 2008; Merkle et al. 2007). In turtles, progenitors contacting the lateral aspects of the CC have a higher proliferative rate than those located on the dorsal and ventral aspects (Fig. 13.2a, b, Russo et al. 2008). Similar to astroglial networks in the brain and spinal cord (Giaume et al. 2010), the progenitors lying in the lateral domains are electrically and metabolically coupled via gap junctions, forming dense clusters of cells that may encompass the whole lateral aspect of the CC (Fig. 13.2c). Gap junction coupling is mediated by connexin 43 (Cx43) expressed selectively on the lateral aspects of the CC at the level of the apical processes of progenitors, being the molecular basis for the location of clusters (Russo et al. 2008). The functional significance of gap junction coupling in the apical processes of CC-contacting progenitors is puzzling. The molecular polarity of RG is thought to have critical functional consequences (Chenn et al. 1998; Götz and Huttner 2005). The strong metabolic coupling of spinal precursors at their apical poles may allow an efficient communication of molecules that regulate key aspects of their biology, like mitotic activity or differentiation (Bruzzone and Dermietzel 2006). In addition, the apical pole of neural stem cells has a set of molecules whose inheritance is determined by the plane of cleavage during cell division, determining the fate of daughter cells (Götz and Huttner 2005). Therefore, Cx43 in CC-contacting progenitors located in the lateral domains is ideally placed to be one of the molecular components whose symmetric or asymmetric inheritance may decide the fate of daughter cells. The progenitors functionally clustered in the lateral domains are heterogeneous in terms of membrane properties. Whereas they are dominated by leak conductances, uncoupling of clustered cells with carbenoxolone showed that a

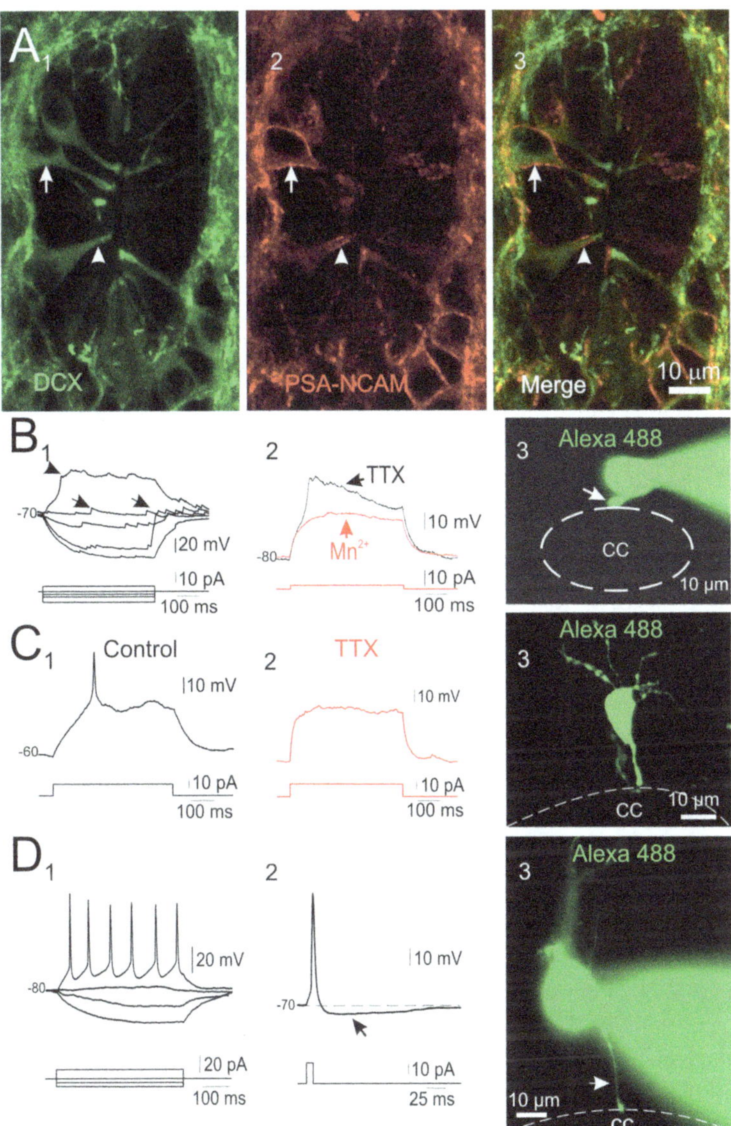

Fig. 13.1 Immature neurons contact the CC of the rat spinal cord. (**a**) Expression of DCX+ (*1, arrow*) and PSA-NCAM (*2* and *3*) in CC-contacting cells. Notice the conspicuous apical processes reaching the CC lumen (*arrowheads*). (**b**) Active response properties in a CC-contacting cell. A depolarizing current pulse produced a slow potential (*1, arrowhead*). Notice the presence of spontaneous synaptic activity (*1, arrows*). The slow response was not sensitive to 1 µM TTX (*2*). Addition of 3 mM Mn²⁺ blocked the slow depolarizing potential (*2*). As revealed by injection of Alexa 488, the recorded cell had a single short process contacting the CC (*3, arrow*). (**c**) Response of a cell in the ependyma to a series of current pulses. A 500 ms depolarizing current pulse applied at the resting membrane potential produced a small single fast spike (*1*). The action potential was blocked by 1 µM TTX (*2*). Confocal image of the cell shown in *1* and *2* (*3*). (**d**) Responses to a series of current pulses of a cell that fired repetitively (*1*). The action potential in this cell had a well-developed sAHP (*2, arrow*). The image in *3* shows the cell was connected to the CC via a thin single process (*arrow*). Modified with permission from the Society for Neuroscience, (Marichal et al. 2009)

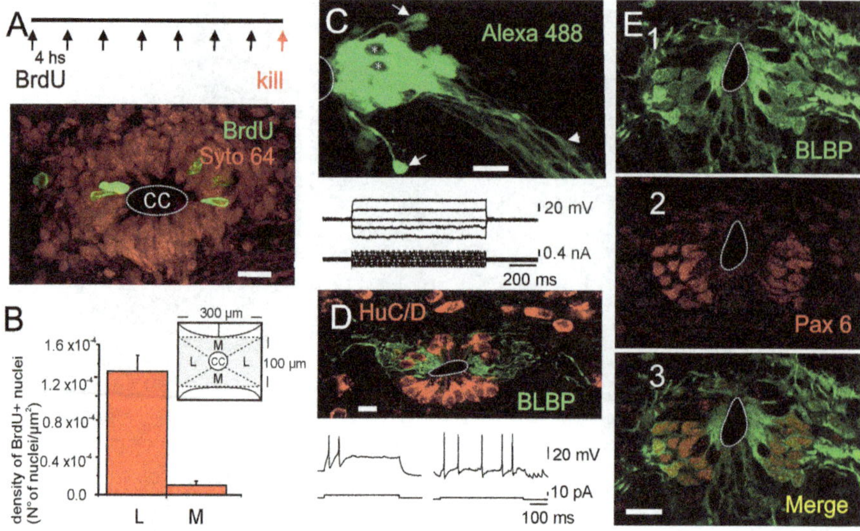

Fig. 13.2 The turtle CC as a neurogenic niche. (**a**) BrdU saturation protocol (*top*). BrdU+ cells (*green*) concentrated on the lateral aspects. Syto 64 (*red*) was used to stain nuclei. (**b**) Densities of BrdU-labeled nuclei in the lateral (*L*) and medial (*M*) aspects of the region surrounding the CC (mean ± SEM). The drawing shows schematically the limits of the regions (*shaded area*). (**c**) A group of dye-coupled cells covering the whole lateral region around the CC. The *asterisk* points to a cell profile that did not belong to the cluster. Notice cell bodies (*arrows*) detached from the main cluster of cells and a cell body intermingled with the bundle of processes (*arrowhead*). The responses to a series of current steps is shown in the *bottom*. (**d**) BLBP and HuC/D immunohistochemistry showing the intimate spatial relationship between BLBP+ precursors and neuroblasts. The responses of two CC-contacting neuroblasts to current steps are shown in the *bottom*. (**e**) BLBP (*1*) and Pax6 (*2*) expression around the CC. The *merged images* show that most BLBP+ cells co-expressed Pax6 (*3*). Scale bars: **a** 10 μm; **b**, **d** and **e** 20 μm. Modified with permission from the Society for Neuroscience, (Russo et al. 2008)

sub-set of cells display a K⁺ delayed rectifier (Russo et al. 2008), resembling precursors in the SVZ of the adult rat (Wang et al. 2003; Liu et al. 2006) and the cortex of the mouse embryo (Bahrey and Moody 2003). It is not clear whether the electrophysiological heterogeneity of clustered cells may reflect functional states related to different phases of the cell cycle, the dynamics of cell maturation or heterogeneity in neural precursors.

Neural progenitors display diverse molecular signatures that correlate with different proliferative potential and lineage differentiation (Campbell and Götz 2002; Pinto and Götz 2007). Interestingly, the expression of BLBP matched the location of clusters of gap junction coupled cells (Fig. 13.2d), forming a tri-dimensional network that enveloped immature neurons (Russo et al. 2004, 2008), in line with the intimate relationship needed for signaling between RG and differentiating neurons (Feng and Heintz 1995). In addition, Pax6 is selectively expressed in the lateral domains of the CC in BLBP+ cells (Fig. 13.2e) (Russo et al. 2008). RG expressing BLBP and Pax6 behave as bi- or multi-potent precursor cells (Hartfuss et al. 2001;

Pinto and Götz 2007), suggesting that clusters in lateral aspects of the CC are discrete neurogenic domains. In line with this possibility, our group has shown that the CC in turtles can give rise to both glial cells and neurons under normal conditions (Fernández et al. 2002).

Central canal-contacting progenitors in the rat The CC in the rat spinal cord has some features in common with its counterpart in freshwater turtles, but also some key differences. As in turtles, the CC of neonatal rats is a heterogeneous stem cell niche organized in lateral and medial domains (Marichal et al. 2012). Although proliferation takes place around all the CC, similarly to the CC of turtles, a higher rate of division takes place in the lateral aspects of the ependyma (Marichal et al. 2012). The dorsal and ventral poles of the ependyma of the rat spinal cord is contacted by cells with the morphological phenotype of RG that is characterized by the expression of nestin and/or vimentin (Fig. 13.3) and bears a single primary cilium (Marichal et al. 2012), typical features of neural stem cells. Although some cells in the lateral aspects of the CC in rats express the stem cell marker vimentin, they lack expression of BLBP and Pax6 (Fig. 13.3) (Marichal et al. 2012), a key difference with turtles (Russo et al. 2008). The absence of RG expressing BLBP/Pax6 may account for the inability to generate new neurons in the post-natal spinal cord of mammals. Progenitor-like cells in the lateral domains are also coupled via Cx43 in clusters that can cover large portions of the lateral CC, with some cells sending distal processes into the parenchyma (Fig. 13.3). These cells, however, have a passive electrophysiological phenotype (Fig. 13.3) similar to RG in the developing cortex of the rat (Noctor et al. 2002). Gap junction coupling has been shown to promote stem cell proliferation (Elias and Kriegstein 2008), and this may be a key factor determining the difference in proliferation capabilities between domains. The spatial profile of Cx43 expression and the clusters of coupled progenitors described in the rat resemble those of turtles (Russo et al. 2008), indicating a phylogenetically conserved functional organization. Indeed, in contrast to cells on the lateral domains, midline RG cells are not electrically coupled (Fig. 13.3).

Unlike neurogenic RG in the developing cortex (Noctor et al. 2002) and cells in the lateral domains of the CC, RG in the dorsal and ventral poles of the ependyma had complex electrophysiological phenotypes displaying various combinations of K^+ currents (I_{KD} and I_A) (Fig. 13.3) and in some cases a low-threshold voltage-gated Ca^{2+} conductance. The presence of I_{KD} is a common feature among adult progenitors in hippocampal nestin + type-2 cells (Filippov et al. 2003) and GFAP+ cells in the SVZ (Liu et al. 2006). On the other hand, I_A is conspicuous in embryonic (Smith et al. 2008) and neonatal (Stewart et al. 1999) SVZ and human stem cells (Schaarschmidt et al. 2009). The electrophysiological phenotype of midline RG is remarkably similar to that of oligodendrocyte progenitors (Chittajallu et al. 2004), raising the possibility that they are bipolar precursors committed to the oligodendrocyte lineage.

The complex repertoire of K^+ currents may regulate fundamental properties of ependymal progenitor-like cells. I_{KD} channels are major regulators of cell

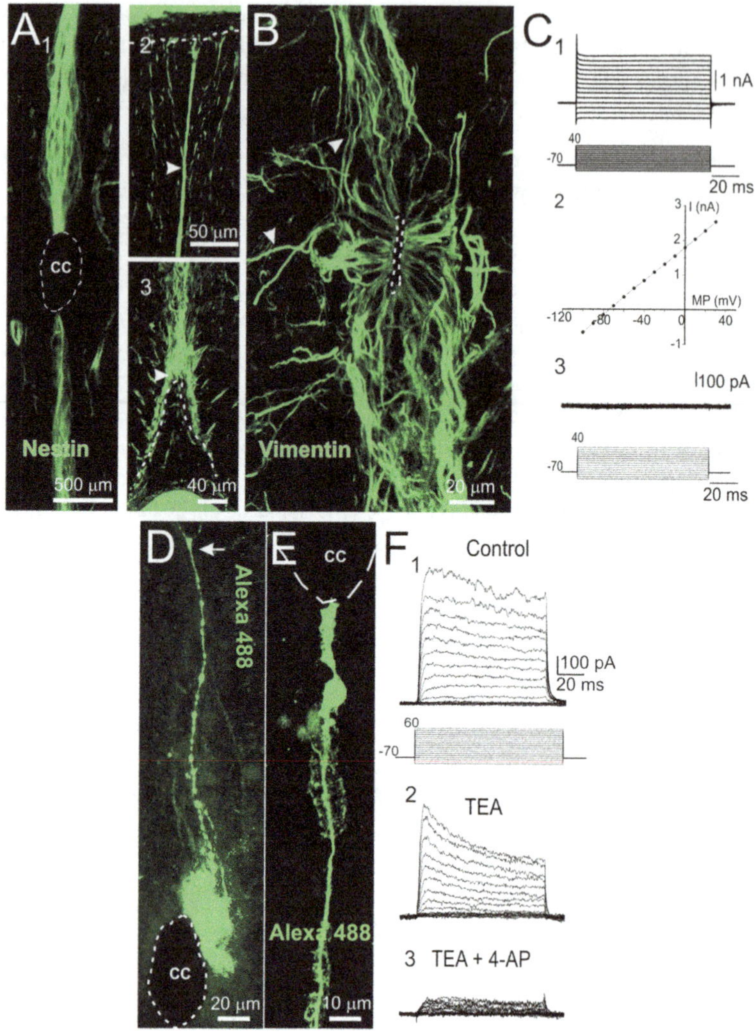

Fig. 13.3 Spatial domains of progenitors in the CC of the rat. (**a**) Nestin immunoreactivity in cells with long radial processes in the dorsal and ventral midline that stretched from the CC lumen (*1*) to the pia on the dorsal (*2, arrowhead*) and ventral (*3, arrowhead*) aspects of the cord. (**b**) Vimentin expressed in cells on the lateral aspects and poles of the ependyma, with some fibers projecting away from the CC (*arrowheads*). (**c**) Responses of a cell recorded on the lateral CC to a series of voltage steps (*1*) that displayed a linear I/V relationship (*2*). This cell lacked voltage-gated currents, as shown in the leak subtracted traces (*3*). (**d**) Gap junction coupled cells covered the whole lateral aspect of the CC with processes projecting toward the pia (*arrow*). (**e**) Alexa 488-filled RG contacting the ventral pole of the CC. F, I_{KD} in RG was blocked by TEA (10 mM, *1* and *2*), whereas the remaining current was blocked by the I_A antagonist 4-AP (2 mM, *2* and *3*). Modified with permission from Stem Cells (Marichal et al. 2012)

proliferation (Ghiani et al. 1999; MacFarlane and Sontheimer 2000a; Chittajallu et al. 2002) and I_A channels are essential for proliferation of multipotent human neural stem cells (Schaarschmidt et al. 2009). Thus, K^+ channels in midline RG may be part of epigenetic mechanisms that regulate proliferation. In addition, I_A has been implied in the differentiation of oligodendrocyte precursors (Sontheimer et al. 1989) and rat spinal cord astrocytes (MacFarlane and Sontheimer 2000b). Thus, another possibility is that K^+ currents participate in the transition from RG to post-mitotic spinal cells. It is not clear whether the molecular heterogeneity and complex electrophysiological phenotypes within midline domains represent different types of progenitors or various functional/developmental stages of a single precursor.

Animal Models with Self-Repair Capability

Primitive vertebrates like cyclostomes, certain fish and tailed amphibians are able to repair their damaged spinal cords and to recover some of the functions lost after SCI (Rovainen 1976; Wood and Cohen 1979; Stensaas 1983; Davis et al. 1990; Armstrong et al. 2003; Dervan and Roberts 2003; Chevallier et al. 2004; McHedlishvili et al. 2007; Shifman et al. 2007; Takeda et al. 2007). A closer relative to mammals, the amniote vertebrate *Trachemys dorbignyi* -a fresh-water turtle- also exhibits outstanding regenerating capabilities after spinal cord transection. Indeed, turtles spontaneously reconnect their severed spinal cords, leading in some cases to substantial recovery of step locomotion (Rehermann et al. 2009).

The strategies for regeneration and functional recovery after SCI in animals with self-repair capabilities are diverse, with the reaction of the ependymal as a main player. Proliferation of ependymal cells near the injury site seems to be a common phenomenon triggered by injury in lampreys (Zhang et al. 2014), tadpoles (Michel and Reier 1979; Gaete et al. 2012), eels (Dervan and Roberts 2003), fish (Reimer et al. 2008; Sîrbulescu and Zupanc 2011) and turtles (Rehermann et al. 2009, 2011). The cells lining the CC in axolotls regenerate the spinal cord after tail amputation by inducing a multipotent blastema, a process regulated by sonic hedgehog signaling (Schnapp et al. 2005; Tanaka and Ferretti 2009).

In anamniotes, ependymal cells migrate from the stumps and form a new CC within the regenerated cord (Dervan and Roberts 2003, Schnapp et al. 2005, Tanaka and Ferretti 2009). In turtles, however, the cellular bridge that reconnects the rostral and caudal stumps of the cord (Fig. 13.4) lacks a distinguishable CC and the typical anatomical features of the spinal cord (Rehermann et al. 2009). Instead, the bridge is formed mostly by glial cells that express BLBP and/or GFAP (Fig. 13.4e–g). In turtles, the BLBP+ progenitors of the ependymal layer appear as the most likely candidate to give rise to the abundant pre-myelinating oligodendrocytes that envelop re-growing axons in the cellular bridge of the regenerating spinal cord (Fig. 13.4) (Rehermann et al. 2009). The ability of BLBP+ RG to contribute to neural repair is supported by the fact that in rodents, transplanted embryonic RG help to bridge the injured spinal cord, promoting functional recovery (Hasegawa et al. 2005). The role

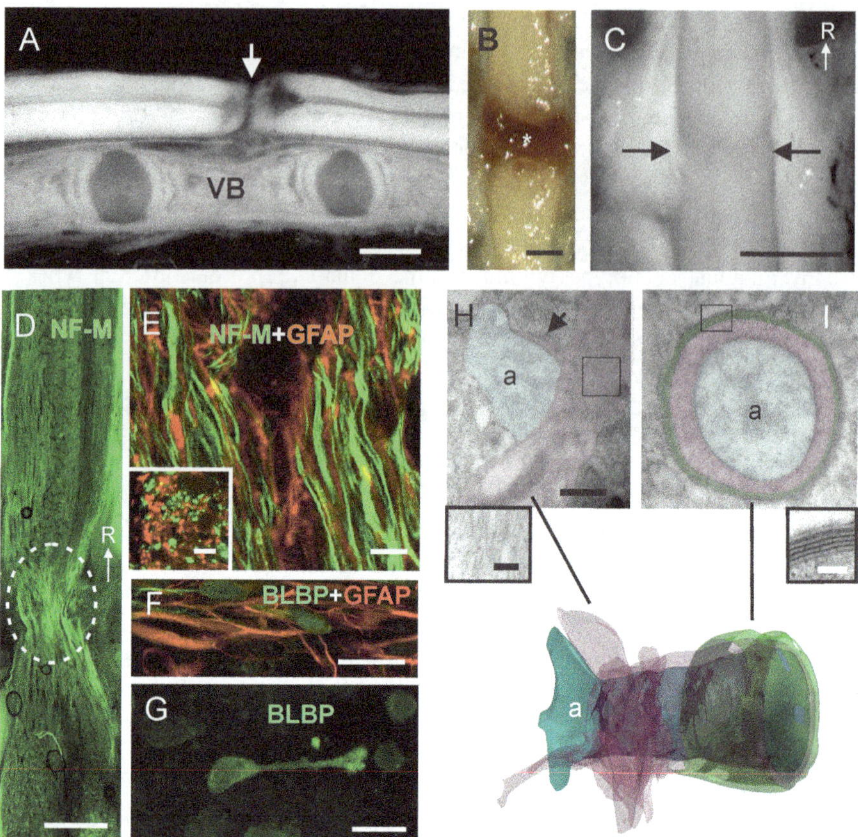

Fig. 13.4 Cellular and molecular basis of regeneration in the turtle spinal cord. (**a**) Longitudinal section of a fixed cord. Note that the section of the cord was complete (*arrow*) from its dorsum to the surface contacting the vertebral bodies (VB). (**b**) Macro photograph showing a compact blood clot 24 h after spinal cord transection (*asterisk*). (**c**) 17 days after surgery the injured site was barely evident when examined at low magnification. The *two arrows* point to the cellular bridge that replaced the primitive clot. The *thin arrow* indicates rostral (*R*). (**d**) Longitudinal section passing through a well-developed bridge (*circled*) containing a high number of unidentified NF-M+ axons. The *thin arrow* indicates rostral (*R*). (**e**) Confocal microscope image showing that GFAP+ processes follow longitudinal pathways paralleling the course of the NF-M+ axons crossing the lesion site. The *inset* shows a transverse section of the same region. (**f**) Cells and processes replacing the clot expressed GFAP or BLBP. (**g**) Many BLBP+ cells bear a single thick process terminating in a conspicuous enlargement. (**h, i**) Example of a 3D reconstruction of an axon sheath (large *double-headed arrows* indicate correspondence between the 3D model and TEM images). The 3D model at the bottom represents a short axon segment with a heterogeneous glial sheath (*blue*, axon; *pink*, glial covering; *green*, myelin). (**h**) The naked axon (ax) begins to be enveloped by a short lamella (*arrow*) stemming from a glial prolongation containing gliofibrils (shown in the *middle left* at higher magnification). (**i**) The enveloping cell produces typical, regularly spaced myelin sheaths. Scale bars: 0.8 mm in **a** and **c**; 350 μm in **b**; 400 μm in **d**; 10 μm in **e**; 20 μm in **f, g**; 0.5 μm **h, i**; 100 nm (*middle right*),80 nm (*middle left*) (**a–g**), modified with permission from the Journal of Comparative Neurology (Rehermann et al. 2009), **h, i** modified with permission from Cell and Tissue Research, Rehermann et al. 2011)

of ependymal cells in early reconnection was originally shown in the tadpole, in which ependymal cells send bundles of radial processes that establish an intimate relationship with axonal sprouts (Michel and Reier 1979). In contrast with the mammalian spinal cord, GFAP+ cells in the regenerating turtle spinal cord do not seem to interfere with the transit of regenerating axons but appear running side by side with axon bundles (Rehermann et al. 2009), suggesting they do not generate the inhibitory signals found in the glial scar in the mammalian spinal cord (Silver and Miller 2004; Thuret et al. 2006).

In anamniotes, the generation of new neurons from endogenous progenitors seems to be part of the regeneration program leading to functional recovery (Tanaka and Ferretti 2009). Indeed, neurogenesis has been reported in lampreys (Zhang et al. 2014), Xenopus (Lin et al. 2007), adult newts (Benraiss et al. 1999) and zebrafish (motoneurons, Reimer et al. 2008). Although in turtles the CC is a neurogenic niche (Fernández et al. 2002; Russo et al. 2004) and SCI triggers a huge increase in proliferation around the lesion epicenter (Rehermann et al. 2011), there is no addition of new neurons in the cellular bridge reconnecting the spinal cord. Thus, in turtles the strategy for endogenous repair is gliogenic, as activated CC progenitors seem to repress the neurogenic program to produce a glial scaffold for the navigation of pre-existing axons (Rehermann et al. 2009, 2011). Therefore, turtles appear to be an unusual amniote occupying a peculiar niche between the anamniotes with "perfect" regenerating capabilities (Tanaka and Ferretti 2009) and mammals, with very limited endogenous repair.

The Modulation of the Ependyma Stem Cell Niche for Spinal Cord Repair

Understanding the mechanisms by which the proliferation, migration and differentiation of CC-contacting progenitors are regulated in the normal and injured spinal cord is of fundamental importance to achieve better endogenous repair in the future. The candidate regulators of the ependymal progenitors are diverse, from growth factor and extracellular matrix components to ion channels and neurotransmitters.

Proliferation The proliferative activity of ependymal cells decreases progressively until the spinal cord stops elongating (Sabourin et al. 2009). The re-activation of cell proliferation by injury seems very specific in terms of the kind of damage, as proliferation increases with a contusion type traumatic injury, but not by chemical- or autoimmune-mediated injury (Lacroix et al. 2014). Furthermore, it has been reported that ependymal cell proliferation in the spinal cord is up-regulated by enhanced physical activity in rats (Cizkova et al. 2009). These studies suggest that the regulation of cell proliferation in the CC stem cell niche is regulated by intricate and multiple pathways.

Many of the molecules involved in the Notch, Wnt, BMP and sonic hedgehog pathways, which regulate the proliferation potential in stem cell niches of the embryo and adult brain, are also present in the CC (Hugnot 2012). Epidermal growth factor (EGF) and fibroblast growth factor 2 (FGF2) promote the proliferation of cells in the CC (Kojima and Tator 2000), and intrathecal application of EGF and FGF2 improves the functional recovery of rats after mild or moderate compression injury of the spinal cord (Kojima and Tator 2002). CSPG, a conspicuous molecular component of the injured spinal cord, has been shown to regulate the proliferation and fate decision of stem/progenitor cells (Sirko et al. 2007).

Membrane receptors operated by neurotransmitters are key regulators of stem/progenitor cells. In the developing brain, for example, GABA plays a central role by controlling various neurogenic steps (Owens and Kriegstein 2002; Ben-Ari 2002; Ben-Ari et al. 2007). Likewise, in the adult mammalian SVZ, GABA released by neuroblasts inhibits the proliferation of GFAP+ progenitors by activating $GABA_A$ receptors in a negative feedback loop (Ge et al. 2006; Bordey 2007). Little is known about GABAergic signaling around the CC, but like in the SVZ, there are GABAergic immature neurons intermingled with ependymal cells both in turtles (Reali et al. 2011) and rats (Stoeckel et al. 2003). In the turtle, a subset of CC-contacting progenitors expresses $GABA_A$ receptors, whereas others express GABA transporter GAT3 to regulate extracellular GABA levels (Reali et al. 2011). GABAergic signaling around the CC has been evolutionarily conserved, since GABA has been recently shown to depolarize ependymal cells via $GABA_A$ receptors in mammals (Corns et al. 2013). The role of GABAergic signaling as a regulator of ependymal cell proliferation remains to be tested.

The voltage gated K^+ channels expressed in a subset of CC-contacting progenitors (Russo et al. 2008; Marichal et al. 2012) are candidates to regulate the proliferative activity of ependymal cells, as they are known to regulate proliferation in various types of cells (Pardo 2004). Kv1.3 and Kv3.1 channels reduce the proliferation of mesencephalic neural progenitors (Liebau et al. 2006). Cultured human embryonic neural stem cells display a variety of K^+ channels, with a predominance of the A-type channel Kv4.2 and a small delayed rectifier (Schaarschmidt et al. 2009). As differentiation proceeds, the Kv4.2 channel is down regulated and blockade of this channel severely impairs proliferation, suggesting a pivotal role for A-type channels in human embryonic stem cells. Oligodendrocyte progenitors display a conspicuous delayed rectifier current, which gets smaller as they differentiate along their lineage (Chittajallu et al. 2002). The expression of Kv1.3 and Kv1.5 seems correlated with cyclin D and increases during the G1 phase of the cell cycle, suggesting a role in the progression to the S phase. As already mentioned, progenitors in the dorsal and ventral poles of the CC have an electrophysiological phenotype that closely resembles that of oligodendrocyte progenitors (Marichal et al. 2012). Thus, it is tempting to speculate that K^+ channels are also key regulators of proliferation in this type of progenitor in the ependyma of the spinal cord. A thorough characterization of the different subunits of K^+ channels expressed by ependymal cells and their selective interference in vivo will allow determining their function.

A common feature of the CC in lower vertebrates and mammals is that cells in the lateral domains are extensively coupled via Cx43 (Russo et al. 2008; Marichal et al. 2012). The fact that this correlates with a higher proliferative rate suggests that Cx43 expression favors proliferation. Indeed, the use of gap junction blockers showed that in the ventricular zone of the developing cortex connexins are needed for cells to enter the S phase (Bittman et al. 1997). Whether this effect is due to interference with gap junction communication or blockade of hemichannels is not clear. It has been proposed that hemichannels are part of the mechanism for the generation of coordinated Ca^{2+} waves that synchronize the cell cycle of a cohort of RG in the developing cortex (Weissman et al. 2004).

Migration A key requirement for healing the spinal cord by endogenous stem cells is the migration of their progeny to the injury site. Using a Cre-recombinase transgenic mouse that labels specifically ependymal cells and their progeny, Meletis et al. (2008) showed that ependyma-derived cells migrated only to the injured site. The mechanisms that guide these cells are not clear. SCI is associated with an inflammatory response and thus the associated signaling molecules are obvious candidates. In line with this, reactive astrocytes and endothelial cells up-regulate the inflammatory chemoattractant stromal cell-derived factor 1 α (SD1-α), and neural stem cells express the corresponding receptor CXC chemokine receptor 4 (CXCR4) (Imitola et al. 2004). The SD1-α/CXCR4 pathway activation promotes chain migration and transmigration of progenitors (Imitola et al. 2004) and the migration of neuroblast towards the damaged zone produced by a stroke in the brain (Thored et al. 2006). These mechanisms could also be at work for ependymal-derived cells, as the SD1-α/CXCR4 pathway is present in the CC (Shechter et al. 2007; Hugnot 2012).

Besides molecular clues, migration of the progeny generated by progenitors in the developing brain and adult neurogenic niches is modulated by the neurotransmitter GABA (Owens and Kriegstein 2002; Ben-Ari 2002; Ben-Ari et al. 2007). GABA reduces the rate of migration of neuroblasts in the SVZ by a poorly understood mechanism that involves intracellular Ca^{2+} independently of membrane depolarization (Bolteus and Bordey 2004). Migration of these cells is also reduced by tonic activation of GluK5 kainate receptors (Platel et al. 2008). It will be interesting to explore whether the cells produced in the ependyma express receptors for these transmitters, as they may transduce the activity of spinal circuits.

Another candidate transmitter to regulate migration is ATP. SCI produces a massive release of ATP (Wang et al. 2004) and may affect ependymal cells, especially those with processes projecting away from the CC. P2Y receptor activation generates Ca^{2+} waves in cortical progenitors, modulating their migration among other cellular events (Ulrich et al. 2012). A recent study from our group shows that cells in both the lateral and medial domains of the rat CC are activated by P2X7 receptors generating Ca^{2+} waves propagated over the whole length of the cells (Marichal et al. 2016). It remains to be explored whether purinergic signaling changes the proliferation and migration of ependymal cells.

Differentiation Once cells have migrated towards the lesion site, they differentiate either in glial cells or neurons. Cell fate selection depends on an intricate interplay of morphogens and transcription factors (Schuurmans and Guillemot 2002).

Whereas in ananmniotes endogenous progenitors generate both kinds of nerve cells during regeneration of the spinal cord, in amniotes (e.g., turtles and rodents) the regeneration strategy is via activation of gliogenesis only (Meletis et al. 2008; Rehermann et al. 2009; Barnabé-Heider et al. 2010). Astrocytes, oligodendrocyte progenitors and ependymal cells in the spinal cord of mice react to injury, but only ependymal cells generate progeny of multiple fates (Barnabé-Heider et al. 2010). Most of the cells derived from the ependyma differentiate in astrocytes in the core of the glial scar (Meletis et al. 2008), whereas the astrocytes in the periphery of the scar derive from astrocyte division (Barnabé-Heider et al. 2010). On the other hand, a small proportion of new oligodendrocytes originate from ependymal cells (Meletis et al. 2008). The mechanisms by which the progeny of ependymal cells differentiate into astrocytes or oligodendrocytes are not known. A recent study showed that ablation of β1-integrin from ependymal cells reduced the number of the progeny of these cells but increased the proportion of cells that differentiated into GFAP+ astrocytes, an effect that would be mediated by a direct interaction of β1-integrin with BMP receptors (North et al. 2015). Modulation of this pathway may be a suitable strategy to manipulate the number of astrocytes produced by ependymal cells and obtain a more favorable balance between ependyma-derived astrocytes/ oligodendrocytes.

Several neurotransmitter systems have been shown to influence lineage choice and differentiation (Tozuka et al. 2005; Ulrich et al. 2012; Jansson and Akerman 2014), but their role in directing the differentiation of the progeny originated by ependymal cells has been ignored. It is necessary to characterize the physiology of ependymal derived cells to determine if there is activity-dependent regulation as described in some stem cell niches of the embryonic and adult CNS.

Conclusions: The Future of SCI Therapeutics

The strategies to treat SCI are multiple, and include the use of natural or artificial scaffold materials to overcome geometrical and molecular constraints for axonal regeneration, modulation of molecules that inhibit axonal growth (e.g. Nogo, CSPG), grafts of embryonic stem cells or induced pluripotent stem cells, modulation of spinal circuits below the lesion, and neuroprosthetics, just to mention some (Horner and Gage 2000; Thuret et al. 2006; Rossignol et al. 2007; Grahn et al. 2014). It is likely that a successful functional recovery after SCI will involve a combination of some of these strategies. Some recent findings are encouraging. A recent study using a combination of surgical removal of the scar, implantation of a natural scaffold and grafting neural stem cells obtained from the olfactory bulb of the same patient, reported a significant improvement of locomotion after SCI (Tabakow et al. 2013).

New data suggest that we will soon be able to intervene to diminish or block the mechanisms that prevent axonal growth. Protein tyrosine phosphatase σ (PTPσ) has a major role in converting growth cones into dystrophic bulbs by stabilizing them with CSPG substrates (Lang et al. 2015). The same study found that interfering with PTPσ relieves CSPG inhibition of axonal growth, promoting the navigation of serotonergic fibers through a spinal cord lesion and leading to some recovery of locomotor and bladder activity.

Stimulation of self-repair of the mammalian spinal cord appears to be a powerful approach for functional recovery (Horner and Gage 2000), and it is now clear that progenitors contacting the CC may have a major part in this therapeutic strategy (Sabelström et al. 2014). In an elegant study, Sabelström et al. (2013) used a FoxJ1-rasless transgenic mouse to delete Ras genes –needed for the transition from G1 to S phase- selectively in ependymal cells. The study showed that when proliferation of ependymal cells is blocked, the extent of the spinal cord lesion is increased because impairment of the formation of the glial scar. Thus, the reaction of ependymal cells to SCI plays an important part in limiting the spread of the

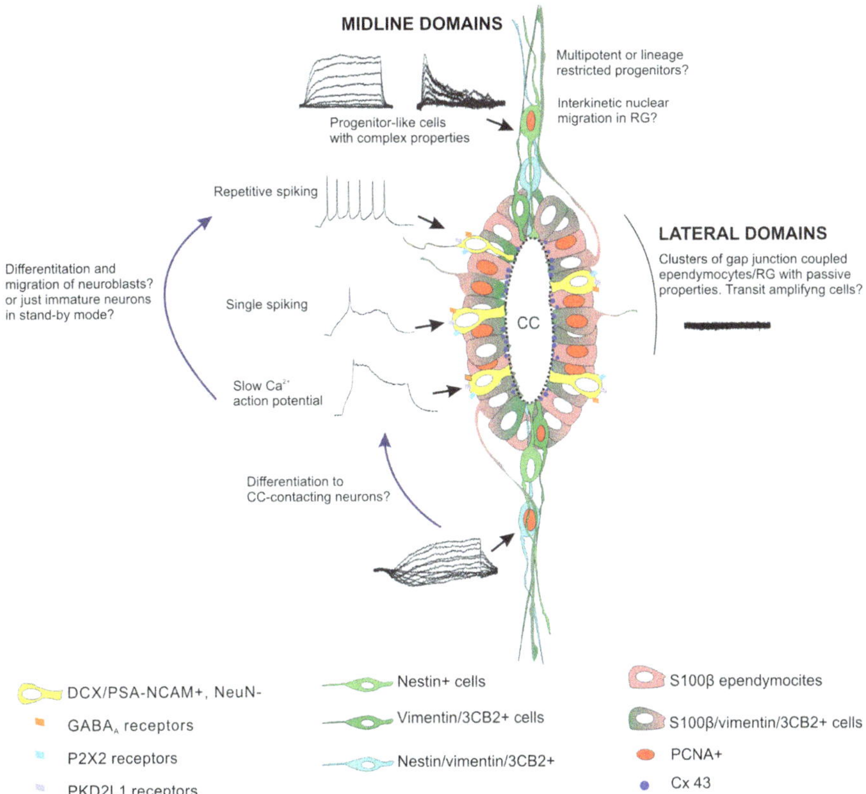

Fig. 13.5 Cartoon depicting the ependyma of the rat spinal cord as a stem cell niche. The heterogeneous molecular and functional phenotypes are *color coded* or illustrated with representative data. Some hypotheses are indicated with *question marks*. *Abbreviations*: *CC* central canal, *RG* radial glia, *PCNA* proliferating cell nuclear antigen, *DCX* doublecortin, *PSA-NCAM* poly-sialylated neural cell adhesion molecule

lesion. In addition, ependyma-derived cells within the lesion secrete neurotrophic factors that contribute to the survival of neurons around the injury site (Sabelström et al. 2013). It would be useful to manipulate the ependymal stem cell niche to maximize the cellular events that are beneficial for endogenous repair and block those that are detrimental. However, caution must be exerted when manipulating stem cells, as exemplified by a study in which stem cells grafted in the injured spinal cord improved motor recovery but induced aberrant axonal growth leading to allodynia (Hofstetter et al. 2005).

The functional complexity of the cells that line the CC is just beginning to unfold (Fig. 13.5). If some of the functional properties of the heterogeneous population of CC-contacting progenitors play a similar role to that of other neural stem cells, they may offer pathways for differential modulation of the ependymal niche. It is hoped that, in the near future we will gain key knowledge to pull the strings that can direct spinal progenitors to generate a successful endogenous repair, thereby changing the long standing statement made by Imothep.

Acknowledgements This work was partly supported by ANII (FCE 3356) and Wings for Life Spinal Cord Research Foundation.

References

Alvarez-Buylla A, García-Verdugo JM, Tramontin AD (2001) A unified hypothesis on the lineage of neural stem cells. Nat Rev Neurosci 2:287–293

Anthony TE, Mason HA, Gridley T, Fishell G, Heintz N (2007) Brain lipid binding protein is a target of notch signaling in radial glial cells. Genes Dev 19:1028–1033

Armstrong J, Zhang L, McClellan AD (2003) Axonal regeneration of descending and ascending spinal projection neurons in spinal cord-transected larval lamprey. Exp Neurol 180:156–166

Bahrey HL, Moody WJ (2003) Voltage-gated currents, dye and electrical coupling in the embryonic mouse neocortex. Cereb Cortex 13:239–251

Barnabé-Heider F, Göritz C, Sabelström H, Takebayashi H, Pfrieger FW, Meletis K, Frisén J (2010) Origin of new glial cells in intact and injured adult spinal cord. Cell Stem Cell 7:470–482

Beattie MS, Bresnahan JC, Komon J, Tovar CA, Van Meter M, Anderson DK, Faden AI, Hsu CY, Noble LJ, Salzman S, Young W (1997) Endogenous repair after spinal cord contusion injuries in the rat. Exp Neurol 148:453–463

Bel-Vialar S, Medevielle F, Pituello F (2007) The on/off of Pax6 controls the tempo of neuronal differentiation in the developing spinal cord. Dev Biol 305:659–673

Ben-Ari Y (2002) Excitatory actions of GABA during development: the nature of the nurture. Nat Rev Neurosci 3:728–739

Ben-Ari Y, Gaiarsa JL, Tyzio R, Khazipov R (2007) GABA: a pioneer transmitter that excites immature neurons and generates primitive oscillations. Physiol Rev 87:1215–1284

Benraiss A, Arsanto JP, Coulon J, Thouveny Y (1999) Neurogenesis during caudal spinal cord regeneration in adult newts. Dev Genes Evol 209:363–369

Bittman K, Owens DF, Kriegstein AR, LoTurco JJ (1997) Cell coupling and uncoupling in the ventricular zone of developing neocortex. J Neurosci 17:7037–7044

Bolteus AJ, Bordey A (2004) GABA release and uptake regulate neuronal precursor migration in the postnatal subventricular zone. J Neurosci 24:7623–7631

Bonfanti L (2006) PSA-NCAM in mammalian structural plasticity and neurogenesis. Prog Neurobiol 80:129–164

Bordey A (2007) Enigmatic GABAergic networks in adult neurogenic zones. Brain Res Rev 53:124–134

Bruzzone R, Dermietzel R (2006) Structure and function of gap junctions in the developing brain. Cell Tissue Res 326:239–248

Campbell K, Götz M (2002) Radial glia: Multi-purpose cells for vertebrate brain development. Trends Neurosci 25:235–238

Carlén M, Meletis K, Göritz C, Darsalia V, Evergren E, Tanigaki K, Amendola M, Barnabé-Heider F, Yeung MS, Naldini L, Honjo T, Kokaia Z, Shupliakov O, Cassidy RM, Lindvall O, Frisén J (2009) Forebrain ependymal cells are Notch-dependent and generate neuroblasts and astrocytes after stroke. Nat Neurosci 12:259–267

Chenn A, Zhang YA, Chang BT, McConnell SK (1998) Intrinsic polarity of mammalian neuroepithelial cells. Mol Cell Neurosci 11:183–193

Chevallier S, Landry M, Nagy F, Cabelguen JM (2004) Recovery of bimodal locomotion in the spinal-transected salamander, Pleurodeles waltlii. Eur J Neurosci 20:1995–2007

Chittajallu R, Chen Y, Wang H, Yuan X, Ghiani CA, Heckman T, McBain CJ, Gallo V (2002) Regulation of Kv1 subunit expression in oligodendrocyte progenitor cells and their role in G1/S phase progression of the cell cycle. Proc Natl Acad Sci U S A 99:2350–2355

Chittajallu R, Aguirre A, Gallo V (2004) NG2-positive cells in the mouse white and grey matter display distinct physiological properties. J Physiol (London) 561:109–122

Cizkova D, Nagyova M, Slovinska L, Novotna I, Radonak J, Cizek M, Mechirova E, Tomori Z, Hlucilova J, Motlik J, Sulla I Jr, Vanicky I (2009) Response of ependymal progenitors to spinal cord injury or enhanced physical activity in adult rat. Cell Mol Neurobiol 29:999–1013

Corns LF, Deuchars J, Deuchars SA (2013) GABAergic responses of mammalian ependymal cells in the central canal neurogenic niche of the postnatal spinal cord. Neurosci Lett 553:57–62

Couillard-Despres S, Winner B, Schaubeck S, Aigner R, Vroemen M, Weidner N, Bogdahn U, Winkler J, Kuhn HG, Aigner L (2005) Doublecortin expression levels in adult brain reflect neurogenesis. Eur J Neurosci 21:1–14

Danilov AI, Covacu R, Moe MC, Langmoen IA, Johansson CB, Olsson T, Brundin L (2006) Neurogenesis in the adult spinal cord in an experimental model of multiple sclerosis. Eur J Neurosci 23:394–400

Davies SJ, Fitch MT, Memberg SP, Hall AK, Raisman G, Silver J (1997) Regeneration of adult axons in white matter tracts of the central nervous system. Nature 390:680–683

Davis BM, Ayers JL, Koran L, Carlson J, Anderson MC, Simpson SB Jr (1990) Time course of salamander spinal cord regeneration and recovery of swimming: HRP retrograde pathway tracing and kinematic analysis. Exp Neurol 108:198–213

Dervan AG, Roberts BL (2003) Reaction of spinal cord central canal cells to cord transaction and their contribution to cord regeneration. J Comp Neurol 458:293–306

Doetsch F, García-Verdugo JM, Alvarez-Buylla A (1997) Cellular composition and three-dimensional organization of the subgerminal zone in the adult mammalian brain. J Neurosci 17:5046–5061

Elias LA, Kriegstein AR (2008) Gap junctions: multifaceted regulators of embryonic cortical development. Trends Neurosci 31:243–250

Elias LA, Wang DD, Kriegstein AR (2007) Gap junction adhesion is necessary for radial migration in the neocortex. Nature 448:901–907

Feng L, Heintz N (1995) Differentiating neurons activate transcription of the brain lipid-binding protein gene in radial glia through a novel regulatory element. Development 121:1719–1730

Fernández A, Radmilovich M, Trujillo-Cenóz O (2002) Neurogenesis and gliogenesis in the spinal cord of turtles. J Comp Neurol 453:131–144

Filippov V, Kronenberg G, Pivneva T, Reuter K, Steiner B, Wang LP, Yamaguchi M, Kettenmann H, Kempermann G (2003) Subpopulation of nestin-expressing progenitor cells in the adult murine hippocampus shows electrophysiological and morphological characteristics of astrocytes. Mol Cell Neurosci 23:373–382

Fu H, Qi Y, Tan M, Cai J, Hu X, Liu Z, Jensen J, Qiu M (2003) Molecular mapping of the origin of postnatal spinal cord ependymal cells: evidence that adult ependymal cells are derived from Nkx6.1+ ventral neural progenitor cells. J Comp Neurol 456:237–244

Gaete M, Muñoz R, Sanchez N, Tampe R, Moreno M, Contreras EG, Lee-Liu D, Larraín J (2012) Spinal cord regeneration in Xenopus tadpoles proceeds through activation of Sox2-positive cells. Neural Dev 7:13–17

Ge S, Goh EL, Sailor KA, Kitabatake Y, Ming GL, Song H (2006) GABA regulates synaptic integration of newly generated neurons in the adult brain. Nature 439:589–593

Ghiani CA, Yuan X, Eisen AM, Knutson PL, DePinho RA, McBain CJ, Gallo V (1999) Voltage-activated Kþ channels and membrane depolarization regulate accumulation of the cyclin-dependent kinase inhibitors p27(Kip1) and p21(CIP1) in glial progenitor cells. J Neurosci 19:5380–5392

Giaume C, Koulakoff A, Roux L, Holcman D, Rouach N (2010) Astroglial networks: a step further in neuroglial and gliovascular interactions. Nat Rev Neurosci 11:87–99

Göritz C, Frisén J (2012) Neural stem cells and neurogenesis in the adult. Cell Stem Cell 10:657–659. doi:10.1016/j.stem.2012.04.005

Götz M, Huttner WB (2005) The cell biology of neurogenesis. Nat Rev Mol Cell Biol 6:777–788

Götz M, Stoykova A, Gruss P (1998) Pax6 controls radial glia differentiation in the cerebral cortex. Neuron 21:1031–1044

Graf T, Stadtfeld M (2008) Heterogeneity of embryonic and adult stem cells. Cell Stem Cell 3:480–483

Grahn PJ, Mallory GW, Berry BM, Hachmann JT, Lobel DA, Lujan JL (2014) Restoration of motor function following spinal cord injury via optimal control of intraspinal microstimulation: toward a next generation closed-loop neural prosthesis. Front Neurosci 8:296

Guillemot F (2007) Spatial and temporal specification of neural fates by transcription factor codes. Development 134:3771–3780

Hamilton LK, Truong MK, Bednarczyk MR, Aumont A, Fernandes KJ (2009) Cellular organization of the central canal ependymal zone, a niche of latent neural stem cells in the adult mammalian spinal cord. Neuroscience 164:1044–1056

Harel NY, Strittmatter SM (2006) Can regenerating axons recapitulate developmental guidance during recovery from spinal cord injury? Nat Rev Neurosci 7:603–616

Hartfuss E, Galli R, Heins N, Götz M (2001) Characterization of CNS precursor subtypes and radial glia. Dev Biol 229:15–30

Hasegawa K, Chang YW, Li H, Berlin Y, Ikeda O, Kane-Goldsmith N, Grumet M (2005) Embryonic radial glia bridge spinal cord lesions and promote functional recovery following spinal cord injury. Exp Neurol 193:394–410

Heins N, Malatesta P, Cecconi F, Nakafuku M, Tucker KL, Hack MA, Chapouton P, Barde YA, Götz M (2002) Glial cells generate neurons: the role of the transcription factor Pax6. Nat Neurosci 5:308–315

Hofstetter CP, Holmström NA, Lilja JA, Schweinhardt P, Hao J, Spenger C, Wiesenfeld-Hallin Z, Kurpad SN, Frisén J, Olson L (2005) Allodynia limits the usefulness of intraspinal neural stem cell grafts; directed differentiation improves outcome. Nat Neurosci 8:346–353

Horner PJ, Gage FH (2000) Regenerating the damaged central nervous system. Nature 407:963–970

Horner PJ, Palmer TD (2003) New roles for astrocytes: the nightlife of an 'astrocyte'. La vida loca! Trends Neurosci 26:597–603

Horner PH, Power AE, Kempermann G, Kuhn HG, Palmer TD, Winkler J, Thal LJ, Gage FH (2000) Proliferation and differentiation of progenitor cells throughout the intact adult rat spinal cord. J Neurosci 20:2218–2228

Hou S, Rabchevsky AG (2014) Autonomic consequences of spinal cord injury. Compr Physiol 4:1419–1453

Hugnot JP (2012) The spinal cord neural stem cell niche. In: Sun T (ed) Neural stem cells and therapy. InTech Europe, Rijeka, pp 71–92

Hugnot JP, Franzen R (2011) The spinal cord ependymal region: a stem cell niche in the caudal central nervous system. Front Biosci 16:1044–1059

Imitola J, Raddassi K, Park KI, Mueller FJ, Nieto M, Teng YD, Frenkel D, Li J, Sidman RL, Walsh CA, Snyder EY, Khoury SJ (2004) Directed migration of neural stem cells to sites of CNS injury by the stromal cell-derived factor 1alpha/CXC chemokine receptor 4 pathway. Proc Natl Acad Sci U S A 101:18117–18122

Jansson LC, Åkerman KE (2014) The role of glutamate and its receptors in the proliferation, migration, differentiation and survival of neural progenitor cells. J Neural Transm 121:819–836

Johansson CB, Momma S, Clarke DL, Risling M, Lendahl U, Frisén J (1999) Identification of a neural stem cell in the adult mammalian central nervous system. Cell 96:25–34

Kohwi M, Osumi N, Rubenstein JL, Alvarez-Buylla A (2005) Pax6 is required for making specific subpopulations of granule and periglomerular neurons in the olfactory bulb. J Neurosci 25:6997–7003

Kojima A, Tator CH (2000) Epidermal growth factor and fibroblast growth factor 2 cause proliferation of ependymal precursor cells in the adult rat spinal cord in vivo. J Neuropathol Exp Neurol 59:687–697

Kojima A, Tator CH (2002) Intrathecal administration of epidermal growth factor and fibroblast growth factor 2 promotes ependymal proliferation and functional recovery after spinal cord injury in adult rats. J Neurotrauma 19:223–238

Kriegstein A, Alvarez-Buylla A (2009) The glial nature of embryonic and adult neural stem cells. Annu Rev Neurosci 32:149–184

Lacroix S, Hamilton LK, Vaugeois A, Beaudoin S, Breault-Dugas C, Pineau I, Lévesque SA, Grégoire CA, Fernandes KJ (2014) Central canal ependymal cells proliferate extensively in response to traumatic spinal cord injury but not demyelinating lesions. PLoS One 9(1):e85916

Lang BT, Cregg JM, DePaul MA (2015) Modulation of the proteoglycan receptor PTPσ promotes recovery after spinal cord injury. Nature 518:404–408

Lee SK, Pfaff SL (2001) Transcriptional networks regulating neuronal identity in the developing spinal cord. Nat Neurosci 4:1183–1191

Liebau S, Pröpper C, Böckers T, Lehmann-Horn F, Storch A, Grissmer S, Wittekindt OH (2006) Selective blockage of Kv1.3 and Kv3.1 channels increases neural progenitor cell proliferation. J Neurochem 99:426–437

Lim DA, Huang Y-C, Alvarez-Buylla A (2008) Adult subventricular zone and olfactory bulb neurogenesis. In: Gage FH, Kempermann G, Song H (eds) Adult neurogenesis. Cold Spring Harbor, New York, pp 175–206

Lin JH, Takano T, Arcuino G, Wang X, Hu F, Darzynkiewicz Z, Nunes M, Goldman SA, Nedergaard M (2007) Purinergic signaling regulates neural progenitor cell expansion and neurogenesis. Dev Biol 302:356–366

Liu X, Bolteus AJ, Balkin DM, Henschel O, Bordey A (2006) GFAP expressing cells in the postnatal subventricular zone display a unique glial phenotype intermediate between radial glia and astrocytes. Glia 54:394–410

Lledo PM, Alonso M, Grubb MS (2006) Adult neurogenesis and functional plasticity in neuronal circuits. Nat Rev Neurosci 7:179–193

Lo Turco JJ, Kriegstein AR (1991) Clusters of coupled neuroblasts in embryonic neocortex. Science 252:563–565

MacFarlane SN, Sontheimer H (2000a) Changes in ion channel expression accompany cell cycle progression of spinal cord astrocytes. Glia 30:39–48

MacFarlane SN, Sontheimer H (2000b) Modulation of Kv1.5 currents by Src tyrosine phosphorylation: potential role in the differentiation of astrocytes. J Neurosci 20:5245–5253

Maekawa M, Takashima N, Arai Y, Nomura T, Inokuchi K, Yuasa S, Osumi N (2005) Pax6 is required for production and maintenance of progenitor cells in postnatal hippocampal neurogenesis. Genes Cells 10:1001–1014

Marichal N, Garcia G, Radmilovich M, Trujillo-Cenóz O, Russo RE (2009) Enigmatic central canal contacting cells: immature neurons in "standby mode"? J Neurosci 29:10010–10024

Marichal N, García G, Radmilovich M, Trujillo-Cenóz O, Russo RE (2012) Spatial domains of progenitor-like cells and functional complexity of a stem cell niche in the neonatal rat spinal cord. Stem Cells 30:2020–2031

Marichal N, Fabbiani G, Trujillo-Cenóz O, Russo RE (2016) Purinergic signalling in a latent stem cell niche of the rat spinal cord. Purinergic Signal 12:331–341

Masahira N, Takebayashi H, Ono K, Watanabe K, Ding L, Furusho M, Ogawa Y, Nabeshima Y, Alvarez-Buylla A, Shimizu K, Ikenaka K (2006) Olig2-positive progenitors in the embryonic spinal cord give rise not only to motoneurons and oligodendrocytes, but also to a subset of astrocytes and ependymal cells. Dev Biol 293:358–369

McHedlishvili L, Epperlein HH, Telzerow A, Tanaka EM (2007) A clonal analysis of neural progenitors during axolotl spinal cord regeneration reveals evidence for both spatially restricted and multipotent progenitors. Development 134:2083–2093

Meletis K, Barnabé-Heider F, Carlén M, Evergren E, Tomilin N, Shupliakov O, Frisén J (2008) Spinal cord injury reveals multilineage differentiation of ependymal cells. PLoS Biol 6:1494–1507

Merkle FT, Mirzadeh Z, Alvarez-Buylla A (2007) Mosaic organization of neural stem cells in the adult brain. Science 317:381–384

Michel ME, Reier PJ (1979) Axonal-ependymal associations during early regeneration of the transected spinal cord in Xenopus laevis tadpoles. J Neurocytol 8:529–548

van Middendorp JJ, Sanchez GM, Burridge AL (2010) The Edwin Smith papyrus: a clinical reappraisal of the oldest known document on spinal injuries. Eur Spine J 19:1815–1823

Ming GL, Song H (2005) Adult neurogenesis in the mammalian central nervous system. Annu Rev Neurosci 28:223–250

Mothe AJ, Tator CH (2005) Proliferation, migration, and differentiation of endogenous ependymal region stem/progenitor cells following minimal spinal cord injury in the adult rat. Neuroscience 131:177–187

Nacher J, Varea E, Blasco-Ibanez JM, Castillo-Gomez E, Crespo C, Martinez-Guijarro FJ, McEwen BS (2005) Expression of the transcription factor Pax 6 in the adult rat dentate gyrus. J Neurosci Res 81:753–761

Noctor SC, Flint AC, Weissman TA, Wong WS, Clinton BK, Kriegstein AR (2002) Dividing precursor cells of the embryonic cortical ventricular zone have morphological and molecular characteristics of radial glia. J Neurosci 22:3161–3173

North HA, Pan L, McGuire TL, Brooker S, Kessler JA (2015) β1-integrin alters ependymal stem cell BMP receptor localization and attenuates astrogliosis after spinal cord injury. J Neurosci 35:3725–3733

Owens DF, Kriegstein AR (2002) Is there more to GABA than synaptic inhibition? Nat Rev Neurosci 3:715–727

Pardo LA (2004) Voltage-gated potassium channels in cell proliferation. Physiology 19:285–292

Pinto L, Götz M (2007) Radial glial cell heterogeneity-the source of diverse progeny in the CNS. Prog Neurobiol 83:2–23

Platel JC, Dave KA, Bordey A (2008) Control of neuroblast production and migration by converging GABA and glutamate signals in the postnatal forebrain. J Physiol 586:3739–3743

Ramón y Cajal S (1914) Estudios sobre la degeneración y regeneración del sistema nervioso. Imprenta de hijos de Moya, Madrid

Reali C, Fernández A, Radmilovich M, Trujillo-Cenóz O, Russo RE (2011) GABAergic signalling in a neurogenic niche of the turtle spinal cord. J Physiol (London) 589:5633–5647

Rehermann MI, Marichal N, Russo RE, Trujillo-Cenóz O (2009) Neural reconnection in the transected spinal cord of the freshwater turtle Trachemys Dorbignyi. J Comp Neurol 515:197–214

Rehermann MI, Santiñaque FF, López-Carro B, Russo RE, Trujillo-Cenóz O (2011) Cell proliferation and cytoarchitectural remodeling during spinal cord reconnection in the fresh-water turtle Trachemys Dorbignyi. Cell Tissue Res 344:415–433

Reimer MM, Sörensen I, Kuscha V, Frank RE, Liu C, Becker CG, Becker T (2008) Motor neuron regeneration in adult zebrafish. J Neurosci 28:8510–8516

Rolls A, Shechter R, Schwartz M (2009) The bright side of the glial scar in CNS repair. Nat Rev Neurosci 10:235–241

Rossignol S, Schwab M, Schwartz M, Fehlings MG (2007) Spinal cord injury: time to move? J Neurosci 27:11782–11792

Rovainen CM (1976) Regeneration of Müller and Mauthner axons after spinal transection in larval lampreys. J Comp Neurol 168:545–554

Rowitch DH (2004) Glial specification in the vertebrate neural tube. Nat Rev Neurosci 5:409–419

Russo RE, Fernández A, Reali C, Radmilovich M, Trujillo-Cenóz O (2004) Functional and molecular clues reveal precursor-like cells and immature neurones in the turtle spinal cord. J Physiol (London) 3:831–838

Russo RE, Reali C, Radmilovich M, Fernández A, Trujillo-Cenóz O (2008) Connexin 43 delimits functional domains of neurogenic precursors in the spinal cord. J Neurosci 28:3298–3309

Sabelström H, Stenudd M, Réu P, Dias DO, Elfineh M, Zdunek S, Damberg P, Göritz C, Frisén J (2013) Resident neural stem cells restrict tissue damage and neuronal loss after spinal cord injury in mice. Science 342:637–640

Sabelström H, Stenudd M, Frisén J (2014) Neural stem cells in the adult spinal cord. Exp Neurol 260:44–49

Sabourin JC, Ackema KB, Ohayon D, Guichet PO, Perrin FE, Garces A, Ripoll C, Charité J, Simonneau L, Kettenmann H, Zine A, Privat A, Valmier J, Pattyn A, Hugnot JP (2009) A mesenchymal-like ZEB1(+) niche harbors dorsal radial glial fibrillary acidic protein-positive stem cells in the spinal cord. Stem Cells 27:2722–2733

Schaarschmidt G, Wegner F, Schwarz SC, Schmidt H, Schwarz J (2009) Characterization of voltage-gated potassium channels in human neural progenitor cells. PLoS One 4:e6168

Schnapp E, Kragl M, Rubin L, Tanaka EM (2005) Hedgehog signaling controls dorsoventral patterning, blastema cell proliferation and cartilage induction during axolotl tail regeneration. Development 132:3243–3253

Schuurmans C, Guillemot F (2002) Molecular mechanisms underlying cell fate specification in the developing telencephalon. Curr Opin Neurobiol 12:26–34

Shechter R, Ziv Y, Schwartz M (2007) New GABAergic interneurons supported by myelin-specific T cells are formed in intact adult spinal cord. Stem Cells 25:2277–2282

Shechter R, Baruch K, Schwartz M, Rolls A (2011) Touch gives new life: mechanosensation modulates spinal cord adult neurogenesis. Mol Psychiatry 16:342–352

Shifman MI, Jin LQ, Selzer M (2007) Regeneration in the lamprey spinal cord. In: Becker CG, Becker T (eds) Model organisms in spinal cord regeneration. Weinheim, Wiley-VCH Verlag, pp 229–262

Shihabuddin LS, Horner PJ, Ray J, Gage FH (2000) Adult spinal cord stem cells generate neurons after transplantation in the adult dentate gyrus. J Neurosci 20:8727–8735

Silver J, Miller JH (2004) Regeneration beyond the glial scar. Nat Rev Neurosci 5:146–156

Sîrbulescu RF, Zupanc GK (2011) Spinal cord repair in regeneration-competent vertebrates: adult teleost fish as a model system. Brain Res Rev 67:73–93

Sirko S, von Holst A, Wizenmann A, Götz M, Faissner A (2007) Chondroitin sulfate glycosaminoglycans control proliferation, radial glia cell differentiation and neurogenesis in neural stem/progenitor cells. Development 134:2727–2738

Smith DO, Rosenheimer JL, Kalil RE (2008) Delayed rectifier and A-type potassium channels associated with Kv 2.1 and Kv 4.3 expression in embryonic rat neural progenitor cells. PLoS One 3:e1604

Sontheimer H, Trotter J, Schachner M, Kettenmann H (1989) Channel expression correlates with differentiation stage during the development of oligodendrocytes from their precursor cells in culture. Neuron 2:1135–1145

Stensaas LJ (1983) Regeneration in the spinal cord of the newt *Notopthalmus* (Triturus) *pyrrhogaster*. In: Kao CC, Bunge RP, Reier PJ (eds) Spinal cord reconstruction. Raven Press, New York, pp 121–149

Stewart RR, Zigova T, Luskin MB (1999) Potassium currents in precursor cells isolated from the anterior subventricular zone of the neonatal rat forebrain. J Neurophysiol 81:95–102

Stoeckel ME, Uhl-Bronner S, Hugel S, Veinante P, Klein MJ, Mutterer J, Freund-Mercier MJ, Schlichter R (2003) Cerebrospinal fluid-contacting neurons in the rat spinal cord, a gamma-aminobutyric acidergic system expressing the P2X2 subunit of purinergic receptors, PSANCAM, GAP-43 immunoreactivities: light and electron microscopic study. J Comp Neurol 457:159–174

Sugimori M, Nagao M, Bertrand N, Parras CM, Guillemot F, Nakafuku M (2007) Combinatorial actions of patterning and HLH transcription factors in the spatiotemporal control of neurogenesis and gliogenesis in the developing spinal cord. Development 134:1617–1629

Tabakow P, Jarmundowicz W, Czapiga B, Fortuna W, Miedzybrodzki R, Czyz M, Huber J, Szarek D, Okurowski S, Szewczyk P, Gorski A, Raisman G (2013) Transplantation of autologous olfactory ensheathing cells in complete human spinal cord injury. Cell Transplant 22:1591–1612

Takeda A, Goris RC, Funakoshi K (2007) Regeneration of descending projections to the spinal motor neurons after spinal hemisection in the goldfish. Brain Res 1155:17–23

Tanaka EM, Ferretti P (2009) Considering the evolution of regeneration in the central nervous system. Nat Rev Neurosci 10:713–723

Thored P, Arvidsson A, Cacci E, Ahlenius H, Kallur T, Darsalia V, Ekdahl CT, Kokaia Z, Lindvall O (2006) Persistent production of neurons from adult brain stem cells during recovery after stroke. Stem Cells 24:739–747

Thuret S, Moon LD, Gage FH (2006) Therapeutic interventions after spinal cord injury. Nat Rev Neurosci 7:628–643

Tozuka Y, Fukuda S, Namba T, Seki T, Hisatsune T (2005) GABAergic excitation promotes neuronal differentiation in adult hippocampal progenitor cells. Neuron 47:803–815

Ulrich H, Abbracchio MP, Burnstock G (2012) Extrinsic purinergic regulation of neural stem/progenitor cells: implications for CNS development and repair. Stem Cell Rev 8:755–767

Wang DD, Krueger DD, Bordey A (2003) Biophysical properties and ionic signature of neuronal progenitors of the postnatal subventricular zone in situ. J Neurophysiol 90:2291–2302

Wang X, Arcuino G, Takano T, Lin J, Peng WG, Wan P, Li P, Xu Q, Liu QS, Goldman SA, Nedergaard M (2004) P2X7 receptor inhibition improves recovery after spinal cord injury. Nat Med 10:821–827

Weissman TA, Riquelme PA, Ivic L, Flint AC, Kriegstein AR (2004) Calcium waves propagate through radial glial cells and modulate proliferation in the developing neocortex. Neuron 43:647–661

Wood MR, Cohen MJ (1979) Synaptic regeneration in identified neurons of the lamprey spinal cords. Science 206:344–347

Yamanaka S (2012) Induced pluripotent stem cells: past, present, and future. Cell Stem Cell 10:678–684

Yiu G, He Z (2006) Glial inhibition of CNS axon regeneration. Nat Rev Neurosci 7:617–627

Zhang G, Vidal Pizarro I, Swain GP, Kang SH, Selzer ME (2014) Neurogenesis in the lamprey central nervous system following spinal cord transection. J Comp Neurol 522:1316–1332

Chapter 14
I_{KD} Current in Cold Transduction and Damage-Triggered Cold Hypersensitivity

Alejandro González, Gaspar Herrera, Gonzalo Ugarte, Carlos Restrepo, Ricardo Piña, María Pertusa, Patricio Orio, and Rodolfo Madrid

Abstract In primary sensory neurons of the spinal and trigeminal somatosensory system, cold-sensitivity is strongly dependent on the functional balance between TRPM8 channels, the main molecular entity responsible for the cold-activated excitatory current, and *Shaker*-like Kv1.1–1.2 potassium channels, the molecular counterpart underlying the excitability brake current I_{KD}. This slow-inactivating outward K^+ current reduces the excitability of cold thermoreceptor neurons increasing their thermal threshold, and prevents unspecific activation by cold of neurons of other somatosensory modalities. Here we examine the main biophysical properties of this current in primary sensory neurons, its central role in cold thermotransduction, and its contribution to alterations in cold sensitivity triggered by peripheral nerve damage.

Keywords Primary sensory neurons • Cold thermotransduction • Kv1 channels • TRPM8 • 4-AP • α-DTx • Cold hypersensitivity

Abbreviations

4-AP	4-AminoPyridine
CCI	Chronic Constriction Injury
CIN	Cold-Insensitive Neuron

Alejandro González, Gaspar Herrera and Gonzalo Ugarte contributed equally to this work.

A. González • G. Ugarte • C. Restrepo • R. Piña • M. Pertusa • R. Madrid (✉)
Departamento de Biología, Facultad de Química y Biología, and Millennium Nucleus of Ion Channels-Associated Diseases (MiNICAD), Universidad de Santiago de Chile, Alameda L. Bdo. O'Higgins 3363, 9160000 Santiago, Chile
e-mail: rodolfo.madrid@usach.cl

G. Herrera • P. Orio
Centro Interdisciplinario de Neurociencia de Valparaíso and Instituto de Neurociencia, Facultad de Ciencias, Universidad de Valparaíso, 2340000 Valparaíso, Chile

© Springer International Publishing AG 2017
R. von Bernhardi et al. (eds.), *The Plastic Brain*, Advances in Experimental Medicine and Biology 1015, DOI 10.1007/978-3-319-62817-2_14

CSN   Cold-Sensitive Neuron
DRG   Dorsal Root Ganglia
DTx-K   Dendrotoxin-K
HT-CSN   High-Threshold Cold-Sensitive Neuron
Kv1.1   Potassium Voltage-gated channel subfamily A member 1
Kv1.2   Potassium Voltage-gated channel subfamily A member 2
LT-CSN   Low-Threshold Cold-Sensitive Neuron
PBMC   1-phenylethyl-4-(benzyloxy)-3-methoxybenzyl(2-aminoethyl)
             carbamate
TEA   Tetraethylamonium
TG   Trigeminal Ganglia
TRPM8   Transient Receptor Potential Melastatin 8 channel
TsTx   Tityustoxin-Kα
TTx   Tetrodotoxin
α-DTx α-Dendrotoxin

I_{KD} Current in Cold Transduction

In primary somatosensory neurons, several transduction and voltage-gated ion channels operate concertedly to give shape to their net excitability in response to thermal stimuli (Vriens et al. 2014; González et al. 2015). In cold thermoreceptors, cold sensitivity is strongly dependent on the counterbalance between the functional expression of TRPM8 channels, the main molecular entity responsible for the cold-activated current (I_{cold}) (McKemy et al. 2002; Peier et al. 2002; Latorre et al. 2011; McCoy et al. 2011; Almaraz et al. 2014; Madrid and Pertusa 2014) and *Shaker*-like Kv1.1 and Kv1.2 channels, the molecular counterpart underlying the inhibitory brake potassium current I_{KD} (Viana et al. 2002; Madrid et al. 2009; McKemy 2013; González et al. 2015; González et al., 2017). In cold-sensitive neurons (CSNs), the inhibitory I_{KD} current works as an excitability brake, contributing to set their thermal threshold (Viana et al. 2002; Madrid et al. 2009). The I_{KD} current is a fast-activating and slow-inactivating voltage-dependent K$^+$ current that dampens the effect of the cold-induced depolarizing TRPM8-dependent current during cold stimulation. This outward current shifts the temperature threshold of CSNs to higher values, retarding and reducing the net response of the neuron to a temperature drop. Thus, CSNs express both I_{KD} and I_{TRPM8}, each of them having opposing actions on temperature threshold. Furthermore, not only thermal threshold but also spontaneous and cold-evoked firing can be modulated as a consequence of Kv1 and TRPM8 function.

The I_{KD} current exerts its action at membrane potentials subthreshold to the action potential firing (Storm 1988; Viana et al. 2002), reducing the neuronal excitability by counterbalancing the effect of cold-induced depolarizing currents. I_{KD} also prevents the unspecific activation by cold of primary sensory neurons of other somatosensory modalities (Viana et al. 2002; Belmonte et al. 2009; Madrid

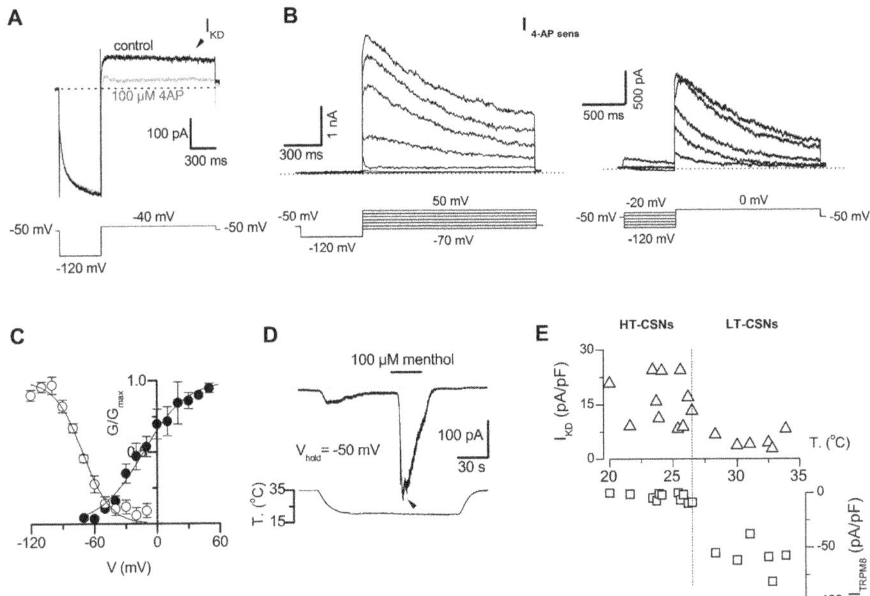

Fig. 14.1 Pharmacological isolation and biophysical properties of I_{KD} in cold-sensitive neurons. (**a**) Whole cell current in a trigeminal CSN during a bipolar voltage step from -120 mV to -40 mV, from a holding potential of -50 mV (*bottom panel*), in extracellular control solution and in the presence of 100 µM 4-AP. (**b**) 4-AP sensitive current (I_{KD}) obtained by digital subtraction of whole cell currents in control conditions and in the presence of 4-AP using standard activation (*left panel*) and inactivation (*right panel*) voltage protocols (*bottom panels*). (**c**) Average activation and steady-state inactivation curves obtained from six trigeminal CSNs. Note the window current around the resting membrane potential. (**d**) Simultaneous recording of membrane current (*top trace*) and bath temperature (*bottom trace*) in a low-threshold CSN during a cooling step combined with application of menthol (100 µM). (**e**) Plot of the excitatory I_{TRPM8} ($I_{cold+menthol}$) and I_{KD} current density in individual CSNs. The abscissa corresponds to the thermal threshold measured using calcium imaging. I_{TRPM8} was measured at the peak of the cold-induced inward current potentiated by menthol (*black arrow head* in **d**). I_{KD} current was measured at -40 mV from a holding potential of -120 mV in the presence of TTx (Modified from Madrid et al. 2009)

et al. 2009; González et al. 2017). Under voltage clamp in whole-cell patch clamp experiments, the I_{KD} current can be revealed by using a bipolar voltage protocol (Fig. 14.1). In primary sensory neurons held at -50 mV, a hyperpolarizing pulse to -120 mV can be used to remove its inactivation. The slow inactivating current revealed after returning to a membrane potential subthreshold for action potential firing (-50 or -40 mV) is taken as I_{KD} (Fig. 14.1a). The large inward current activated by the hyperpolarizing pulse in CSNs corresponds to I_h, depending on HCN channels (Orio et al. 2009). The I_{KD} current is sensitive to micromolar concentrations of 4-AP (Fig. 14.1a). The activation and inactivation properties of I_{KD} can be obtained by using standard activation and inactivation protocols, repeated in the presence and in the absence of 100 µM 4-AP. In CSNs, series of depolarizing voltage

Table 14.1 Activation and inactivation properties of I_{KD}

	Activation		Inactivation	
	$V_{1/2}$ (mV)	Slope factor s (mV)	$V_{1/2}$ (mV)	Slope factor s (mV)
TG CSNs	-13 ± 2	21 ± 2	-71 ± 2	-13 ± 1
DRG CSNs	-12 ± 2	12 ± 1	-68 ± 4	-10 ± 2

Modified from Madrid et al. (2009) and González et al. (2017)

TG trigeminal ganglia, *DRG* dorsal root ganglia, *CSNs* cold sensitive neurons. $V_{1/2}$: voltage for half-maximal activation and slope factor s were obtained from the fittings of normalized conductances to a Boltzmann function $G = G_{max} / (1 + exp. [(V_{1/2} - V)/s])$. Average activation and steady-state inactivation curves were obtained in six (TG) and five (DRG) CSNs from adult mice

steps between -70 and $+50$ mV (1.5 s duration) are used to obtain the I-V relationship for the total K^+ current, in the presence of TTx. In the inactivation protocol, a fixed pulse to 0 mV (1.4 s) is preceded by a variable prepulse (0.5 s) between -120 and -10 mV. Digital subtraction of both sets of currents before and after application of 4-AP corresponds to I_{KD} (Fig. 14.1b).

The biophysical properties of I_{KD} obtained by using these stimulation protocols and the pharmacological isolation strategy, reveal that this current starts to activate at membrane potentials near the resting potential of CSNs of trigeminal ganglia (TG). Fittings of the steady-state activation and inactivation data using the Boltzmann function reveal the voltage for half-maximal activation, inactivation, and the slope factor of this current (Fig. 14.1c). These values are very similar to those observed in CSNs from dorsal root ganglia (DRG) in adult mice (Table 14.1). In both cases, activation and inactivation curves show a significant overlap, giving rise to a window current around the resting membrane potential. The low activation threshold, fast activation, and slow inactivation of I_{KD} bring on a damping effect on the depolarization induced by TRPM8 activation, shifting the thermal threshold of the neuron to lower temperatures.

In cold thermoreceptor neurons, differential functional expression of TRPM8 channels and the molecular counterpart of I_{KD} in low- (LT-) and high-threshold (HT-) CSNs it is intimately linked to their thermosensitive phenotype. LT-CSNs have higher sensitivity to menthol than HT-CSNs. The higher density of cold + menthol-sensitive currents (I_{TRPM8}) (Fig. 14.1d) in LT-CSNs suggests that they express higher levels of TRPM8 than HT-CSNs. In the case of I_{KD}, there is a negative correlation between temperature threshold and I_{KD} current density. The expression of I_{KD} in HT-CSNs is twice as great as that in LT-CSNs (Madrid et al. 2009). Thus, LT-CSNs express higher levels of TRPM8 and lower levels of I_{KD} than HT-CSNs, and vice versa (Fig. 14.1e). In trigeminal and dorsal root ganglia, LT-CSNs correspond to about 75% of CSNs, which detect innocuous cold. HT-CSNs on the other hand represent approximately 25% of total population of CSNs, which are involved in the detection of unpleasant cold stimuli. Both populations are indistinguishable in terms of their basic electrophysiological properties (Madrid et al. 2009), and they are functionally separated from polymodal- and mechano-nociceptors.

In order to determine the molecular nature of I_{KD} in CSNs, Madrid et al. (2009) tested several K^+ channel blockers and determined their effect on I_{KD} and on the thermal threshold of CSNs, previously identified by calcium imaging

(Madrid et al. 2009). They found that α-DTx, a blocker of Kv1.1, Kv1.2 and Kv1.6 *Shaker* potassium channels subunits, had a strong effect on I_{KD} and shifted their thermal threshold to higher temperatures. This fully reversible effect is very similar to the result observed using 4-AP (Fig. 14.2). DTx-K, a polypeptide that selectively blocks Kv1.1 containing channels had identical effects on thermal threshold, and TsTx, a blocker of Kv1.2 and Kv1.3 containing channels also shifted the cold threshold to higher values. Homotetrameric Kv1.1 and Kv1.2 present different sensitivities to TEA (IC_{50} < 0.5 mM versus IC_{50} > 100 mM, respectively) and heteromeric channels fall in between. 1 mM TEA had no effect on thermal threshold of CSNs, but 10 mM was enough to obtain a significant shift to higher temperatures. Since Kv1.1 and Kv1.2 homomeric channels are almost absent in neurons (Koschak et al. 1998; Manganas and Trimmer 2000), the molecular substrate of I_{KD} corresponds, most probably, to heteromers of Kv1.1-Kv1.2 channels.

The pharmacological suppression of I_{KD} using 4-AP (or α-DTx) not only induces a shift in the thermal threshold of cold-induced responses to higher temperatures in CSNs, but also induces cold sensitivity in neurons from other sensory modalities that express high levels of I_{KD} (Viana et al. 2002; González et al. 2017). Figure 14.2c shows two representative trigeminal neurons of this latter case, identified using calcium imaging. The intracellular calcium increase in response to cold in these cells is the result of the action potential firing induced by cold (Viana et al. 2002; Madrid et al. 2006; González et al. 2015; Gonzalez et al., 2017). Figures 14.2d, e show the effect of 100 µM 4-AP on the thermal threshold of the action potential firing of a HT-CSN, and the appearance of cold-induced discharges in a cold-insensitive neuron (CIN) due to the I_{KD} blockage by 4-AP, respectively. Thus, a large percentage of HT-CSNs become low-threshold under pharmacological suppression of I_{KD} (Fig. 14.2f), and a subpopulation of CIN become cold-sensitive into the mild cold range (Fig. 14.2). The I_{KD} current density in these transformed neurons is larger than in CSNs (Fig. 14.2).

I_{KD} Current in Painful Hypersensitivity to Innocuous Cold in Response to Chronic Nerve Damage

Several ion channels have been proposed in the last years as important elements in developing and maintaining painful hypersensitivity to cold in different models of nerve damage. These include the cold-activated channels TRPM8 and TRPA1, voltage-gated and background K^+ channels, voltage-gated Na^+ channels, and HCN channels, among others (for reviews see Abrahamsen et al. 2008; Patapoutian et al. 2009; Emery et al. 2012; Du and Gamper 2013; Yin et al. 2015; Lolignier et al. 2016). Changes in the functional expression level of several voltage-gated K^+ channels in response to different forms of nerve injury have been reported, especially in murine models, including Kv1 channels (Rasband et al. 2001; Kim et al. 2002; Yang et al. 2004; Cao et al. 2010; Duan et al. 2012; Zhao et al. 2013; Fan et al. 2014; Li et al. 2015; Wang et al. 2015; Calvo et al. 2016; for reviews see

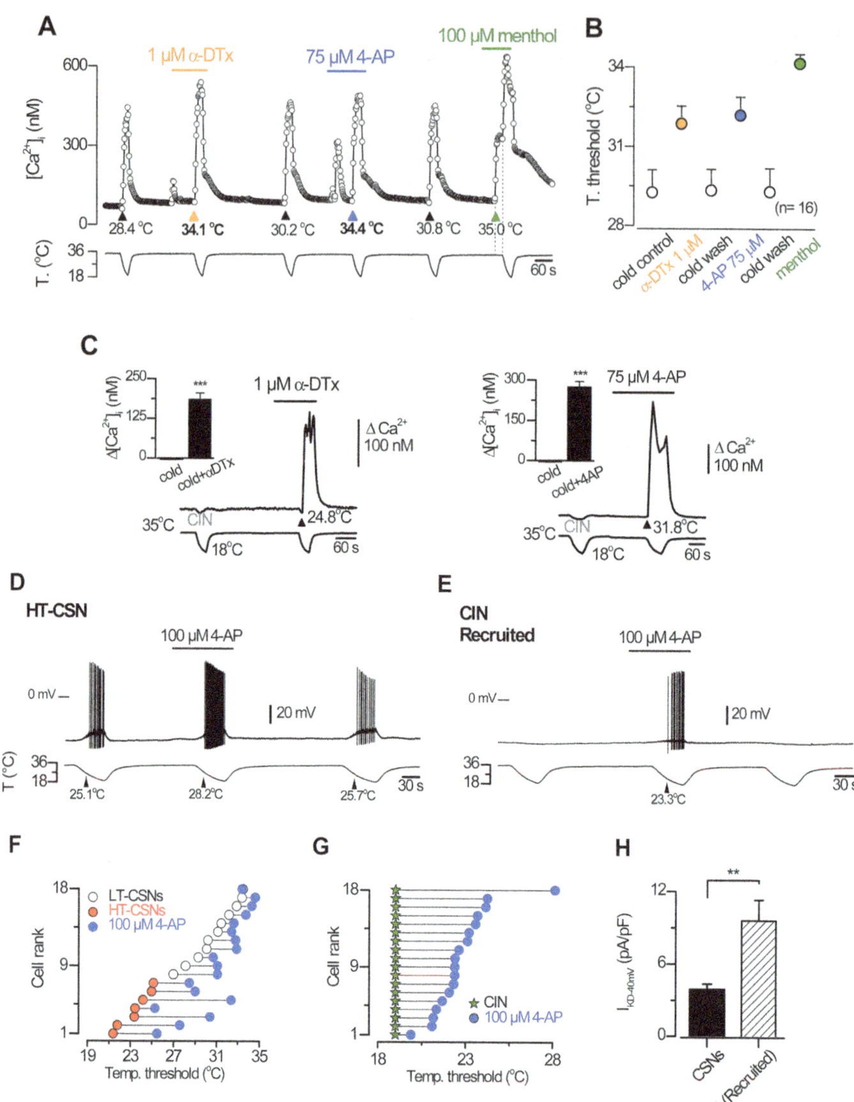

Fig. 14.2 Thermal modulation of primary sensory neurons by pharmacological suppression of I_{KD}. (**a**) Time course of $[Ca^{2+}]_i$ elevation in response to six consecutive cooling ramps, and the effect of 1 µM α-DTX and 75 µM 4-AP on thermal threshold, in a trigeminal CSN. Note the reversibility of the blockers and the sensitization of cold-induced response by the TRPM8-activator menthol. (**b**) Mean thermal threshold of 16 CSNs studied in the same conditions. (**c**) Simultaneous recording of $[Ca^{2+}]_i$ and bath temperature in two trigeminal CINs. In the neuron at the *left panel*, the application of 1 µM α-DTx unmasked a cold-induced $[Ca^{2+}]_i$ response during the second cooling ramp. In the neuron at the *right panel*, the application of 75 µM 4-AP also revealed a cold-induced $[Ca^{2+}]_i$ response during the second cooling step. *Insets* in both panels correspond to a summary of the effect of these I_{KD} blockers on $[Ca^{2+}]_i$ responses during cooling. (**d**) Simultaneous recording of membrane potential (*top trace*) and bath temperature (*bottom trace*) during three

Takeda et al. 2011; Du and Gamper 2013; Tsantoulas and McMahon 2014), and it has been proposed that Kv1.1 acts as an important brake in mechanical and pain sensitivity (Hao et al. 2013). Nevertheless, the possibility that axonal damage may affect I_{KD} current, thereby inducing exacerbated responses to cold in response to axonal damage has received less attention. Peripheral application of Kv1 channel blockers such as 4-AP and α-DTx increase cold-induced nocifensive behavior in mice (Madrid et al. 2009), suggesting that suppression of I_{KD} in intact animals can induce cold-evoked nocifensive behaviors, similar to those observed in mice with chronically injured nerves (González et al. 2017). Consistent with this view, we have recently found that painful sensitivity to innocuous cold in response to axonal damage by chronic constriction of the sciatic nerve (Fig. 14.3a), is linked to a reduction of the I_{KD} current density (Fig. 14.3) in both high-threshold CSNs and in a subpopulation of nociceptive neurons normally insensitive to mild cold temperatures (González et al. 2017). These neurons became sensitive to innocuous and mild cold by the functional reduction of I_{KD}, increasing the percentage of CSNs in response to axonal damage and shifting the mean temperature threshold of the entire population of CSNs to higher values. Although no major differences in TRPM8-dependent currents were found (Fig. 14.3c), TRPM8 channels appear to be critical to the cold responses in CSNs from both control and injured mice, since the TRPM8 channel blocker PBMC is able to prevent cold-induced responses in neurons from both groups of animals (González et al. 2017).

Based in our findings, we propose that the transformation of normally silent (cold-insensitive) nociceptive neurons into neurons sensitive to innocuous or mild cold, due to a reduction of the functional expression of I_{KD} induced by chronic constriction injury, is part of the main neural and molecular mechanisms that underpin the painful hypersensitivity to innocuous cold in this form of chronic nerve damage.

Fig. 14.2 (continued) consecutive cooling ramps in a cold-sensitive DRG neuron recorded under current-clamp (I_{hold} = 0 pA), using conventional patch clamp whole-cell recording. Application of 100 μM 4-AP reversely shifted the temperature threshold ~4 °C to higher temperatures, and greatly enhanced the action potential firing during the cooling ramp. (**e**) Simultaneous recording of membrane potential and bath temperature in a cold-insensitive dorsal root ganglia neuron recorded under current-clamp mode (I_{hold} = 0 pA), in the same conditions that the neuron in **d**. The pharmacological suppression of I_{KD} using 4-AP induces cold-sensitivity in this particular CIN. (**f**) Summary of effect of 100 μM 4-AP on cold-evoked temperature threshold in 18 CSNs from dorsal root ganglia. Note that the effect on thermal threshold in larger in HT neurons. (**g**) Thermal thresholds (*blue dots*) of 18 CINs transformed in CSNs by pharmacological suppression of I_{KD} using 4-AP. *Green stars* represent the lower temperature during the first cooling ramp before 4-AP. Note that several neurons respond at temperatures into the mild cold range. (**h**) Mean amplitude of normalized I_{KD} current in a group of 34 CSNs and 23 transformed neurons from dorsal root ganglia. The I_{KD} current was measured at −40 mV, 1 s after a 500 ms conditioning pulse to −120 mV, as in Fig. 14.1 (Panels **a–c**, modified from Madrid et al. 2009; Panels **d–h**, modified from González et al. 2017)

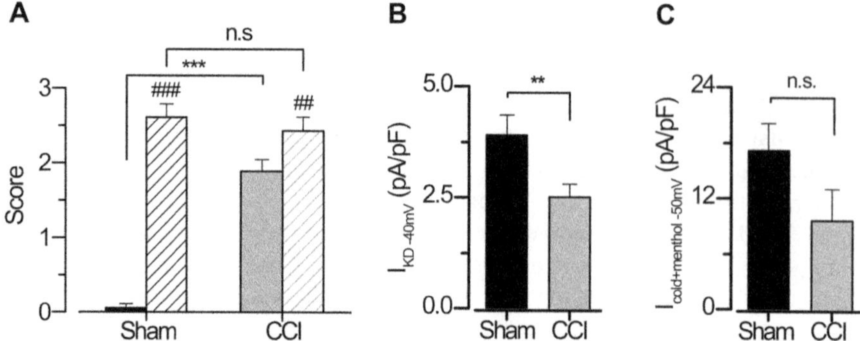

Fig. 14.3 Cold-evoked nocifensive behavior in sham and injured mice would be related to a differential expression of I_{KD} and I_{TRPM8} in CSNs. (**a**) Cold-evoked nocifensive behavior assessed by acetone response score, evaluated 7 days after surgery. We used chronic constriction injury (CCI) of the sciatic nerve as a model of nerve damage manifesting cold allodynia in mice hindpaw. *Bar graph* in **a** shows the nocifensive score before and after application of acetone to the plantar surface of the hindpaw in sham (*black*) and CCI (*grey*) mice, before (*filled bars*) and after (*dashed bars*) pharmacological suppression of I_{KD} by intraplantar injection of 4-AP (n = 6). Intergroup analyses were performed by means of two-way analysis of variance (ANOVA) followed by the Bonferroni post hoc multiple comparisons test (** p < 0.01; *** p < 0.001). Intragroup analyses were performed by means of paired *t* test (## p < 0.01; ### p < 0.001); n = 6 for each group. (**b**) Bar graph of the mean I_{KD} current density at −40 mV in sham (n = 34) and CCI (n = 36) CSNs. (**c**) Bar graph summarizing the mean $I_{cold+menthol}$ (I_{TRPM8}) current density (measured at the pick of the cold-induced current potentiated by 100 μM menthol) in 26 CSNs from sham and 20 CSNs from CCI animals (V_{hold} = −50 mV). Electrophysiological experiments were performed 7 days after surgery. (** p < 0.01; *** p < 0.001, unpaired *t* test) (Modified from González et al. 2017)

Using Mathematical Models to Study the Contribution of I_{KD} to Cold Sensing and Cold Hypersensitivity

In order to further study the contribution of I_{KD} to cold transduction in normal and pathological conditions, we have developed a mathematical model of cold-sensitive primary sensory neurons, based on the work of Olivares et al. (Olivares and Orio 2015; Olivares et al. 2015), that includes the I_{KD} (González et al. 2017). After the addition of this brake potassium current, the equation for the membrane potential is described by

$$C_m \frac{dV}{dt} = -I_{sd} - I_{sr} - I_d - I_r - I_{M8} - I_{KD} - I_l + I_{wn}$$

where C_m is the membrane capacitance, I_{sd} and I_{sr} are the slow depolarizing and repolarizing currents, respectively. These currents create an intrinsic oscillation of the membrane potential. I_d and I_r are general Hodgkin and Huxley-type depolarizing and repolarizing currents necessary for action potential firing. I_{M8} is the cold-activated current depending on TRPM8, I_l is an ohmic leakage current and I_{wn} is a noise term (Olivares et al. 2015; González et al. 2017). The brake potassium current I_{KD} has been described by the equation

$$I_{KD} = \rho(T) g_{KD} a_{KD} c_{KD} \left(V_m - E_K\right)$$

where $\rho(T)$ is a function of temperature to adjust the conductance ($Q_{10} = 1.3$), and g_{KD} the maximal conductance density of 4-AP-sensitive channels. a_{KD} and c_{KD} are variables for the fast activation and slow inactivation respectively, with values for s and $V_{1/2}$ fixed to the mean values shown in Table 14.1. V_m is the membrane potential and E_K is the reversal potential for K$^+$ currents set to -90 mV. Activation time course of the conductance follows the differential equation

$$\frac{da}{dt} = \varphi(T) \frac{a^\infty(V) - a}{\tau_a(V)}$$

where $\varphi(T)$ is a function of temperature to adjust the channel kinetics ($Q_{10} = 3$), $a^\infty(V)$ and $\tau_a(V)$ are functions of voltage for the steady state activation level and the relaxation time constant, respectively.

Consistent with our experimental findings, our model shows that a reduction in I_{KD} current density shifts the thermal threshold to higher temperatures, in neurons where the excitatory current depends on TRPM8 channels (Fig. 14.4). Only by reducing I_{KD}, a cold-insensitive model neuron (a) is turned into cold sensitive (b), while a high-threshold CS model neuron (c) is turned to low threshold (d). Moreover, our simulations using this model have revealed that although the ratio of I_{KD}/I_{TRPM8} expression is important to set the threshold of CSNs, and the brake potassium current is a critical factor to determine it (González et al. 2017). Thus, low expression levels of I_{KD} yield low thermal thresholds even in neurons with very low functional expression of excitatory TRPM8-dependent current.

A schematic representation of the functional role of I_{KD} in cold transduction and cold-induced pain in response to nerve damage is shown in Fig. 14.5.

Concluding Remarks

In primary somatosensory neurons, cold-sensitivity is tightly linked to the functional expression of *Shaker*-like Kv1.1–1.2 potassium channels, the molecular counterpart underlying the excitability brake current I_{KD}. This slow-inactivating outward K$^+$ current reduces the excitability of cold-sensitive neurons, and prevents the unspecific activation by cold of sensory neurons of other somatosensory modalities, including nociceptors. We have recently found that a reduction of the I_{KD} density in response to axonal damage would be related to the exacerbated cold sensitivity that appears in response to nerve damage. Moreover, our mathematical of model of primary sensory neurons supports the role of I_{KD} not only in normal cold sensing, but also in damage-triggered cold hypersensitivity. The reduction of I_{KD} in response to nerve damage would occur in both high-threshold CSNs and in a subpopulation of nociceptors, which become sensitive to innocuous and mild cold after axonal damage, unveiling their contribution to the molecular and neural mechanisms underlying this sensory alteration.

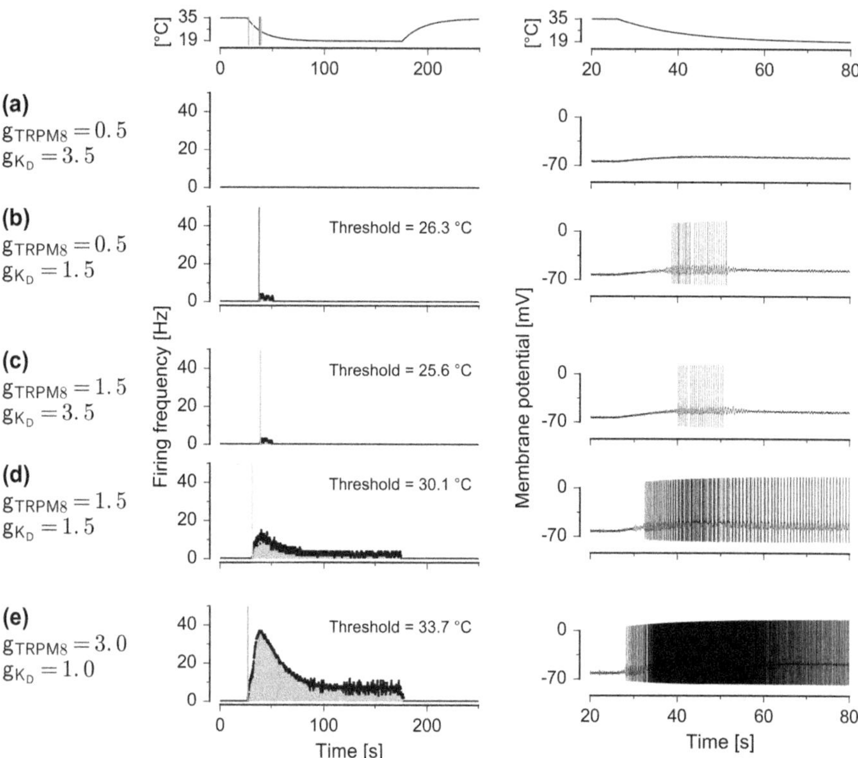

Fig. 14.4 Mathematical model of CSNs with different densities of I_{KD} and I_{TRPM8}. (**a-e**) Firing rate (*left*) and voltage trace (*right*) of model CSNs with different densities of I_{KD} current and I_{TRPM8} in response to the cooling stimulus depicted at the *top*. The Kv1.1–1.2 (g_{KD}) and TRPM8 (g_{TRPM8}) maximum conductance configurations are indicated at left in mS/cm². Note the shift of the thermal threshold to higher temperatures induced by a reduction of I_{KD} density (**d** versus **c**), and the response to cold induced in a cold-insensitive neuron when I_{KD} density is reduced (**b** versus **a**), even with a very low density of I_{TRPM8} (Modified from González et al. 2017)

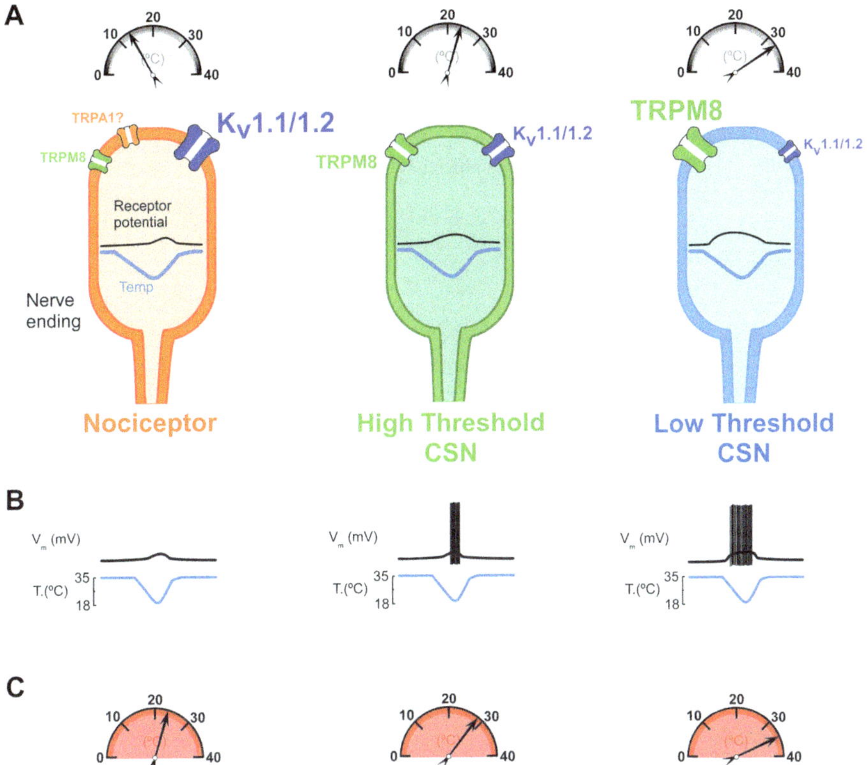

Fig. 14.5 Cold detection mechanisms and role of I_{KD} in thermal sensitivity of primary sensory neurons. (**a**) Schematic representation of cold detection thresholds and mechanisms in primary somatosensory neurons under physiological conditions. Nerve endings depict TRPM8 and I_{KD} as the main molecular mechanisms involved in the detection of cold stimuli in three functional subtypes of peripheral sensory neurons. The temperature thresholds at which these nerve endings would be excited by temperature drops are shown by the thermometers at the *top*. The size of labels reflects the relative density of the channels responsible for I_{cold} (TRPM8) and I_{KD} (Kv1.1–1.2) in the different subclasses of sensory neurons. (**b**) Schematic representation of simultaneous recording of membrane potential (*top trace*) and temperature (*bottom trace*) during cooling ramps in one CIN (*left panel*) and two CSNs (*middle* and *right panel*). (**c**) Schematic representation of the thermal threshold in these sensory neurons in injured mice, shifted to higher temperatures by the reduction of I_{KD}. After axonal damage, the reduction in the functional expression of I_{KD} would recruit a subpopulation of polymodal nociceptors normally activated by noxious cold temperatures that will respond to mild cold. HT-CSNs will respond to innocuous cold temperatures signaling cold discomfort. LT-CSNs expressing low levels of I_{KD} would remain largely unaffected. We propose that, after injury, the reduction in the functional expression of I_{KD} increases the cold sensitivity of HT-CSNs signaling cold discomfort and recruits polymodal nociceptors normally activated by extremely cold temperatures that cause pain

Acknowledgements Supported by Grants FONDECYT 1161733 and 1131064 (RM), CONICYT ACT-1113 (RM, PO, MP, GU), FONDECYT 1130862 (PO), FONDECYT 11130144 (MP) and FONDECYT 3150431 (AG). GH holds a CONICYT PhD fellowship. RM thanks Dr. F. Viana and VRIDEI-USACH. We also thank Dr. J. Gómez-Sánchez for his contribution. RM&MP Lab thanks R. Pino and M. Campos for excellent technical assistance. The Centro Interdisciplinario de Neurociencia de Valparaíso is a Millennium Science Institute funded by the Ministry of Economy, Chile. MiNICAD is a Millennium Science Nucleus funded by the Ministry of Economy, Chile.

References

Abrahamsen B, Zhao J, Asante CO, Cendan CM, Marsh S, Martinez-Barbera JP, Nassar MA, Dickenson AH, Wood JN (2008) The cell and molecular basis of mechanical, cold, and inflammatory pain. Science 321:702–705

Almaraz L, Manenschijn JA, de la Pena E, Viana F (2014) Trpm8. Handb Exp Pharmacol 222:547–579

Belmonte C, Brock JA, Viana F (2009) Converting cold into pain. Exp Brain Res 196:13–30

Calvo M, Richards N, Schmid AB, Barroso A, Zhu L, Ivulic D, Zhu N, Anwandter P, Bhat MA, Court FA, McMahon SB, Bennett DL (2016) Altered potassium channel distribution and composition in myelinated axons suppresses hyperexcitability following injury. Elife 5:e12661

Cao XH, Byun HS, Chen SR, Cai YQ, Pan HL (2010) Reduction in voltage-gated K+ channel activity in primary sensory neurons in painful diabetic neuropathy: role of brain-derived neurotrophic factor. J Neurochem 114:1460–1475

Du X, Gamper N (2013) Potassium channels in peripheral pain pathways: expression, function and therapeutic potential. Curr Neuropharmacol 11:621–640

Duan KZ, Xu Q, Zhang XM, Zhao ZQ, Mei YA, Zhang YQ (2012) Targeting A-type K(+) channels in primary sensory neurons for bone cancer pain in a rat model. Pain 153:562–574

Emery EC, Young GT, McNaughton PA (2012) HCN2 ion channels: an emerging role as the pacemakers of pain. Trends Pharmacol Sci 33(8):456–463

Fan L, Guan X, Wang W, Zhao JY, Zhang H, Tiwari V, Hoffman PN, Li M, Tao YX (2014) Impaired neuropathic pain and preserved acute pain in rats overexpressing voltage-gated potassium channel subunit Kv1.2 in primary afferent neurons. Mol Pain 10:8

González A, Ugarte G, Piña R, Pertusa M, Madrid R (2015) TRP channels in cold transduction. In: Madrid R, Bacigalupo J (eds) TRP channels in sensory transduction. Springer, Cham, pp 187–209

González A, Ugarte G, Restrepo C, Herrera G, Piña R, Gómez-Sánchez J, Pertusa M, Orio P, Madrid R (2017) Role of the excitability brake potassium current I_{KD} in cold allodynia induced by chronic peripheral nerve injury. J Neurosci 37(12):3109–3126

Hao J, Padilla F, Dandonneau M, Lavebratt C, Lesage F, Noel J, Delmas P (2013) Kv1.1 channels act as mechanical brake in the senses of touch and pain. Neuron 77:899–914

Kim DS, Choi JO, Rim HD, Cho HJ (2002) Downregulation of voltage-gated potassium channel alpha gene expression in dorsal root ganglia following chronic constriction injury of the rat sciatic nerve. Brain Res Mol Brain Res 105:146–152

Koschak A, Bugianesi RM, Mitterdorfer J, Kaczorowski GJ, Garcia ML, Knaus HG (1998) Subunit composition of brain voltage-gated potassium channels determined by hongotoxin-1, a novel peptide derived from Centruroides limbatus venom. J Biol Chem 273:2639–2644

Latorre R, Brauchi S, Madrid R, Orio P (2011) A cool channel in cold transduction. Physiology (Bethesda) 26:273–285

Li Z, Gu X, Sun L, Wu S, Liang L, Cao J, Lutz BM, Bekker A, Zhang W, Tao YX (2015) Dorsal root ganglion myeloid zinc finger protein 1 contributes to neuropathic pain after peripheral nerve trauma. Pain 156:711–721

Lolignier S, Gkika D, Andersson D, Leipold E, Vetter I, Viana F, Noel J, Busserolles J (2016) New insight in cold pain: role of ion channels, modulation, and clinical perspectives. J Neurosci 36:11435–11439

Madrid R, Pertusa M (2014) Intimacies and physiological role of the polymodal cold-sensitive ion channel TRPM8. Curr Top Membr 74:293–324

Madrid R, Donovan-Rodriguez T, Meseguer V, Acosta MC, Belmonte C, Viana F (2006) Contribution of TRPM8 channels to cold transduction in primary sensory neurons and peripheral nerve terminals. J Neurosci 26:12512–12525

Madrid R, de la Peña E, Donovan-Rodriguez T, Belmonte C, Viana F (2009) Variable threshold of trigeminal cold-thermosensitive neurons is determined by a balance between TRPM8 and Kv1 potassium channels. J Neurosci 29:3120–3131

Manganas LN, Trimmer JS (2000) Subunit composition determines Kv1 potassium channel surface expression. J Biol Chem 275:29685–29693

McCoy DD, Knowlton WM, McKemy DD (2011) Scraping through the ice: uncovering the role of TRPM8 in cold transduction. Am J Physiol Regul Integr Comp Physiol 300:R1278–R1287

McKemy DD (2013) The molecular and cellular basis of cold sensation. ACS Chem Neurosci 4:238–247

McKemy DD, Neuhausser WM, Julius D (2002) Identification of a cold receptor reveals a general role for TRP channels in thermosensation. Nature 416:52–58

Olivares E, Orio P (2015) Mathematical modeling of TRPM8 and the cold thermoreceptors. In: Madrid R, Bacigalupo J (eds) TRP channels in sensory transduction. Springer, Cham, pp 210–223

Olivares E, Salgado S, Maidana JP, Herrera G, Campos M, Madrid R, Orio P (2015) TRPM8-dependent dynamic response in a mathematical model of cold thermoreceptor. PLoS One 10:e0139314

Orio P, Madrid R, de la Peña E, Parra A, Meseguer V, Bayliss DA, Belmonte C, Viana F (2009) Characteristics and physiological role of hyperpolarization activated currents in mouse cold thermoreceptors. J Physiol 587:1961–1976

Patapoutian A, Tate S, Woolf CJ (2009) Transient receptor potential channels: targeting pain at the source. Nat Rev Drug Discov 8:55–68

Peier AM, Moqrich A, Hergarden AC, Reeve AJ, Andersson DA, Story GM, Earley TJ, Dragoni I, McIntyre P, Bevan S, Patapoutian A (2002) A TRP channel that senses cold stimuli and menthol. Cell 108:705–715

Rasband MN, Park EW, Vanderah TW, Lai J, Porreca F, Trimmer JS (2001) Distinct potassium channels on pain-sensing neurons. Proc Natl Acad Sci U S A 98:13373–13378

Storm JF (1988) Temporal integration by a slowly inactivating K+ current in hippocampal neurons. Nature 336:379–381

Takeda M, Tsuboi Y, Kitagawa J, Nakagawa K, Iwata K, Matsumoto S (2011) Potassium channels as a potential therapeutic target for trigeminal neuropathic and inflammatory pain. Mol Pain 7:5

Tsantoulas C, McMahon SB (2014) Opening paths to novel analgesics: the role of potassium channels in chronic pain. Trends Neurosci 37:146–158

Viana F, de la Pena E, Belmonte C (2002) Specificity of cold thermotransduction is determined by differential ionic channel expression. Nat Neurosci 5:254–260

Vriens J, Nilius B, Voets T (2014) Peripheral thermosensation in mammals. Nat Rev Neurosci 15:573–589

Wang XC, Wang S, Zhang M, Gao F, Yin C, Li H, Zhang Y, Hu S, Duan JH (2015) Alpha-dendrotoxin sensitive Kv1 channels contribute to conduction failure of polymodal nociceptive C-fibers from rat coccygeal nerve. J Neurophysiol 115(2):947–957

Yang EK, Takimoto K, Hayashi Y, de Groat WC, Yoshimura N (2004) Altered expression of potassium channel subunit mRNA and alpha-dendrotoxin sensitivity of potassium currents in rat dorsal root ganglion neurons after axotomy. Neuroscience 123:867–874

Yin K, Zimmermann K, Vetter I, Lewis RJ (2015) Therapeutic opportunities for targeting cold pain pathways. Biochem Pharmacol 93:125–140

Zhao X, Tang Z, Zhang H, Atianjoh FE, Zhao JY, Liang L, Wang W, Guan X, Kao SC, Tiwari V, Gao YJ, Hoffman PN, Cui H, Li M, Dong X, Tao YX (2013) A long noncoding RNA contributes to neuropathic pain by silencing Kcna2 in primary afferent neurons. Nat Neurosci 16:1024–1031

© Springer International Publishing AG 2017
R. von Bernhardi et al. (eds.), *The Plastic Brain*, Advances in Experimental
Medicine and Biology 1015, DOI 10.1007/978-3-319-62817-2

CPSIA information can be obtained
at www.ICGtesting.com
Printed in the USA
LVHW082003110119
603608LV00001B/19/P